ROUTLEDGE LIBRARY EDITIONS: DEVELOPMENT

URBANISATION AND PLANNING IN THE 3RD WORLD

AF695272

URBANISATION AND PLANNING IN THE 3RD WORLD

Spatial Perceptions and Public Participation

ROBERT B. POTTER

Volume 77

Routledge
Taylor & Francis Group
LONDON AND NEW YORK

First published in 1985

This edition first published in 2011
by Routledge
2 Park Square, Milton Park, Abingdon, Oxfordshire OX14 4RN

Simultaneously published in the USA and Canada
by Routledge
711 Third Avenue, New York, NY 10017

Routledge is an imprint of the Taylor & Francis Group, an informa business

First issued in paperback 2013

© 1985 Robert Potter

All rights reserved. No part of this book may be reprinted or reproduced or utilised in any form or by any electronic, mechanical, or other means, now known or hereafter invented, including photocopying and recording, or in any information storage or retrieval system, without permission in writing from the publishers.

British Library Cataloguing in Publication Data
A catalogue record for this book is available from the British Library

ISBN 13: 978-0-415-59672-5 (hbk)
ISBN 13: 978-0-415-585327-9 (pbk)
ISBN 13: 978-0-415-58414-2 (Set)

eISBN 13: 978-0-203-84035-1 (Set)
eISBN 13: 978-0-203-83622-4 (Vol 77)

Publisher's Note
The publisher has gone to great lengths to ensure the quality of this reprint but points out that some imperfections in the original copies may be apparent.

Disclaimer
The publisher has made every effort to trace copyright holders and welcomes correspondence from those they have been unable to contact.

Urbanisation and Planning in the 3rd World

Spatial Perceptions and Public Participation

Robert B. Potter

CROOM HELM
London & Sydney

©1985 Robert Potter
Croom Helm Ltd, Provident House, Burrell Row,
Beckenham, Kent BR3 1AT

Croom Helm Australia Pty Ltd, Suite 4, 6th Floor,
64-76 Kippax Street, Surry Hills, NSW 2010, Australia

British Library Cataloguing in Publication Data

Potter, Robert B.
Urbanisation and planning in the Third World:
spatial perceptions and public participation.
1. Urbanisation – Developing countries
I. Title
307.7'6'091724 HT153

ISBN 0-7-99-2055-5

Contents

Figures

Figures

Tables

Plates

Preface

I started writing this book for two main reasons. First, in teaching an urban geography course which has come to include more and more material on Third World urbanisation and development, I have found it difficult to recommend a text to students that covers adequately urban development and planning at the global, national and local scales. Regrettably, and quite astonishingly given the scale and magnitude of the urbanisation process in the Third World, standard textbooks on urban geography are generally almost exclusively Western in orientation. On the other hand, the numerous fine books which have by now been published on Third World urban processes tend, by and large, to focus in isolation either on particular facets of urbanisation, or on particular levels of analysis. Thus, there are several excellent texts on Third World urban housing, spontaneous settlements and self-help imperatives. Equally, there are recognised classics on the political economy of urbanisation and on urban and regional planning, as well as those which focus on particular regions and even those providing a city by city approach. The present book tries to help fill this gap and is thereby presented as a contribution to the urban geography of the Third World. In particular, Chapters 1 to 4 inclusive have been written with this goal in mind. The second reason for writing was to present in wider terms, some of the results and ideas stemming from field research I have carried out in the Caribbean region. This work, which started in 1980, has involved the examination of urban planning practices in a number of Caribbean territories, along with assessments of the utility of public participation in the urban planning process, and the potential role of designative and appraisive perception studies in such contexts. Thus, Chapters 5 to 7 are somewhat more research-oriented, although by no means exclusively so. It is hoped that this dual approach will be seen as an advantage, rather than as a limitation of the work.

The book has been written for all those concerned with the role of cities and the urbanisation process in the poor countries of the world. It is hoped that it will be of interest not only to geography students, but also to those in the closely-related fields of regional and urban planning, development studies, regional science, economics, sociology and environmental psychology. The term 'student' is used intentionally, but not merely to refer to those who are engaged in full time education at recognised institutions of higher education. Thus, it is hoped that as well as finding a niche for sixth-formers and undergraduates, the book may also be of relevance to interested lay readers and other non-specialists.

Whilst fully recognising and exploring the strong underpinnings of the processes of urbanisation, urban growth and urbanism in the realms of international social, economic and political organisation, this book also takes a strongly behavioural-cognitive view of the Third World urban process. Thus, as is emphasised in the title, exploring this theme is a major aim of the book. The theme of the volume serves to stress that urbanisation and urban growth are fundamental contemporary processes affecting humanity, and that reacting to the challenges presented by them will undoubtedly tax severely the capabilities, resources and inventiveness of governments and communities in the last quarter of the twentieth century and beyond. A further conclusion that is reached is that dealing with these processes will entail much more than merely responding in a piecemeal fashion to past trends. It will involve an ideological commitment to promoting a more balanced and equitable pattern and process of development. In this regard, it is argued that the need to take into account the environmental and more general perceptions and cognitions of ordinary people cannot be over-stressed. All over the world the pressing need is for the evolution of systems of planning and development that recognise the salience of, and which seek to act upon the aspirations and cognitions of the general populace. Thus, however difficult it may be to achieve, the present volume argues that inherent to efficacious planning and development is the need for the effective participation of the public. Together these are the themes of the book: global urbanisation, planning, spatial perceptions and the role of public involvement in the planning process.

I should like to acknowledge the generosity of the British Academy, the Central Research Fund of the University of London and the 20th International Geographical Congress Fund for grants which enabled me to work in Barbados, St Lucia and Trinidad and Tobago, and which indirectly also afforded me the opportunity of visiting Venezuela. In the Caribbean, I received much assistance from colleagues, friends and others too numerous to mention by name. Special thanks, however, must go to the 540 or so people who spent time giving their views and reactions in interviews. In Caracas, Cristina Ferrin and Alexis provided genuine hospitality and helped us visit parts of the city that otherwise we should not have been able to see. Closer to home, Susannah Hall and Ron Halfhide drew the majority of the figures making up the book, whilst Simon Barker photographically reproduced most of the artwork. Of particular assistance were John Coshall, Derek Diamond, John Gold, Elizabeth Higgins, David Hilling, Alan Mountjoy, John Parr and Tim Unwin, all of whom kindly read and commented on particular chapters. They are, of course absolved from any sins either of omission or commission which remain, by virtue of the simple fact that none read more than two or three chapters, and several only one. Once again, thanks are due to Rosemary

Dawe, who typed the draft and final texts.

Finally, Virginia, as always, read each chapter as it appeared and offered useful guidance and encouragement, despite her own busy work load. Wisely, she reciprocated by allowing me to help her from time to time. I hope that soon the scorecard will have been equalised!

London

Robert B. Potter

Acknowledgements

I am grateful to the authors and/or publishers for permission to reproduce in revised or redrawn form, the following copyright materials:

Figures 6.1 and 6.2 from P. R. Gould and R. White (1974): Mental Maps. Penguin (reproduced here as Figures 1.2 and 1.3).
Figure 1.3 from S. D. Brunn and J. F. Williams (Eds.) (1983): Cities of the World: World Regional Urban Development. Harper and Row (reproduced here as Figure 2.3).
Figure 7 from C. A. Doxiadis and J. G. Papaioannou (1974): Ecumenopolis: the Inevitable City of the Future. W. W. Norton (reproduced here as Figure 2.5).
Figure 18 from J. E. Vance (1970): The Merchant's World: the geography of Wholesaling. Prentice-Hall (reproduced here as Figure 3.1).
Figure 2.1 from J. Friedmann (1966): Regional Development Policy: a case study of Venezuela. M.I.T. Press (reproduced here as Figure 3.2).
Figure 1 from D. Conway (1980): "Step-wise migration: toward a clarification of the mechanism", International Migration Research, 14, 3-14 (reproduced here as Figure 3.11).
Figures 1.7 and 1.8 from D. J. Dwyer (1975): People and Housing in Third World Cities: Perspectives on problems of Spontaneous Settlements. Longman (reproduced here as Figures 3.12 and 3.13).
Figure 2 from L. A. Eyre (1972): "The shantytowns of Montego Bay, Jamaica", Geographical Review, 62, 394-412 (reproduced here as Figure 3.14).
Figure 5 from E. Jones (1964): "Aspects of urbanisation in Venezuela", Ekistics, 18, 420-5 (reproduced here as Figure 3.16).
Figure 14 from H. W. Richardson (1978): Regional and Urban Economics. Penguin (reproduced here as Figure 4.1).
Figure 2 from J. B. Parr (1980): "Frequency distribution of central places in Southern Germany", Economic Geography, 56, 141-54, and Figures 1 and 4 from J. B. Parr (1981): "Temporal change in a central place system", Environment and Planning, 13, 97-118 (reproduced here as Figures 4.3 and 4.4).
Figure 9.1 from D. Pocock and R. Hudson (1978): Images of the Urban Environment. Macmillan (reproduced here as Figure 5.1).

Chapter 1

Introduction

However we choose to define the word 'urbanisation', and whether we are inclined to regard it as referring to a primarily economic, demographic or social-behavioural process, there can be little doubting the fact that it signifies one of the most problematical processes confronting mankind in the last two decades of the twentieth century and beyond. It is hardly surprising, therefore, that during the post-war period, a vast literature has developed within the social sciences which specifically seeks to examine the manifold historical, contemporary and prospective aspects of the global trend of urbanisation. This corpus of writing obviously reflects the simple truism that urbanisation involves fundamental shifts and changes in the whole fabric of nations, regions and continents, not least in their social, economic and demographic structuring. Further, it is hard to refute the assertion that along with efforts to alleviate global poverty and hunger, to promote greater equality of opportunity by careful social development, and to enhance world peace and individual security, the tackling of problems associated with rapid urban development will test the ingenuity of mankind to the utmost, and may yet determine the capacity for sustained global development in the future.

The pressing nature of these three major developmental issues of world poverty, peace and human settlements, together with that of rapid total population growth has been clearly exemplified in the document North-South: A Programme for Survival, wherein it is argued that:

> There is a real danger that in the year 2000 a large part of the world's population will still be living in poverty. The world may become overpopulated and will certainly be overurbanized. Mass starvation and the dangers of destruction may be growing steadily - if a new major war has not already shaken the foundations of what we call world civilization (Brandt, 1980,p.11).

Whilst the twin threats of poverty and war are discussed in detail in the Brandt report, the problems posed by 'overurbanisation' and human settlements are not fully considered, an omission strongly criticised by many reviewers of the document. This lack of attention is surprising, for at the present juncture, the course of world urban development should hold considerable fascination for all those concerned with global developmental issues. Although historically towns, cities and predominantly urban modes of living have all been synonymous with 'development', 'industrialisation' and 'modernisation', many of the urban areas of the Western world are today increasingly facing problems of atrophy, decay and economic regression in the guise of inner city problems, fiscal bankruptcy and counter-urbanisation (Berry, 1976).

In fact, the problems and challenges afforded by the processes of rapid urban growth and urbanisation are coming increasingly to be associated with the 'less economically developed' or 'poor' nations of the world, and questions as to whether massive urban agglomerations should be allowed to continue growing or whether decentralisation should be encouraged, what urban form should be envisaged and how the problems of cities can be combated are vexed ones. Thus, the urban containment and decentralisation ideologies which had their origins in Western-industrialised urban systems in the late nineteenth and early twentieth centuries are now increasingly being accepted as the major planning guidelines in many contemporary less developed countries. However, at the same time, views as to the relative social-economic benefits and costs of the growth of large urban agglomerations and urban primacy are far from unequivocal. Similarly, the actual role of cities in the spatial development process is an important, albeit a highly contentious issue. Thereby, the need for sound research and planning is crucial, for even if decanting and spatial dispersion are seen as desirable, the precise configuration of settlements to be promoted and their relative sizes remain as thorny, if not intractable issues. Such topical and fundamental issues underlie the content and orientation of the present volume.

URBAN GEOGRAPHY AND THE THIRD WORLD

The present volume is primarily addressed to students, academics and professionals in the fields of geography, planning and development studies, and is regarded as a contribution to the study of the urban geography of developing nations. This orientation must be stressed for, despite the manifest importance and relevance of the contemporary processes of urbanisation and urban growth in Third World countries, it is true to say that surprisingly few urban geographers have explicitly turned their attentions to them. This is certainly witnessed by the negligible space devoted to the phenomenon of Third World urban development in most student text-books on urban geography, a comment which by and large stands for established classics, as well as recent additions to the literature (Johnson, 1967; Northam, 1979; Hartshorn, 1980; Carter, 1981; Herbert and Thomas, 1982; Clark, D., 1982; Knox, 1982; Short, 1984).

This rather parochial focus of much mainstream urban geography is also exemplified by the content of published research registers covering the field. The Urban Geography Study Group of the Institute of British Geographers (IBG), for instance, compiled a listing of some 227 projects being carried out by its members in 1982. Of these, only ten were classified by the present author as dealing with Third World topics, representing a mere 4.41 per cent of the total. Notwithstanding the fact that geographers who do not profess to work under the prefix 'urban' also study urban based topics, the figure is worryingly small. However, of a total of 120 projects listed in the 1982 Research Register of the Developing Areas Study Group of the IBG, again only 19 or 15.82 per cent were on explicitly urban related themes. In fact, it may well be the case that urban geographical interest in Third World urbanisation is at present on the wane rather than the increase. Thus, Warnes (1978) reviewed the contents of an earlier, 1976, IBG Urban Study Group research register and found that 21 or 12 per cent of the total indexed projects were concerned with "urbanization, urban growth", putting the topic in joint fourth place among members. However, although most of these projects were international or comparative in character, some were entirely Western in focus. Warnes also noted that only 4 projects involved the study of "squatter settlements". This can be compared with the 1973 Register consisting of 72 projects, where a total of 10, or 21 per cent were on "urbanization, urban growth", this comprising the first ranking research topic.

This already relatively low and apparently declining level of interest shown in Third World urbanisation can undoubtedly be explained, at least in great part, by changes in the wider economic and academic environments in the post-war period (Farmer, 1983). At the present time of economic recession, and with mounting socio-economic problems in the urban areas of British and other Western countries, it is perhaps inevitable that increasing attention will be given to these pressing domestic issues. However, there is the ever-present danger that this will lead to a form of academic insularity if the interdependent connections existing between global environmental and development problems are ignored, or at best neglected (Brookfield, 1975).

The apparently mounting malaise in Third World urban geographical study can also be viewed historically. The hallmark of human geography in the 1960s was the adoption of an increasingly quantitative approach, an orientation that was closely allied with a commitment on the part of practitioners to theory construction, model building and the search for empirical generalisations. Urban and economic geographers were at the forefront of these developments but their focus was almost exclusively on developed regions and their metropolitan areas. Since the early 1970s, it is fair to say that the topics dealt with by human geographers and the associated methods and philosophies employed have broadened considerably, rendering a buoyant, if not always overtly coherent field of academic teaching and research. An early call in the 1970s was for increasing 'social relevance', and an enhanced concern with the potential policy-related implications of human geographical research. Another theme has emphasised the importance of the concepts of social justice and social welfare in geographical analyses. Both approaches have given a fresh impetus to urban, economic, social and political geography, but again seem not to have been translated into a direct focus on Third World urban problems.

At the same time, human geographers have been showing a greater interest in political processes as they operate in space from the national to micro-scales. In particular, the provision of welfare and social services has been studied, and again the theme is clearly germane to Third World urban studies, but has largely been pursued in a developed world context. Also, a strong radical approach, often employing Marxist-based ideas has emerged (Peet, 1977), and is reflected in some very significant texts on the general theme of urbanism, particularly those by Harvey (1973) and Castells (1977). A further recent major growth area within human geography has been the study of environmental perceptions and cognitions under the general banner 'behavioural geography' (see, for example, Walmsley and Lewis, 1984). This is not to be seen as an entirely new and separate sub-discipline, but rather a fresh paradigm that has been applied in most of the traditional areas of geographical concern. Here too, although much of this work has taken the planning relevance of such an approach as its major theme (Saarinen, 1976; Pocock and Hudson, 1978), and despite some early tentative applications in developing countries (Gould and White, 1974), its full potential has not as yet been assessed in the Third World context.

In summary, therefore, it can be argued that more work has been carried out on Third World urban processes by non-urban geographers and indeed by non-geographers such as sociologists, demographers, historians, economists, social anthropologists and regional scientists than by urban geographers. However, there are some indications that a substantive reorientation is occurring in that human geographers are supplementing their traditional interest in the nature of spatial patterns with a revitalised concern for the aggregate structural and societal processes that give rise to those patterns. Further, there is a current associated trend for the rejection of logical positivism in favour of structuralist, and humanistic approaches such as phenomenology and idealism. As a direct consequence, geography's inter-disciplinary links are becoming more numerous, and in some instances, much stronger. Hopefully, the important topic of Third World urbanisation will thereby receive greater attention in the not too distant future as these methodological ripples spread, and indeed there are already some encouraging signs in this respect (Johnston, 1980; Gilbert and Gugler, 1982; Brunn and Williams, 1983).

SETTLEMENT, PREFERENTIAL, MIGRATORY AND POLICY SPACE IN THIRD WORLD COUNTRIES

The present volume pursues two broad themes. First, it stresses the pressing need for the identification of soundly based and appropriate systems of urban and regional planning in Third World countries, particularly frameworks which explicitly acknowledge the importance of harnessing the aspirations, initiatives and perceptions of individuals and groups, rather than those which are premised on alien models and modes of planning. Secondly, the volume stresses the generally neglected theme that, viewed at one level, the process of rural to urban migration, and the associated phenomena of urbanisation, urban growth and urbanism in Third World countries are the tangible outcome of individuals' perceptions of environmental and socioeconomic opportunities. Thus, a central argument is that the study of individual and group environmental perceptions stands as a potentially rewarding theme in the analysis of Third World urban and regional patterns; and further, should therefore be regarded as a major potential input into the planning process at the survey, analysis and plan stages (see also Potter, 1983d,1984b,1984c).

It may be argued that the principle of public participation in planning is as vital in Third World countries as it is in Western nations, albeit in a somewhat modified form (Franklin, G.H., 1979; Potter, 1984b, 1984c; Conyers, 1982). This is, for example, reflected in the mounting awareness that community-based planning programmes (Kent, 1981) and self-help imperatives must be recognised by Third World planners and politicians, for they express the aspirations and desires of the mass of the populace, and thereby represent a major cultural resource for change. It is all too easy for planners and policy-makers, many of whom have been trained abroad, to assume without question, that tried and tested modes of planning which have proved useful in developed nations will be equally suited to meeting the formidable problems faced by less developed countries (Zetter, 1981).

The various strands of this overall argument are developed in the following chapters of this book. It should be stressed that this wider argument concerning the need to open up a genuine and effective dialogue between planners and the planned, applies equally well at both the intra-urban and inter-urban levels. However, as an introductory example, some of the primarily spatial connotations of this suggestion are briefly illustrated here by means of a simple graphical model drawn up at the inter-urban scale (Figure 1.1), and then subsequently by the real world example of Tanzania, also mainly couched at the inter-urban scale.

The urban settlement pattern of a given country or region, that is its configuration of urban places by size and location, is the product of the historical processes of urbanisation and urban growth (Figure 1.1a). It is highly unlikely that urban development will have proceeded evenly and uniformly through space. Typically, certain regions of the national space, in this example the north and western coastal areas, will have become the focus of growth associated with a process of cumulative causation. The corollary will be the lack of development of the peripheral regions, in the example here, the east and south. Decision-makers will inevitably respond to the actual and perceived socio-economic opportunities

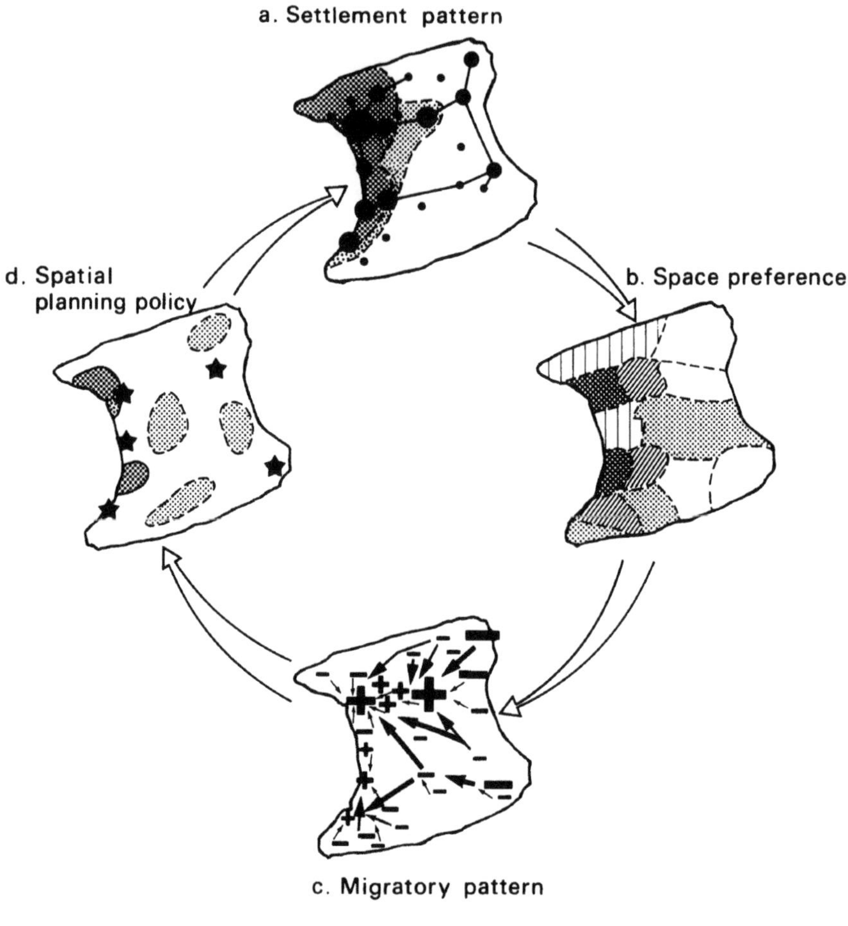

Figure 1.1: Settlement, preferential, migratory and policy space

offered by different localities, that is inter-regional inequalities, so that areas where average wage rates, employment and housing opportunities, medical and social services are believed to be favourable, will be regarded as more desirable than the others. Individuals, corporations, and indeed government organisations and the state itself will all be characterised by their own distinctive national space preference surfaces (Figure 1.1b). Although these will contain many unique elements, it is likely that there will be a generalised or consensus regional patterning.

However, by definition, space preferences whether individual or corporate are based on past experiences and conditions. Additionally, they are the outcome of far from perfect information gathering and processing abilities. As a consequence, therefore, spatial imagery is likely to show evidence of bias, inertia and misinformation. Thus, space preferences may effectively become stereotypes, showing a poor correspondence with actual environmental conditions. This, of course, is particularly likely if rapid socio-economic change is occurring. For example, labour and capital may continue to migrate toward a particular urban region long after it has reached its social and economic optimum. Similarly, areas of potential growth may remain under-perceived. Some have referred to such disparities between space preferences and actual conditions as constituting a "myth" map.

The present book argues that such perceptions and misperceptions, and the structural circumstances and constraints which underlie them need to be examined and understood if efficacious urban and regional planning is to be promoted in Third World countries. The links existing between settlement patterns and group/individual and institutional space preferences offer valuable insights for planners and environmental policy-makers. The collection of data on environmental perceptions can be regarded as a useful, albeit a specialised method of encouraging public involvement in the planning sequence. Although it must be accepted that spatial migration is basically a structural phenomenon (Lipton, 1980; Thapa and Conway, 1983), these social and economic structures are reflected in individuals' reasons for spatial migration. Thus, at this level, migration can be regarded as a behavioural response that is based on space preferences, and environmental images of complex socio-economic structures (Wolpert, 1966).

Whether we are talking of direct or step-migration, temporary or permanent moves, there must be a difference in evaluations between the present location and the proposed destination for migration to occur. The only exception to this is in situations where political-military force and coercion are used to move people, unfortunately, a not altogether infrequent situation in developing countries. However, other things being equal, migration occurs from areas of negative or low space preference levels to those of positive or high levels (Figure 1.1c). It can be argued, therefore, that to affect patterns of rural to urban migration, to promote deconcentration from primate cities, or to promote rural development, public images and space preferences must be understood and modified. Hopefully, this will be achieved by transforming the structure of the actual environment, but of course, it must be recognised that politicians and planners sometimes attempt this by means of counter-image campaigns (Burgess, J.A., 1982; Hoare, 1981; Pocock and Hudson, 1978), which verge on environmental propaganda and brainwashing. Following from this, it must also be recognised that certain vested interests may seek to retain the sharply divided structural conditions and perceptions which underlie spatial inequalities. All of these factors will influence spatial planning policies (Figure 1.1d).

But it must be stressed that the overall argument presented in this volume is not solely concerned with inequalities and migrations at the national scale. A similar, or at least parallel line of argument can be adopted at the local, or intra-urban level. Here the massive social and spatial inequalities that exist within the cities of the Third World, reflect the lack of power which the majority of the population have in the decision-making processes that influence, and as frequently determine, the circumstances of their daily lives. Here again it can be argued that a vital prerequisite for efficacious planning and development is the successful articulation of the voices of the ordinary people, their perceptions and aspirations. In short, what is needed is the participation of the public in the planning process, instead of all forms of change emanating from 'above'. This theme constitutes the focus of the present book.

THE EXAMPLE OF TANZANIA

As will be seen from subsequent chapters, the diagrams in Figure 1.1 are loosely based on the case of Trinidad in the West Indies (Potter, 1983c). However, a brief but more specific example which broadly illustrates some of these points can profitably be considered, that of Tanzania (see Figures 1.2 and 1.3). One of the world's 25 poorest countries, Tanzania became independent in 1961 after a German and British colonial history (Hoyle, 1979). At the 1978 Census, Tanzania recorded a total population of 17.5 million, of which 15.1 million lived in rural areas. A principal feature of the country is the highly uneven character of its population distribution, with heavy concentrations in the northern areas and the coastal region, and paucity in the south and central west. The settlement pattern has traditionally comprised dispersed villages, although the colonial period witnessed greater spatial concentration, especially around Dar es Salaam, which Hoyle (1979) has recently described as an "hypertrophic cityport" (see also Hoyle, 1983; O'Connor, 1983). In respect of size, Dar es Salaam on the Indian Ocean coast is still the principal city and in 1978, housed a total population of 870,020 (Siebolds and Steinberg, 1982). In fact, Dar es Salaam had become the capital and seat of government in 1891.

On the basis of research carried out in the 1960s, Gould (1969a; see also Gould and White, 1974) showed that there was a strong urban bias in the space preferences of a sample of Tanzanian nationals, so that there was an associated "problem of getting well-qualified people to serve in places that need them desperately, but are perceived as low valleys on a mental map ..." (Gould and White, 1974, p. 160). The essential correspondence implied in this quotation between settlement, preferential and migratory space is reflected in Figures 1.2 and 1.3, which have been adapted from Gould's research. In order to investigate preferential space, a group of Tanzanian university students were requested to rank order their residential preferences for the constituent districts of their national territory, assuming that on graduation they were to enter government service as a career. The aggregate space preferences of Tanzania thereby revealed are shown in Figure 1.2 by means of an isopleth map, wherein values have been standardised on a scale running from 0 to 100. The most preferred zones, those recording values in excess of 40 are shown shaded. The 'mental map' or 'residential desirability surface' pinpoints all of the major urban areas as peaks to the space preference surface. In particular, Morogoro, Iringa and Mbeya constitute a corridor of very high desirability running inland in a south-easterly direction from Dar es Salaam. Notably, however, Dar es Salaam is itself only moderately rated. The other major administrative and urban nodes of the northern half of the country also stand out, particularly Arusha-Moshi, Shinyanga, Tabora and Kigoma (Figure 1.2). In contrast, the rural areas are generally held in relatively low esteem, especially the southern and west central areas. In fact, the key to examining these patterns is the realisation that the perception map matches almost perfectly the spatial configuration of 'modernisation' within the country, and provides an excellent example of the somewhat hypothetical case shown in Figure 1.1. In further research, Gould (1970) employed multivariate statistical techniques to analyse 41 variables relating to the provision of modern amenities and services, measured for some 289 hexagonal cells making up the entire Tanzanian national space. The pattern revealed by Gould's analysis is shown in Figure 1.3 and is the characteristic one consisting of 'islands' of modernisation which are linked by major lines of transport (see Gould, 1969a,1970; Hoyle, 1979; Safier, 1969).

Tanzania has attracted the attention of many academics concerned with settlement planning and development issues due to its national planning policies since independence (Hirst, 1978; Lundqvist, 1981; Nyerere, 1982; Kulaba, 1982; Briggs, 1983). The Tanzanian case is useful in that it demonstrates how existing environmental conditions, both real and perceived, together comprise the context for planning policies and initiatives. Lundqvist (1981) has identified four major phases of planning in Tanzania since independence. The period from 1961 to 1966 is thereby seen as the legacy of the colonial period. The planning carried out at this time was largely sectoral in nature rather than regional, as a result of which social and physical infrastructure remained concentrated in the principal towns and the rural-urban dichotomy was perpetuated.

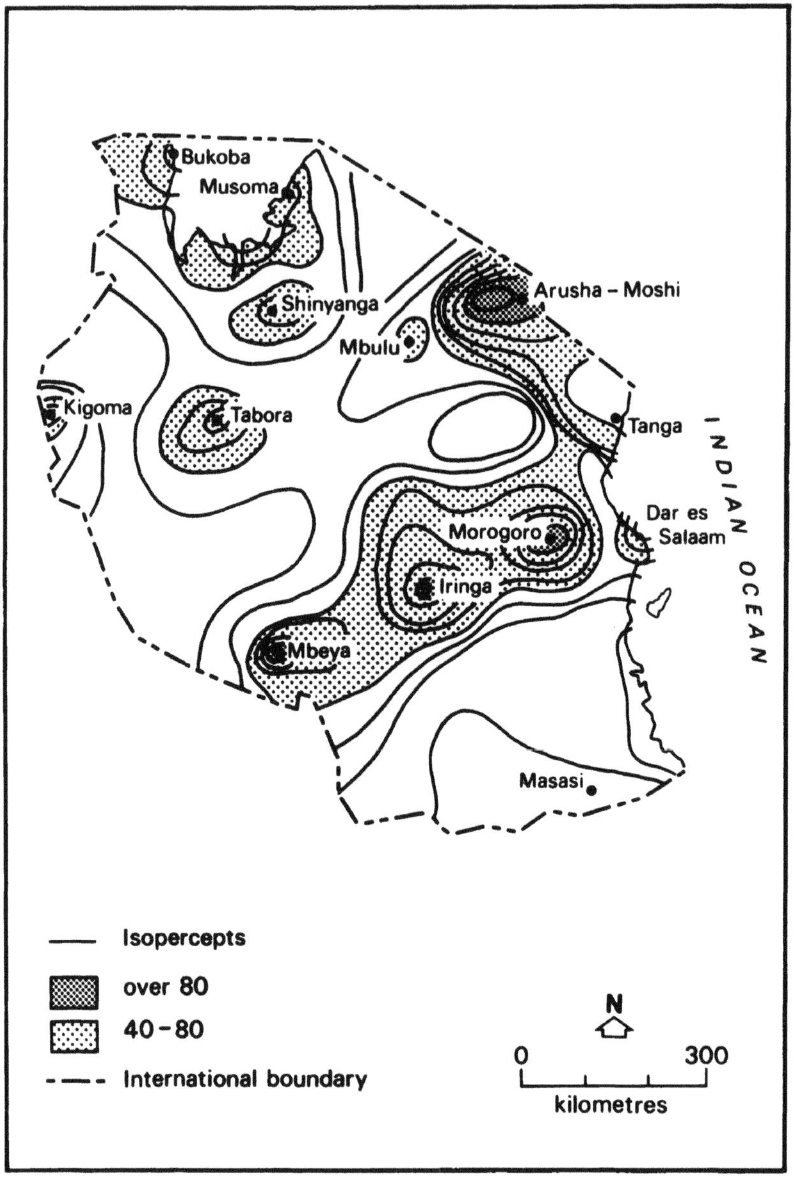

Figure 1.2: Aggregate space preferences in Tanzania (Source: Gould and White, 1974)

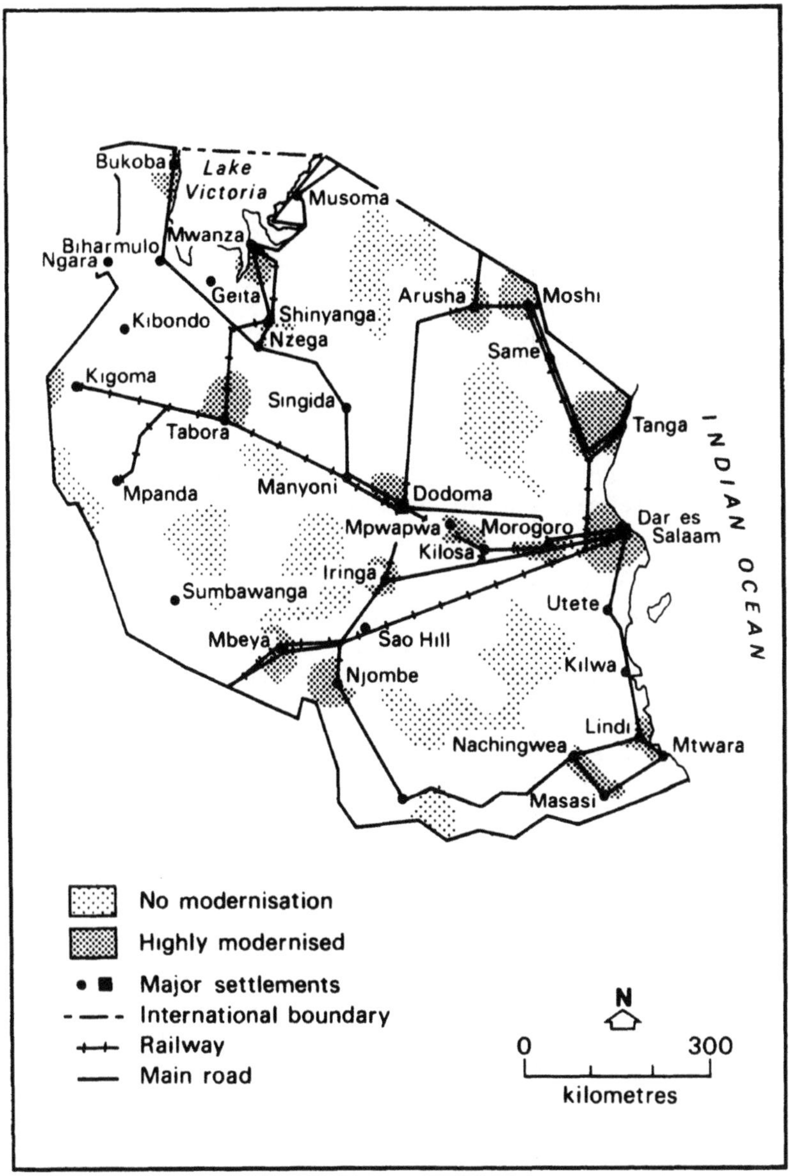

Figure 1.3: Settlements and modernisation in Tanzania (Source: Gould and White, 1974)

The main policy efforts to reduce the urban-rural gap represent the second and most important phase, extending from 1967 to 1972, which witnessed the emergence of a commitment to a rural based development linked to African socialism. These policies were based on the Arusha Declaration of 1967 which attacked privilege and placed strong emphasis on equality, co-operation, self-reliance and nationalism. Such ideals were largely put into practice in the Second Five Year Plan, 1969-74. The major policy imperative was the Ujamaa villages which were regarded as the expression of "modern traditionalism", that is, a twentieth century version of traditional African village life. The basic idea was to concentrate scattered rural populations, and by such a process of villagisation to provide the services requisite for viable settlements. The reduction of rural-to-urban migration was the principal aim of the policy, along with the lessening of dependence on major cities such as Dar es Salaam. Ujamaa villages were envisaged as co-operative ventures by means of which initiative and self-reliance would be fostered, along socialist lines of planning. An avowed intention, therefore, was to encourage participatory democracy at the local level, as summarised by Hirst (1978, p.124):

> In the first phase, where not less than 250 households have settled and made their home in a defined area, a village may be registered and a village council elected to foster social and economic development and encourage communal enterprises. In the second phase, when the authorities are satisfied that a substantial portion of the economic activities of the village are undertaken and carried out on a cooperative and communal basis, the village may be designated as an ujamaa or socialist village.

In addition, efforts were also made to spread urban development away from Dar es Salaam towards nine selected regional centres. In overall terms, President Nyerere regarded these policies as a distinct move away from a slavish imitation of Western-style planning and development programmes.

Lundqvist recognises the period 1973-78 as constituting a third phase, described as 'villagisation by order', during which compulsory movement to development villages occurred. Hirst (1978) attributes this to the speed envisaged for the programme after 1970, which he argues almost inevitably meant that coercion would have to be used to compel people to move to new villages. Obviously, antagonism has resulted and some commentators have argued that this has led to an ironic lack of commitment and involvement of the peasantry in implementing the policy - the very antithesis of its original aims (Hirst, 1978; Briggs, 1983). At the beginning of the phase, in 1973, the centrally located town of Dodoma was selected as the new national capital in place of the peripherally located Dar es Salaam. However, in spite of a lack of investment in urban renewal and development, Dar es Salaam has continued to grow, primarily due to rural in-migration and port-related economic developments (Hoyle, 1979). In fact, the population of the former capital city has continued to grow at an alarming rate of around 7 per cent per annum according to recent estimates (Siebolds and Steinberg, 1982).

Taken together these policies represent a concerted effort to change the map of perceived spatial attractions in Tanzania, and thereby to influence patterns of migration and settlement. Although having received much praise from certain quarters, the policies of Tanzania have been viewed with considerable scepticism by others, especially those from a committed Marxist viewpoint (Lundqvist, 1981). In fact, it has been argued that a fourth stage of planning policy, starting in 1978 must be viewed as one in which industry and urban development seem likely to be upgraded at the expense of Ujamaa villages and rural progress, partly as a reaction to the near-disaster brought about by the 1973-78 phase. Thus, the Fourth Five Year Plan 1981/2-1985/6 notably gave top priority to industrial development. Briggs (1983, p. 68) has commented that "By 1979/80 ujamaa and villagisation had become little more than a memory for government and planners", and that in broader terms, "It remains to be seen whether ujamaa will be revived some time in the future, or whether it should be interpreted as a bold experiment which did not work out and which has subsequently been discarded". Whatever the long-term outcome, Tanzania since 1961 offers a useful example of the ways in which settlement patterns, migrations, perceptions, principles of participatory democracy and planning strategies are inextricably linked together.

AIMS AND ORIENTATION OF THE PRESENT WORK

Although containing quite substantial review sections, particularly in Chapters 2, 3, 4 and 5, the present book should be regarded perhaps as much as a research-oriented volume as a straightforward textbook. Thus, whilst it is hoped that the book will be of interest to undergraduates interested in the fields of urban geography, urban and regional planning, development studies and behavioural geography, it does not endeavour to chronicle all aspects of urbanisation and planning practice in the Third World. Rather, its contribution is seen as being considerably more specific than this.

Although there is now almost a surfeit of books dealing with the general topic of Third World urbanisation, the majority focus exclusively either on the provision of housing at the intra-urban scale (Dwyer, 1975; Drakakis-Smith, 1981; Lloyd, 1979, 1980), or alternatively, provide a macro-scale and therefore highly generalised account of the political economy of urban development (Roberts, 1978; Santos, 1979; Safa, 1982). Some other texts have recently followed a basically city-by-city approach (Pacione, 1981; Brunn and Williams, 1983). Surprisingly, with the recent exception of Gilbert and Gugler (1982), there is no text that examines in detail both the inter-urban (i.e. urban systems) and intra-urban (i.e. local) dimensions of Third world urbanisation. Further, at present there is relatively little published on spatial-geographical planning procedures and policies in less developed countries, although interest in wider aspects of Third World planning is clearly on the increase (Abu-Lughod and Hay, 1977; Stretton, 1978; Taylor and Williams, 1982; Conyers, 1982). Likewise, no text presently available stresses the importance of the behavioural-cognitive approach to spatial change in the Third World and its potential as a medium for mobilising public participation in the planning process. The strong accent placed on these specific themes in the present book will hopefully ensure that it complements existing texts in the field, rather than competing directly with them.

The book draws heavily on the author's own empirical research, notably in Chapters 6 and 7, but also in connection with Chapters 3, 4 and 5. This material derives from data collected as part of a project entitled 'urban development, physical planning and environmental perceptions in the Southern Caribbean', which has involved field research and social surveys in three West Indian countries, namely Trinidad and Tobago, Barbados and St. Lucia. However, in sections which are not specifically oriented to reporting the results of this research work, every effort has been made to draw examples from other world regions. In this regard, Venezuela, Cuba, Tanzania, China and Hong Kong, for example, are used as detailed case studies. It is inevitable, however, that considerable emphasis should be placed on the area best known to the author, that is the Caribbean basin.

In Chapters 2 and 3, historical and contemporary patterns of world urbanisation, and the urbanisation process in the Third World are viewed in detail. Chapter 2 seeks to provide an overview by examining and accounting for the development of urbanism at the global scale. In particular, it seeks to establish the factors and processes that have been involved in world urbanisation and the degree to which these are linked to environmental, economic, technical, social and political transformations. Chapter 3 examines the general conditions that occur in association with contemporary urbanism in the Third World at both the intra-urban and inter-urban scales. Issues relating to spontaneous settlements, shanty towns, housing, employment provision, the role of the state, the movement of low income groups, the development and structure of urban systems and city size differentiation are all stressed. A strong emphasis is placed on the behavioural viewpoint, and the links existing between the formative processes of urbanisation and environmental perceptions-cognition are elaborated, as yet a relatively unexplored approach in Third World urban studies. The actual and potential roles of urban and regional planning strategies are examined in Chapter 4, which covers physical development, social and economic planning at scales ranging from the national to the local levels. Particular emphasis is placed on the evolution of sound national urban development strategies in Third World countries. As already suggested, relatively little geographical work has been completed on the theme of the potential role of public participation in Third World planning, despite many instances of attempted implementation. The issue is specifically tackled in Chapter 5. The possible deleterious influence of the unquestioning adoption of Western-style values and methods of planning is the principal theme, along with the importance of planner-public relationships and their political foundations. In Chapter 6, this theme is taken further. It is argued from an environmental standpoint that a major input of public involvement in planning could be effected through studies of nationals' spatial perceptions and wider environmental cognitions. Techniques which may be employed to realise this suggested potential, such as semantic differentials, personal construct theory and repertory grids, residential desirability surveys, myth map analyses, free-recall mental maps, cloze procedures and others are reviewed and assessed by means of examples. More substantial case studies of planning-related perception and cognition research are provided in Chapter 7, these being based on the author's own research in Trinidad and Barbados. The emphasis given to practical techniques and methodologies is specifically designed to make these two chapters particularly useful to planners and policy-makers actually involved in Third World environmental change and development. Conclusions and wider points of debate are finally presented in Chapter 8.

THE THIRD WORLD, THE INDIVIDUAL AND THE STATE

Finally in this introductory chapter, it is necessary to devote attention to some of the definitional and procedural problems involved in a volume of this type. The first and most pressing issue is that of the meaning and geographical definition given to the expression the 'Third World'.

The adoption of the term the Third World in this book should be regarded as no more than a convenient, albeit a potentially troublesome form of shorthand. Despite having been coined in the Cold War period of the 1950s and its entry into general parlance, heated debate has recently raged concerning usage of the term (O'Connor, 1976; Auty, 1979; Wolf-Phillips, 1979; Mountjoy, 1980; Ward, 1980). The expression originally had, in fact, a political connotation, serving to denote nations not committed to either the Western free-market bloc nor the Eastern socialist centrally-planned group of countries - that is, essentially the newly independent and largely non-aligned nations (O'Connor, 1976). Today, however, the orientation of the term has changed to a primarily economic one and it is used to signify the poor countries of the world. Frequently in the past, the descriptions 'underdeveloped', 'less-developed', 'developing' have been used, but they are all problematical in so far as they imply that development is a simple unidirectional-linear process, the foundations of which are basically economic rather than socio-cultural - in short, the terms are all basically Eurocentric (Cannon, 1975; Jackson, 1975; Brookfield, 1973; Unwin, 1983).

O'Connor, however, among others has raised several objections to use of the term 'Third World'. First, it implies the existence of distinct and enduring entities, whereas in reality there is obviously a wide variety of countries which are better regarded as making-up a continuum. Further, there is the point that in terms of original usage, countries such as Cuba, Vietnam and China must all be seen as part of the second world. In fact, in the present volume, the term Third World is employed in the current sense of referring to the poorer, or low-income, countries of the world; as a general guide, those which had per capita incomes of less than $ US 3,000 per annum in 1978, and a life expectancy lower than 70 years (Gilbert and Gugler, 1982). This largely pragmatic adoption of the phrase is reflected in that other terms are used almost interchangeably in the text. What is broadly being implied is the 'Fourth World' as described by Ward (1980), consisting of the poorest of the poor countries, notably those of South Asia, tropical Africa, the Caribbean and parts of Latin America, plus what Ward referred to as the NICs, or newly industrialising countries such as Brazil, Mexico, Singapore.

A further issue should be raised at the outset to be returned to later in the text. Given this undoubted diversity among poor countries in different parts of the world, is it desirable or even possible to generalise about extant conditions and future solutions to environmental and social problems? Certainly, it is tempting to suggest that cultural diversity alone is likely to negate any efforts to derive neat universal solutions. We must heed Gilbert and Gugler's (1982) apposite caution that "Too often policies useful in one country at one specific time are turned into panacea for all countries at all times" (p. 162), and that "instant solutions taken from the latest vogue generalization have wrought havoc in the planning field" (p. 163).

In this connection, attention should also be addressed to the emphasis which is placed on the policy and planning relevance of inventories of environmental perceptions and cognitions in this volume. Some might venture that this is merely the latest in a line of 'vogue' but 'inappropriate' generalised panacea. However, the cognitive approach, coupled with the call for the examination of the most efficacious means of establishing participatory democracy in all types and at all spatial scales of planning is, in fact, premised on this very need to explicitly recognise the country-specific and people-specific nature of sound and efficacious Third World urban planning. If planning specifically seeks to take account of local views, conditions and imperatives, it should at least be in a better position to avoid the wholesale adoption of fashionable but inappropriate exogenously derived solutions to social and environmental problems. Thus, as argued elsewhere, such methods do not constitute a panacea for urban planning problems in Third World countries, but rather a general context for their appropriate discussion in, and by the public (Potter, 1983d, 1984b, 1984c).

Following from this, it is obvious that calls for increased public involvement and participation in planning have strong political connotations. This theme will be fully explored in Chapter 5, but the major dilemmas involved are worth noting here, having been well-summarised by Stretton (1978, p.210-11) when he observed that:

> If people cannot participate effectively in planning and local government, many interests will be treated unfairly. If, on the other hand, there are strict and universal requirements for consultation and participation, planning will be slow and conservative - there will be more ways to delay or veto change, and few chances of forcing changes to occur ... Hence the paradox that direct, detailed participation in policy-making is usually fought for by radicals, but often conservatizes the systems which adopt it.

The role of the individual vis-a-vis the state in Third World urban development is thus a crucial issue. Some might contend, for example, that the perception approach suggested in this volume is essentially reductionist and further, that there is the danger that if the individual's role in the planning process is seen as omnipotent, the state may regard this as an excuse to abdicate its responsibility. A good example of this is afforded in the sphere of housing provision. Whilst many would now follow both Turner (1967, 1968) and Mangin (1967) in regarding self-help housing with approbation, it is possible that some governments may see it as a means of cheap labour reproduction and thereby of avoiding all responsibility for becoming involved in the housing market (see Ward, 1982). The present book does, in fact, accept the veracity of the argument that the solution of many urban problems has much to do with the political, economic and ideological aspirations of states, and this theme is pursued in Chapters 2 and 3. As noted by Gilbert and Gugler (1982) the rise of 'dependency' and 'neo-Marxist' interpretations of development have witnessed a shifting of analysis away from the individual, toward a class-based interpretation of society under the rise of monopoly Capitalism. It may be argued by some that this reorientation contravenes the focus of the present book, based as it is on the need to take into account the aspirations, perceptions and cognitions of individuals. An extended quotation taken from the work of a social anthropologist who has recently worked in shanty towns in Lima, Peru, will hopefully help to set this vital issue in context:

> It thus becomes important to see how men view the world within which they live, how their behaviour is governed by their image of society ... we should try to understand how the shanty-town dweller sees his society ... Some may retort that the shanty-town people have so little control over their destiny that it is not worthwhile inquiring into their views and attitudes; and indeed there is much justification in this, for economic policies are certainly not decided by them. Yet they do have a potential for action and if this potential is to be harnessed or developed their view of their world is certainly important (Lloyd, 1979, p.10-11).

Elsewhere the same author had observed that in many Third World countries, the "attitude of people in the dominant groups towards the poor tends to be negative and pejorative whatever their political persuasion ..." so that an "appraisal of the capacity of the poor to act positively is more likely to be made by those completely outside the society studied" (Lloyd, 1979, p. 49). Taken together these quotations summarise the raison d'être of the present study, that is the examination of urbanisation and urban planning from the standpoint of the views, attitudes and perceptions of denizens of Third World countries.

Chapter 2

The Course of World Urbanisation

In outline terms at least, the course of global urbanisation can be summarised succinctly by a few distractingly simple statistics. Humans have lived in settlement clusters of sufficient size to occasion use of the label "urban" for at least 5,500 years, and perhaps longer. But viewed historically, this represents only a relatively short period in relation to the 30,000 or so years that modern humans are thought to have inhabited the world. However, as will be exemplified later, once established, the spread of urban settlements and predominantly urban modes of life was at first a relatively slow process. For example, by 1800, nearly five and a half millennia after the development of the first true cities, only an estimated 3 per cent of the total world population was to be found living in towns and cities. The fact that when the change came it did so with unprecedented speed is attested by the observation that at the present time, approximately 40 per cent of the world's fast increasing total population resides in large urban agglomerations. It is projected that this proportion will rise to 50 per cent by the turn of the twentieth century and 90 per cent by the year 2050. Thus, in a mere 250 year period, the world will have been transformed from a predominantly rural mode of life to an almost exclusively urban one. Accounting for the processes underlying these ostensibly straightforward facts concerning the temporal march of world urbanisation is an altogether much more difficult task, and one that has fascinated academic social scientists for some considerable time.

This chapter seeks to provide an overview by charting the course of world urbanisation and endeavouring to chronicle the processes involved. For example, what circumstances and factors led to, or at least were correlates of, the emergence of the first cities? From these incipient stages, why did it take so long for the idea of urban living to become widely established? Subsequently, in the past 200 years, why and how has rapid urbanisation come about? It is only by seeking at least a partial understanding of these historical facets and processes of urbanisation that a realistic appreciation of today's urban processes in the developing and developed realms can be gained. This has been neatly summarised by Bird (1977, p. 27) when he comments that "A study of city origins throws a great search light on the march forward of human society". Although archaeologists, anthropologists and historians have all been involved in the study of urban origins and subsequent urban evolution, much spatial-geographical work and writing in the general field of urbanisation, has tended to be disappointingly ahistorical. Thus, Friedmann and Wulff (1976, p.10) have recently warned of:

> ... the facile generalizations of those social scientists who are inclined to think that the start of urbanization in Third World countries coincided more or less with the beginning of their own interest in the study of this process.

The need for an explicitly historical perspective in studies is clearly exemplified by the fact that urbanisation is by no means a new process in the contemporary Third World. In fact, it may be forgotten by many that the first regions of significant urban development were all located in what are today classified as Third World countries.

THE CHARACTERISTICS OF URBAN SETTLEMENTS

Any discussion, however, of the origins of urban settlements and the subsequent history of world urbanisation is somewhat premature. Quite simply, the question 'when and how did towns first develop?' begs the question 'what is a town?'. Similarly, it is also important to define what is meant by the term 'the process of urbanisation'.

The common recognition of urban status is quite clear in that it is associated with the built environment or 'concrete jungle' of bricks, mortar and tarmac. In more scholarly terms, towns and cities constitute but a part of the continuum of settlements which represents different ecological relationships between human groups and their environments. Dansereau (1978) has stressed the high density of population, high relative coverage of buildings and the transformation of wild landscapes associated with urban settlements. A basic fact, however, is that there is not and can never be a clear and universal dividing line between urban and non-urban settlements. In this sense, Glass (1976) stresses that the question 'what is urban?' is unanswered and unanswerable. However, there is some agreement as to the criteria that may be used in different local situations to define urban settlements. Frequently, of course, population size is used, but the minima vary greatly. Thus, a threshold as low as 200 people is taken as an official indication of urbanity in Denmark, but it is 2,000 in Cuba, 2,500 in Mexico and 10,000 in Senegal. Density of population is also often regarded as critical, so that India, for example, requires that an urban place consist of at least 5,000 people living at a density of not less than 1,000 persons per square mile. Another hallmark of urban status is seen as the practising of non-agricultural activities. Thus, in India, a further requirement is that at least 75 per cent of the adult male population must be engaged in non-agricultural occupations. Similarly, in Israel, any place with 2,000 or more inhabitants is accepted as urban, save where more than 33% of household heads are engaged in agriculture. In yet other situations, urban places are defined by either administrative designation or by the presence of infrastructure regarded as indicating urban status. In terms of the latter, for instance, in the Philippines a town is defined as a network of streets which possesses at least six commercial and/or recreational establishments and some of the amenities of a city, such as a town hall, church, public plaza, market place, school or hospital. It is here that the essentially arbitrary, subjective and circular nature of urban definitions can best be appreciated. Thus, a town becomes that which the state is prepared to call a town, so, as Jones (1966) observes, a city is a city is a city. This is nowhere better exemplified than in the Demographic Yearbook, where the definition of urban for Guinea is officially listed as "urban centres"! (United Nations, 1977, p. 161).

Therefore, from a statistical angle at least, an urban settlement may be regarded as a large, dense population of mainly non-agricultural workers who reside in an area with definite urban fuctions and administrative status. This accords well with Wirth's (1938, p. 8) argument in his seminal paper 'Urbanism as a way of life' that:

> For sociological purposes a city may be defined as a relatively large, dense, and permanent settlement of socially heterogeneous individuals.

The latter part of Wirth's definition introduces the idea that from a sociological-behavioural point of view, the characteristic of social heterogeneity is vital to urban status. Some would even argue along Wirthian lines that urbanisation is as much a matter of distinct values, behaviour and personality as of physical form and economic structure.

Whatever criteria are employed, however, we can agree with Tuan (1978, p.1) that "It is futile to seek a definition of the city that commands universal assent". Certainly, the fact that definitions vary from country to country and that all turn a continuum into a dichotomy as a necessary statistical evil must be recognised. The most sensible approach is to follow that taken by the United Nations, whereby the definition employed by the statistical service of each country is respected. A vital point here is of course that there is good reason for urban definitions to vary from country to country or from one time to another. For example, in a basically agricultural economy, a settlement cluster of even 7,000 may exhibit the attributes of rurality in all other respects. The common consensus, however, is that a population of 5,000 leaves hardly any doubt that we are dealing with an urban settlement and 10,000 none at all.

The most important reason for considering urban definitions is not, however, statistical but rather substantive, for in striving to recognise the essential attributes of towns and cities we are indirectly identifying the key dimensions of change that are associated with the urbanisation process. This contention is exemplified by Tuan (1978, p.1) when he argues that:

> cities are artifacts and worlds of artifice placed at varying distances from human conditions close to nature. I assume that a life close to nature is bound to food production and to the needs of survival, that it follows closely the natural rhythms of day and night and of the seasons.

Thus, Tuan regards cities as mechanisms whereby human groups have successfully distanced themselves from natural conditions by severing agricultural ties, civilizing the effects of winter and conquering the dark of night. Perhaps we should rest content with this type of broad and admittedly subjective definition of urban settlements, rather than striving for an ostensibly precise, but inevitably spurious statistical one.

THE URBANISATION PROCESS: AN OVERVIEW

It is tempting therefore to define urbanisation as the process which leads to an increasing number of people living in large, dense and basically non-agricultural settlements. Thus, it is sometimes argued that the essence of urbanisation is the increasing concentration of both people, and secondary and tertiary economic activities in space (Yeates and Garner, 1980). However, although this is undoubtedly true as a description, considered in process terms it is as circular an argument as stating that a city is a city.

In a sense, it is perhaps safest to start with a purely statistical definition of urbanisation. Thus, in demographic analysis, the word 'urbanisation' is normally used to describe the proportion of the total population of a region or nation that lives in urban settlements, however the latter are defined. Thus, the degree or level of urbanisation (U) is normally measured by an index that is some kind of ratio between the urban population (Pu) and the total population (Pt):

$$U\% = \frac{Pu}{Pt} \times 100$$

The 'process of urbanisation' as ordinarily used refers to an increase in the degree to which the proportion of a nation or region's population is to be found concentrated in towns and cities. It is thus a measure of the relative growth of urban populations within a given territory.

But there are other terms that are frequently used, by some as if they are directly interchangeable with the word 'urbanisation'. Foremost among these are 'urban growth', 'urbanism' and 'urban development'. It is necessary to clearly distinguish between these terms which are "frequently used and almost as often abused within the literature" (Potter and Potter, 1978, p.350). The term 'urban growth' is best reserved to describe either the absolute growth of urban populations, or the physical expansion of urban fabric. Confusion between the terms 'urbanisation' and 'urban growth' arises because the two processes to which they refer have tended historically to occur hand in hand. But as Davis (1972, p. 47-8) has recorded, "to treat a growth of the urban population as if it were equivalent to an equal rise of the urban fraction is to commit a blunder". Quite simply, the two need not occur together and in this sense refer to two quite separate processes. For example, urban growth may continue, whilst urbanisation actually declines, if the total population of a region grows at a faster rate than its urban population. As an illustration, data contained in the United Nations Demographic Yearbook 1976 show that the United Kingdom in 1967 had an urbanisation level of 79.1 per cent and a total urban population of some 38.05 million. However, by 1973 the urbanisation level had fallen to 77.7 per cent, but the aggregate urban population had increased marginally to 38.21 million.

The word 'urbanism' on the other hand is normally used to describe the extent to which a given population conforms to what is deemed to be an urban lifestyle. As such, it is closely associated with the sociological concept of the rural-urban dichotomy and owes much to Wirth and his claim that a distinct urban personality exists, so that behavioural change characterises the urban in-migrant. The argument that urbanisation is fundamentally a behavioural process is at once both fascinating and highly contentious. At this point, suffice it to say that given technological change, especially in both transport and communications, it is more realistic to couch any such hypothesised differences in terms of a rural-urban continuum rather than a dichotomy. However, once again, it must be stressed that urbanism should be seen as a process quite separate from urbanisation and urban growth. As an illustration of this, it is frequently asserted that current Third World urbanisation is witnessing the increasing ruralisation of the towns, so that 'urbanism' is decreasing. Similarly, in advanced countries, it is sometimes argued that the entire population, regardless of location, leads what may be regarded as basically an urban lifestyle (Pahl, 1965; Lewis and Maund, 1976). Given the precise meanings of these three terms, 'urban development' remains as a useful catchall phrase covering aspects of urbanisation, urban growth and urbanism.

The search for clear and unambiguous definitions has again provided insights into processes. Thus, urbanisation can be regarded as first, involving changes in the size, density and composition of populations in different areas. Secondly, it involves fundamental changes in the economic structure of a society and thirdly, and perhaps more contentiously, changes in human behaviour. The salience of these three sets of processes involved in urbanisation has been recognised by Lampard (1965, p. 519) in his important essay on 'historical aspects of urbanisation':

> Broadly speaking, three conceptions of urbanisation have currency in the social sciences: the behavioral, the structural, and the demographic.

The structural perspective basically focusses attention on the patterning of economic activities of entire populations. Urbanisation is thereby seen to involve the movement of people out from agricultural communities and into larger non-agricultural ones. In the simplest terms, it is argued that urbanisation can only occur when agricultural productivity is sufficiently high to enable farmers to produce over and above their requirements. This surplus affords the opportunity for occupational specialisation to occur as some sections of the society are released from the land. As a result, a distinctly differentiated ordering of occupations occurs within urban areas. This framework is the basis for all economic models of urbanisation. The approach also involves recognising that fundamental social and political changes accompany urbanisation. Quite simply, in the scenario outlined above, some form of social control and the redistribution of the surplus are required once directly non-productive occupational specialisations are included within the evolving urban society.

The demographic concept, on the other hand, postulates that urbanisation is primarily a process of population concentration, an argument briefly mentioned at the outset of this section. Thus, consideration is only given to two main variables, those of population numbers and space. Finally, under the behavioural formulation, urbanisation is viewed as an adjustment of personal behaviour and the focus is therefore placed on the individual within society. As previously discussed in defining the word urbanism, certain patterns of thought, behaviour and action are thereby considered to be characteristically urban.

It is necessary to stress, however, that these three approaches and the processes upon which they focus must not be regarded as mutually exclusive. It is unfortunate, but perhaps inevitable that work on world patterns of urbanisation employing one perspective has often involved neglect of the others. Realistically, it can only be posited that urbanisation is a complex process that simultaneously involves all three types of change, with their relative importance varying from time to time and from place to place, and inescapably, from one researcher to another.

URBAN ORIGINS AND THE EVOLUTION OF URBAN LIVING

What were the circumstances whereby humans commenced living in large, dense and permanent agglomerations associated with fundamental structural, demographic and behavioural transformations? Such changes were so far reaching that they are known as the "second" or "urban revolution", following after the first or agricultural revolution, which is also referred to as the Neolithic period (Childe, 1950, 1951; Adams, 1960).

An account of these changes must start with a cautionary note in that as reliable statistics are not available even today for many areas of the world, in examining urban origins, commentators are forced to make use of whatever credible evidence is available. Thus,certain archaeological evidence is fragmentary and contention surrounds its interpretation. Further, recent finds have continued to influence previous conclusions, thereby leading to some notable revisions of ideas. The reason for examining urban origins is to cast light on the historical and contemporary growth of cities, for as Davis (1973, p. 9) has noted, "As information has grown, the theory of how cities began is beginning to be integrated with the theory of how they operate in modern society". There are really two linked questions concerning urban origins that must be addressed: first, where and when did cities initially appear and second, what were the processes involved?

The current view is that there were at least seven regions of "primary urban generation", that is areas of apparently independent urban development (Wheatley, 1971). These areas are: (1) Mesopotamia, (2) Egypt, (3) the Indus Valley, (4) the North China Plain, (5) Mesoamerica, (6) Central Andes and (7) South-western Nigeria. Together these areas constitute the pre-industrial civilizations. The very first true cities are thought to have developed in the "cradles of civilization" or the "fertile crescent", made up by the Tigris, Euphrates and Nile riverine valleys of the Middle East. According to Sjoberg (1960, 1965), the rise of urban civilization dates from 3500-3000 BC in Mesopotamia; among the earliest cities being Eridu, Ur, Lagash, Larsa, Kish, Jemdet Nasr and Uruk. With regard to the Nile Valley, the main period of urban development dates from 3100 BC and was associated with the dynasties of the Pharaohs centred on cities such as Memphis and Thebes. These were largely mortuary cities associated with pyramids and temples and some argue that the settlements were quite small, perhaps reflecting the practice of changing the site of the capital with the ascendancy of each new Pharaoh. Initially, archaeologists such as Childe (1951) suggested that the so-called urban revolution occurred only in the ancient near East and that later developments in other areas represented a form of spatial diffusion of urban life from the single area of invention. But the more recent view is that whilst specific items of technology may have been diffused, the actual development of urban life arose independently in these other areas. Thus, cities in the Indus Valley appear to have developed around 2500 BC, two well documented examples being Mohenjo-Daro and Harappa in modern day Pakistan. Both settlements show evidence of irrigation and were well-planned on a regular rectangular grid basis. However, urban life in this area appears to have atrophied and ceased after 1500 BC. In China, urban development started around 1500 BC on the alluvial plains of the Yellow River, with Anyang and Chengchow being two such sites. In the case of meso-America, ceremonial complexes such as that at Teotihuacan in Mexico seem to date from around 1000 BC. In the Central Andes, settlements based on maize and shifting cultivation are dated from 500 BC. However, some archaeologists doubt that these were true cities and argue that they were merely ceremonial foci for low density rural populations. Finally, there is evidence that the Yoruba territories of present day Nigeria constitute a somewhat later region of primary urban generation. It is suggested that ceremonial centres may have appeared as early as the end of the first millenium AD, the first such development being at present day Ife. It is believed that other subsequent areas of urban development such as Crete, Southeast Asia and Etruria were all secondary, or derived from these seven primary ones (Wheatley, 1971).

Even this brief resume of pre-industrial urban development gives some hints as to the hypotheses that may be advanced to explain urban genesis. Widening the discussion somewhat, Carter (1977, 1983) argues that four main explanations for the initial emergence of towns may be recognised. First, hydraulic (or environmental/ecological) theses suggest that cities occurred due to the presence of a favourable physical environment, which allowed the extraction of an agricultural surplus. Economic theories on the other hand imply that the city was a product of the articulation of long-distance trade and regional market functions. Thirdly, it may be posited that towns grew for military purposes at defensive strong points. Lastly, religious theories envisage urban development occurring about the foci offered by shrines and temples.

But before turning to consider these explanations in greater detail, the precursors to the urban revolution must be considered briefly. During the Palaeolithic or Old Stone Age, a period generally equivalent to the geologists' Pleistocene, human groups relied entirely on hunting, fishing and gathering. Then, some 8,000-10,000 BC, the Neolithic Revolution commenced in the Middle East. The term is used as a shorthand to describe the period when humans first began to domesticate animals and to plant, cultivate and improve edible grasses and roots. In other words, humans started to modify the environment rather than merely adapting to it, so that the possibility of establishing permanent settlements came about (Carol, 1964). A seemingly universal attribute of Neolithic cultures was the cultivation of wheat and barley, whilst a notable development in the realm of material culture was the manufacture of pots. These, and a whole series of subsequent inventions and discoveries such as the plough, the wheeled cart, the sailing boat and the chemical processes of smelting, equipped humans for urban life (Childe, 1951).

The emphasis, however, must be placed squarely on the word 'equipped'. Naturally, the transition from food gathering to food producing and the attendant increase in food production, both per head and per unit of land must have increased greatly the carrying capacity of the land. But these developments alone cannot explain the emergence of the first urban settlements. Indeed, with regard to Egypt, Mesopotamia and the Indus Valley, there is much evidence that initially these conditions did not lead to the growing size of individual villages, but rather, resulted in the development of an increasing number of relatively small self-sufficient villages.

Some of these early trends are charted in Figure 2.1. The graph depicted in Figure 2.1a is adapted from Doxiadis and Papaioannou (1974) and shows that only after 3500 BC did urban living start to evolve. Below, in Figure 2.1b, adapted from Kolars and Nystuen (1974), changes in the physical form of human settlements from 10, 000 BC to 2000 AD are summarised. In the Prehistoric period settlements were first non-permanent and later small and isolated.

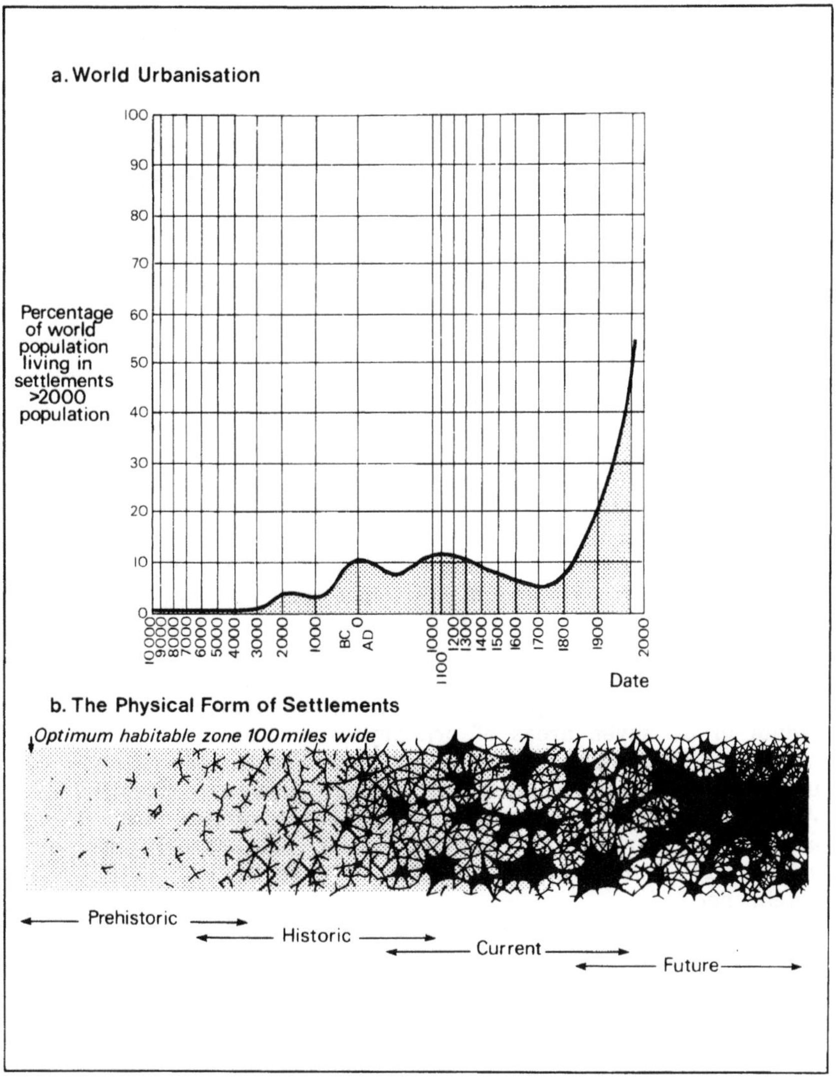

Figure 2.1: World urbanisation trends and settlement form 10 000BC - 2 000AD

The number of settlements increased with the Neolithic Revolution, but they remained small and self-sufficient until the urban revolution (Figure 2.1).

It was Childe (1950) who argued strongly that the direct link between Neolithic villages and later city life was the production of a food surplus. Hence, it was posited that the first cities were the outcome of favourable hydrological and ecological circumstances, such as the existence of periodically flooded river valleys in the 'fertile crescent' and elsewhere. However, such an ecological base can only be seen as a necessary, but not sufficient stimulus to urbanism. For example, we are forced to ask why agriculturalists would be prepared to give up some of their produce to support directly nonproductive groups and a ruling elite. Clearly, for the first occurrence of occupational specialisation and the establishment of priests, traders, craftsmen and other groups, further catalysts were required.

The movement away from an entirely ecological-hydrological explanation of the origin of cities had much to do with the work of Adams (1966). Against conventional wisdom, he argued that social factors were paramount in the creation of the first cities. More specifically, he maintained that it was social change that precipitated agricultural and technical changes and not the other way around. This argument has since gained general currency and has been developed by a number of different scholars (Johnston, 1977). Thus, both the military and religious theories of urban genesis argue that an elite group emerged and extracted the food surplus as tribute in return for either physical protection or spiritual security respectively. Such ideas do, of course, suggest reasons why the 'economically exploited group' might have been prepared to give up some of its surplus product. Further, it is salient that the settlements in all seven areas of primary urban generation were indeed associated with religio-military temples and forts. The role of religion in urban generation has been most strongly argued by Wheatley (1971). Warfare and military explanations of urban origins seem credible, although as Carter (1977) concludes, they are probably best viewed as factors that intensified urbanism once it had actually been established.

These socially oriented theories do provide suggestions as to how egalitarian societies became both occupationally specialised and socially differentiated once the potential for surplus production had arisen. Some other commentators have advanced the economic thesis that trade and marketing functions were the prime progenitors of the first cities, Jacobs (1969) being notable among them. But as Carter (1977) concludes, as with military explanations, although trade was clearly an important element in the intensification of early cities, it is probably best regarded as a by-product of urbanism, not as its creator. This view is also exemplified by Johnston (1980) when he argues that part of the economic benefits of trade were used as further rewards for the 'exploited' segment of society if and when scepticism grew about the benefits of the spiritual or temporal security afforded by urban living.

Although we can thereby logically assert that the prime factors in urban genesis were social, it seems sensible to conclude that no single causal factors were operative. This viewpoint is well-demonstrated by Sjoberg (1960, 1965), who argues that the course of pre-industrial development was intimately related to three preconditions: a favourable ecological base, allied to a relatively advanced technology, in turn linked with a new type of social organisation in the form of the emergence of political and economic power. Such impetuses would inevitably become cumulative and mutually reinforcing and would subsequently be strengthened by military and trade functions. Similarly, it can be ventured that the self-sustaining nature of the urban process would have been bolstered by increasing social heterogeneity. Once the advantages of wealth, power and status had become vested with particular groups, it is likely that they would seek to preserve and strengthen the existing system. Presumably, the continued development of urban living was smoothest where the quality of life of the economically exploited group increased in absolute terms, despite any mounting relative disparity. Thus, the relationship between the rulers and the workers was at once both exploitative and protective.

Closely linked with these ideas on cumulative causation is the thesis that continued urban development was a response to the psychological factors of human perception and behaviour. For example, in his book Centrality and Cities, Bird (1977) argues that a fixed point or centre was necessary for human orientation, especially in connection with religion. He continues by arguing that once there are central villages and non-central villages, even if only by virtue of their order of establishment, other factors become effective for:

> ... once a location acquires some centrality because it embodies conscious or subconscious memories of a group, it is at once perceived or held to be more central than it is by people who do not live in that central place.

In this connection, Bird cites Helson's (1964) adaptation level theory which stresses the subjective and relative bases of perceived assessments of objects. Thus, almost by definition, once designated as such, centres become more central. Bird also talks of what he describes as the "proclaimed impulse" for, if a city is an agreed centre, the agreement must be proclaimed back to all those concerned. Certainly there is no denying that a strong link exists between nodal centres and human perceptions of space (Haggett, 1965; Bird, 1977; Potter, 1982). One is almost tempted to the view that as centres did not exist, it was necessary for humans to create them.

The intention in the above account has been to stress process explanations, for questions concerning the precise descriptive details of urban origins abound. For example, the issue of the size of such centres is frequently raised and it is asked whether these settlements were truly urban in character. In this respect, estimates do vary;

with regard to Mesopotamian cities, for example, Sjoberg talks of populations from 5,000 to 10,000, Childe of 7,000 to 20,000. However, Childe surveyed the features of all known cities of antiquity and concluded that they were all truly urban. This was not just a reflection of their size, but related to their density, occupational specialisation, class structure, their role in the collection and storage of a surplus, their monumental structure and artistic achievements. Notably, such criteria can be broadly aligned with those employed today to designate urban status. Further, there is as yet no universal acceptance of the designation of the seven areas mentioned as being the only possible sites of primary urban generation. For example, archaeological excavations on the Anatolian plateau in Turkey at Catal Huyuk show evidence of ceremonial shrines and graveyards dating back to 6500 BC. Thus, it has been argued that prior to 6000 BC, some three or four millenia before the famous cities of Mesopotamia, Catal Huyuk was a fully-fledged and thriving urban centre (Mellaart, 1964, 1967).

However, we may conclude that the unravelling of the complex of forces associated with urban genesis is as important as the precise historical details involved. Clearly, the causal factors were extremely complex and the overall process one of slow evolution to an urban base. Thus, in temporal terms at least, we can express some sympathy with Friedmann and Wulff's (1876, p. 7) observation that:

> ... compact cities are not just suddenly invented like New Towns but represent the culmination of a long period of evolutionary change in the spatial organization of society. To call this 'revolution' as Gordon Childe has done (1950;1951,Chapter 7) is simply verbal sorcery.

HISTORICAL PATTERNS AND PROCESSES OF URBANISATION

It is clearly beyond the scope of this chapter to provide a fully detailed account of the evolution of urbanisation from the so called urban revolution right through to the modern period, although a broad outline of trends is attempted. It can, however, be summarised that for many centuries after the original development of cities, the overall level of world urbanisation rose only very slowly and the urban areas that did exist were relatively small and, in effect, represented local phenomena (see Figure 2.1).

After the urban revolution, the subsequent history of urbanisation in the Middle East and Europe was complex and probably involved elements of both independent invention and spatial diffusion. However, it is generally recognised that it was with the rise of the great empires of the Greeks and the Romans and to a lesser extent the Muslims that urban life spread widely in Europe. The primary diffusion of the city that occurred under the Greeks was intensified under the Roman Empire in the first three centuries AD (Pounds, 1969). However, city life declined with the fall of the Roman Empire in the fifth century AD. In fact, it was not until the tenth and eleventh centuries that city development started to become important once more and it was the twelfth and thirteenth centuries that saw the rise of the Medieval city based on increasing local and long-distance trade. By the end of the Middle Ages, most of the major European cities of today were already in existence.

These developments and subsequent ones can best be summarised by considering Johnston's (1980) argument that the history of urbanisation reflected fundamental changes in social organisation. More particularly, Johnston has recognised five broad changes in the structure of society. In fact, the transition from the first to second stages has already been discussed at length above. The initial stage is that of the 'Reciprocal Society', representing the first small societies that were of limited territorial extent. These groupings were fully egalitarian and were based on consensus and democratic forms of decision-making. In them, exchange was based on reciprocal principles and no power or elite group existed. Such societies were pre-urban in all respects. Where productivity increased and a surplus product was first stored, the egalitarian structure broke down, being replaced by a rank-ordering of the society's members. At the same time, goods and labour were redistributed among members, so that the stage is referred to as that of 'Rank Redistribution'. This is of course synonymous with the urban revolution and the first emergence of religious and military power. Subsequently, trade was required in order to enhance economic growth and this was met by small scale territorial expansion. This was the precursor to the third societal stage identified by Johnston, the 'Money-Exchange System', for trade necessitated the establishment of a common exchange or money system. At this point, monetary rents could start to be demanded by the owners of land for its use, rather than payment in either goods or labour. Johnston argues that this led to a clear threefold division of society into an elite, a subject group and an intermediate set of administrators and military personnel.

However, in regard to global organisation, it was with the emergence of the fourth and fifth societal stages, those of 'Mercantilism' and 'Capitalism' respectively that major developments occurred. The main hallmark of Mercantile Societies was the expansion of trade and an outcome was the establishment of merchants whose job it was to articulate trade by buying and selling commodities, mainly over long distances. As the role of merchants demands that they buy goods first and then sell them, this required that they had capital before starting operations. Surplus capital was now held by landlords who could offer loans for merchants to establish themselves, so that a close interdependence emerged between these two groups. Thus, some of the surplus previously resting entirely in the hands of the ruling elite was now shared with the merchant group. These early developments of mercantile societies occurred in Europe around the fourteenth-fifteenth centuries and were associated with much more complex and interlinked settlement and economic patterns; as witnessed by the development of periodic or travelling marketing systems and fairs in Britain during the Medieval period (Fox, 1970).

The main development, however, came with colonialism, for continued mercantile growth required greater land resources. Historically, this occurred during the sixteenth to eighteenth centuries, first by means of trading expeditions, and then by distant colonialism, a trans-oceanic version of local colonial expansion, mainly involving areas of low socio-economic development. If an established population already existed, colonial power could be enforced by administrative elites on tours of duty. In extreme cases, local populations were entirely destroyed, as in many Caribbean islands, where they were replaced by African slaves and indentured labourers (Lowenthal, 1972; Gilbert and Gugler, 1982). In formerly unoccupied lands, a colony could be established and under such conditions, ports came to dominate the urban system, acting as gateways to the new colony. Frequently, a coastal-linear settlement pattern emerged, often characterised by the beginnings of strong primacy and spatial polarisation. Thus, the settlement systems of both the colony and the colonial power developed symbiotically, as is shown in Vance's (1970) mercantile model of settlement evolution (see Chapter 3).

Whilst some stress that this process of European colonisation of the traditional world by the Spanish, Portuguese, British, Germans, Belgians, French, Dutch and Italians represented the spread of economic growth and development, others would argue that it represented a form of exploitation and expropriation. Certainly the process involved the flow of surplus value, often hailing from the production of a staple agricultural product such as sugar, first from rural areas to the colony's gateway primate city and thence to the major cities of the colonial power. These developments witnessed the increasing economic specialisation of countries and the international division of labour. What is undeniable, therefore, is that from this time onward, the economic evolution of the developed and developing world has been closely interrelated. Thus, Roberts (1978, p. 1) states that the:

> ... form of economic expansion that occurred in Europe from the sixteenth century onwards has shaped the patterns of growth in under-developed areas of the world.

The fifth historical societal stage identified by Johnston, capitalism, represents a complex development, but its principal outcome was that the scale of employment and production expanded dramatically with the development of the factory system. However, Johnston (1980, p. 37) comments that "although industrial capitalism is inextricably bound up with the factory system" its most important feature was "the alienation of labour power which enforces the working for wages". Thus, those with capital gained control over the means of production, whilst workers could only sell their labour. This first stage is known as industrial capitalism; as a subsequent stage, late or monopoly capitalism occurs when the potential market for goods becomes saturated. Under such circumstances, firms can only expand sales by capturing more of the existing market, so that successive mergers lead to a process of increasing industrial concentration.

It is the rise of industrial capitalism in the eighteenth century that brings us to the onset of the modern period of world urbanisation. Thus, the exponential growth of world urbanisation started only around 1800 (Figure 2.1a). As previously noted, at this point, only 3 per cent of world population was to be found living in urban settlements. The great growth in urbanisation came only with the Industrial Revolution in the late eighteenth and early nineteenth centuries in Europe. Davis (1965) has stressed the importance of the enormous growth in productivity which came with the use of inanimate sources of energy and machinery. Similarly, Sjoberg (1960) has emphasised the importance of the scientific revolution as the basis for the industrial revolution. Thus, he notes that the advent of industrialisation brought large-scale production, improvements in agricultural implements and farming techniques, improvements in food preservation and better transport and communications. All of these technical developments in the initial stage of the Industrial Revolution up to 1830 contributed to both increasing productivity in agriculture and the parallel development of industry. The steam engine was undoubtedly the key invention, leading to the factory system, whereupon mechanisation and mass production took hold. Taken together these fundamental changes allowed people to congregate in the industrial city.

During the eighteenth and nineteenth centuries economic development was still essentially an interdependent process, involving both industrial and non-industrial countries (Brookfield, 1975). This close relationship is expressed by dependency theory, which is also sometimes referred to as neo-colonialism or dependent capitalism. The approach was largely derived from the Latin American context and suggested that the development of the South American continent had been dictated by its integration into the capitalistic mode of production (Frank, 1969). The theory has clear implications for urban development in the form of "dependent urbanisation" (Castells, 1977; Gilbert and Gugler, 1982), the basic components of which have been stated by Friedmann and Wulff (1976). They summarise that it basically involves the notion of developed countries establishing outposts in Third World cities for three related reasons. First, to extract a surplus by way of primary products; secondly, to expand the market for goods developed under advanced monopoly capitalism and third, to ensure the continued stability of indigenous political systems that will most willingly support the capitalist system.

Whether or not one agrees with the central tenets of dependency theory, it is indisputable that the close interrelation existing between developed and underdeveloped countries from the sixteenth century onwards has had a clear bearing on patterns of world urbanisation. The epochs of mercantilism and industrial capitalism set the stage for the rapid urbanisation trends that were shortly to follow. Johnston (1980) observes that early industrial capitalism led to the accentuation of the primacy of gateway cities in the developed realm and to the bolstering of the divisions existing between levels of the settlement hierarchy. At the same time, the seeds for spatially polarised development and rapid urbanisation were being laid in the countries of the Third World, so that;

> ... it is possible to argue that without the intrusion of industrial capitalism and imperialism some Third World societies would still lack major cities. In major parts of America and Africa urban development was superimposed by capitalism on essentially rural societies (Gilbert and Gugler, 1982, p.13).

CONTEMPORARY PATTERNS AND PROCESSES OF URBANISATION

The rapid urbanisation that occurred in the developed countries of Western Europe and North America during the late eighteenth and early nineteenth centuries was associated with a process of industrialisation. During this period, a rapid increase in demand for labour occurred in the towns, whilst at the same time, technical developments in agriculture allowed for a declining rural population. Thus, a steady stream of migration occurred from the rural to urban areas. Health and sanitary conditions in the towns were very poor. As a result, there was little natural increase in urban populations: cities were the death traps portrayed in the novels of Dickens.

Thus, the course of rapid urbanisation in the developed world followed what may be described as a smooth progression, matching gradual changes in industrialisation and demographic structure. This is probably best discussed in relation to Kingsley Davis' (1965) classic "cycle of urbanisation" model. By examining the experience of developed countries over the past 150 years, especially Britain, Davis argued that the progress of urbanisation in a given territory can best be represented by a curve in the form of an attenuated "S". Such a curve, which is characteristic of many growth processes over time, is described as a logistic curve. The cycle of urbanisation envisaged for developed countries is shown in Figure 2.2 at bottom left. For example, in the case of the United Kingdom, the swiftest rise in urbanisation came between 1811 and 1851. As the level of urbanisation climbed over 50 per cent, the curve began to flatten. The curve levels out completely or even declines after reaching 75 per cent. Thus, urbanisation is regarded as a finite process in that a saturation or ceiling point is assumed to exist. Thus, Davis was effectively recognising the occurrence of suburbanisation of population or counter-urbanisation in the later stages of the cycle in developed countries. This relatively gradual process of urbanisation over a 150 year period was paralleled by slow demographic change in developed countries. Thus, the cycle of urbanisation and demographic transition can be compared directly for developed countries (Figure 2.2). At the beginning of the nineteenth century, urbanisation was low and so was the rate of population growth. The period of most rapid urbanisation came in the early part of the nineteenth century, a time when the death rate dropped sharply and the population expanded quickly. As Great Britain moved into the late expanding stage of the demographic transition, urbanisation had

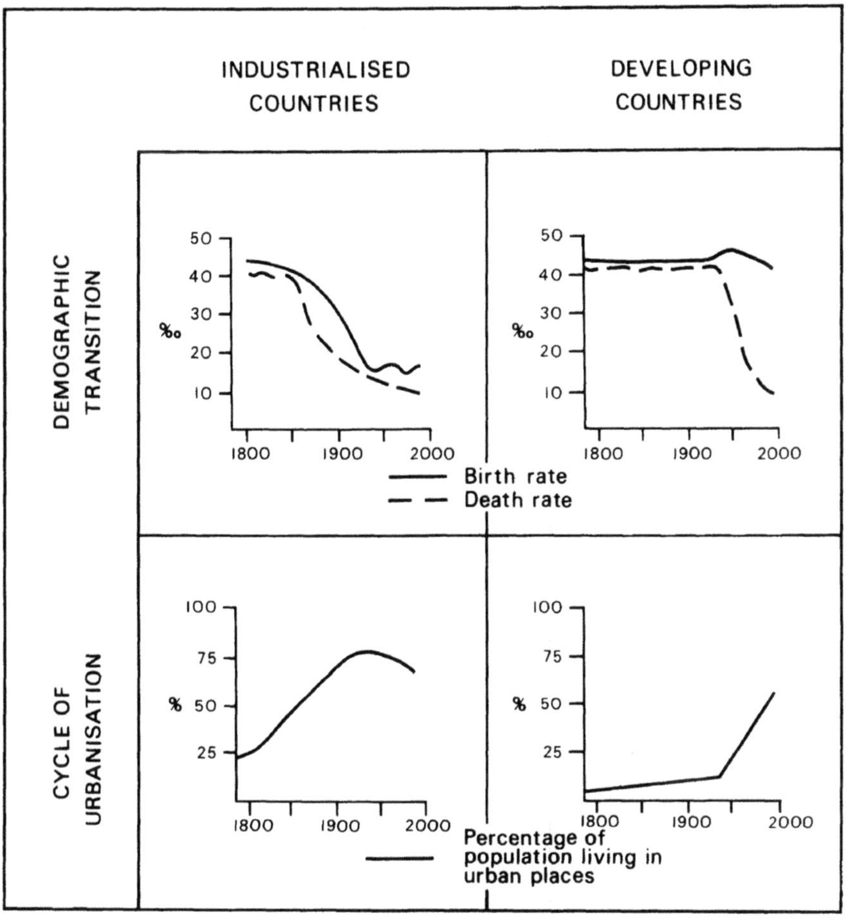

Figure 2.2: The demographic transition and the cycle of urbanisation

slowed down noticeably and, on entry into the low fluctuating demographic stage, urbanisation had largely run its course. Thus, the process of urbanisation was a gradual one and Davis notes that it took England and Wales 79 years to move from a situation where 10 per cent of its population was urban, to one where 30 per cent lived in towns and cites. In this regard, Davis has observed a general trend, for the later each country has become industrialised, the faster it has urbanised. Hence, the same change in urbanisation took 66 years in America, 36 years in Japan, but only 26 years in Australia.

The rapid rate of urbanisation that currently characterises the Third World is the product of the post-war period. Thus, the majority of less developed countries are not following the cycle of urbanisation associated with early economic growth. They are urbanising more rapidly than the industrial nations did in their heydays. Davis (1965), for example, calculated that for the underdeveloped countries for which data were available in the 1940s and 1950s, the average annual gain in urban population was 4.5 per cent. In comparison, for nine European countries during their periods of fastest urbanisation, the average annual gain per year had been 2.1 per cent. Thus, as shown in the graph at bottom right in Figure 2.2, most less developed countries displayed a sharp and almost immediate urbanisation trend after the Second World War.

Just as the developing countries have telescoped the cycle of urbanisation, so they have foreshortened the demographic transition. Thus, a major feature of such countries has been a very rapid growth of total population. In the period 1970–75, for example, the population of the world grew at an overall rate of 1.9 per cent per annum. However, this overall rate comprised a growth of 2.4 per cent per annum in the less developed realm and one of 0.9 per cent per annum in the developed world. This reflects another important contrast between contemporary Third World urbanisation and that of the developed world in the past, and that is the huge absolute numbers involved in the Third World. This is mainly the outcome of a sharp decline in the death rate due to better medical and health care facilities in the post-war period whilst birth rates have remained at traditionally high levels (Figure 2.2, top right). Hence, there is massive growth of total population in many Third World countries, both rural and urban.

Thus, even if there were no relative shift in the proportion of many Third World countries' populations living in urban areas (that is urbanisation _per se_), rapid urban growth would still occur due to the very high rates of natural increase of population involved. As health and social welfare facilities generally tend to be much better in the cities than in the rural areas, Third World cities exemplify _par excellence_ the combination of pre-industrial fertility with post-industrial mortality. Thus, the contemporary Third World city has the highest rate of natural increase of population ever found in urban areas (Dwyer, 1975), a fact that is frequently understressed by commentators (Davis, 1965). For example, in Caracas, Venezuela, from 1960 to 1966, 52 per cent of the population increase was accounted for by in-migrants; and an identical figure was recorded for Bombay,

India for the period 1951-61. MacGregor and Valverde (1975) in a study of Mexican dryland cities note that 31 per cent of the total population of Monterrey and 48 per cent of Tijuana are in-migrants. In fact, very often, natural increase and in-migration contributed in broadly equal proportions to the total growth of urban populations.

But relative rural-to-urban shifts in population are occurring too, so that Third World cities are growing at extremely rapid rates. Generally, the percentage annual growth of urban population is running in excess of 6 per cent per annum, and frequently over 10 per cent. At the latter rate, the absolute number of urban dwellers will double every seven years, and at the former every ten years. During the decade 1960-70, the percentage annual growth of urban population for Malawi was 10.1, Tanzania 8.6, Malaysia 5.9, Nigeria 6.0, Venezuela 5.6 and Colombia 5.0, for example.

Why has rapid urbanisation become so characteristic of contemporary Third World countries? Customarily, this is explained by a combination of 'push' and 'pull' factors. With regard to pull factors, the enhanced health care facilities in cities has already been mentioned. After the Second World War, many Third World countries were encouraged to seek economic growth and prosperity by means of industrialisation especially programmes based on import substitution. Historically, markets and infrastructure have normally been centred on the primate cities of less developed countries. Thus, post-war development planning and the legacies of colonial and capitalist penetration have led to the increasing concentration of industrial job opportunities in the major urban areas. Thus, Roberts (1978) argues that it is the form of industrialisation and not inertia or traditionalism that is holding back agrarian development in much of the Third World. Others such as Lipton (1977) have argued that Third World poverty is largely the product of this urban bias in world development. The industries established were often of the 'final touch' type and hence provide relatively few jobs. However, it is undeniable that the jobs that are available offer relatively high wages in comparison with rural incomes. Hence, cities in the Third World have come to offer the hopes of employment and high wages. Added to this is the growth of tertiary jobs, especially in the informal sector of the economy (Santos, 1979), which also offer the chance of gainful employment. Further, some would argue that in the past, inappropriate curricula in schools have oriented children toward white collar urban-based jobs (Mabogunje, 1980). Hence, the Third World city is essentially a set of perceived opportunities, so that "perhaps they do no more than promise the hope for work and settlement; but to obtain even crumbs, one must be near the table" (Jones and Eyles, 1977, p. 210). Some may observe that the overt poverty, crowding and unemployment of Third World cities should act as deterrents to migrants, but this is to totally ignore the massive unemployment and poverty which frequently exist in rural areas (Gilbert and Gugler, 1982). Such conditions represent the push factors stimulating urban growth.

Thus, rates of urbanisation in the world are increasingly becoming inversely related to levels of economic development, altering in dramatic fashion a relationship that has held previously for some five and a half millenia. It is the less developed countries that account for the extremely rapid current rate of world urbanisation, whilst the industrialised nations are exhibiting declining rates of urbanisation, and some even declining levels. Dwyer (1975, p.13) has observed that:

> In all probability we have reached the end of an era of association of urbanization with Western style industrialization and socio-economic characteristics.

But if a map showing the spatial patterning of urbanisation levels in the world today is examined (Figure 2.3) the point is clearly made that, overall, levels of urbanisation are still closely related to levels of economic development. This relationship is also substantiated if urbanisation levels are graphed against the gross national product per capita of individual countries (Figure 2.4). Thus, it is the developed countries that still show the highest levels of urbanisation. United Nations' (1969) estimates suggested that by 1980, the world as a whole would be 46 per cent urban, this varying between 70 per cent in the developed countries and 32 per cent in the less developed countries. North America is the most completely urbanised continent with an estimated 81 per cent urbanisation level in 1980. It is followed by Oceania (75 per cent), the Soviet Union (68 per cent) and Europe (65 per cent). The countries of the less developed areas of the world stand out by virtue of their as yet relatively low urban proportions: South Asia (25 per cent), Africa (28 per cent), and East Asia (31 per cent). Latin America on the other hand is highly urbanised already (60 per cent). Clearly, however, statistics and maps of relative urban proportions must not blind us to the absolute numbers and rates of increase involved in different world regions at present and in the future.

ECUMENOPOLIS: THE GLOBAL PROSPECT?

The rapid progress of urban development in the Third World has meant that the world is at present experiencing its highest ever rate of urbanisation. As Davis (1972) observes, the speed of this process is such that it cannot have existed long in the past, nor can it endure long in the future. During the decades 1950-60 and 1960-70, world urban population increased by 16.8 and 16.9 per cent respectively. Quite simply, if these rates were to continue unabated, then the entire world population would be totally urban by the year 2031.

More realistic United Nations estimates of world urbanisation levels of 50 per cent by 2000 and 90 per cent by 2050 have already been cited in this chapter. Detailed estimates of urbanisation levels between 1920 and 2000 for the world and its major regions are shown in Table 2.1. Many of the trends described in the second half of the present chapter are statistically summarised by this table. In

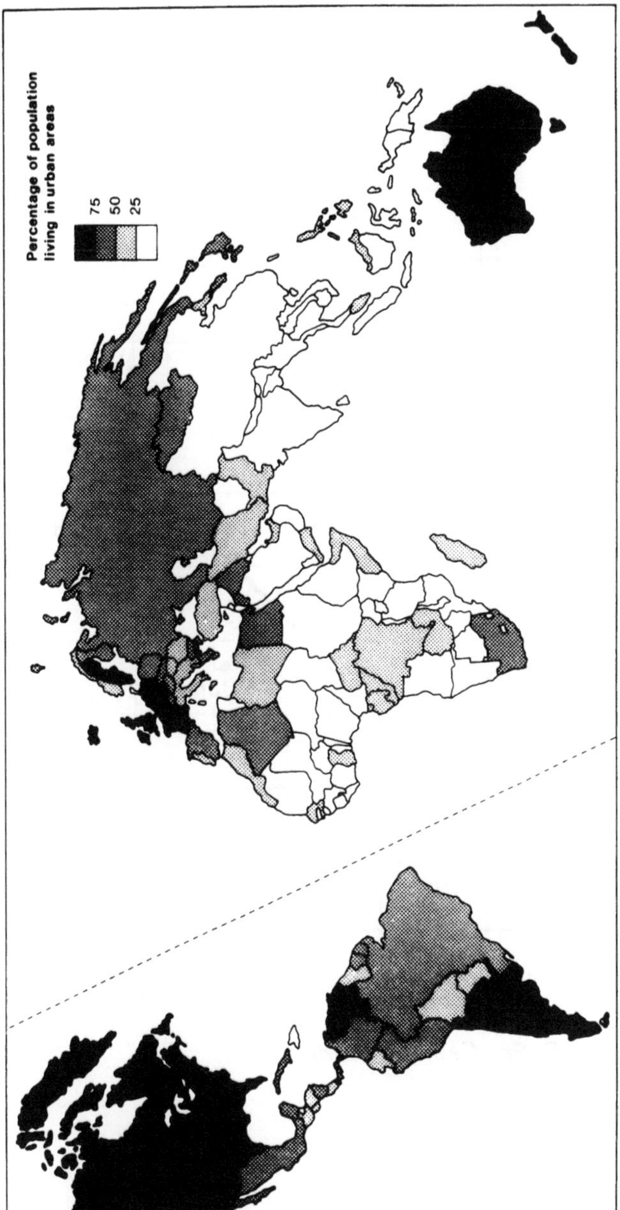

Figure 2.3: The contemporary pattern of world urbanisation (Source: Brunn and Williams, 1983)

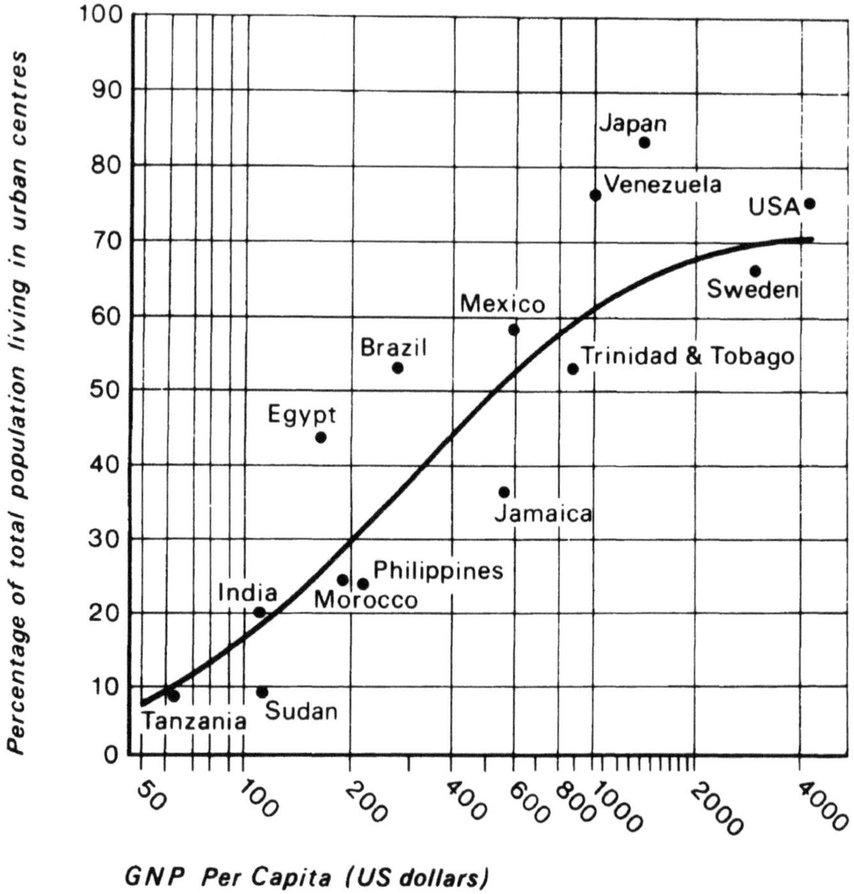

Figure 2.4: Relationship between urbanisation and gross national product

Table 2.1: Estimated level of urbanisation 1920-2000 for the world and its major areas

Area	Percentage of total population urban:				
	1920	1940	1960	1980	2000
WORLD	19	25	33	46	51
Europe	46	53	58	65	71
North America	52	59	70	81	87
Soviet Union	15	32	49	68	85
Oceania	47	53	64	75	80
More Developed Areas	40	48	59	70	80
East Asia	9	13	23	31	40
South Asia	9	12	18	25	35
Latin America	22	31	49	60	80
Africa	7	11	18	28	39
Less Developed Areas	10	14	23	32	43

Source: United Nations (1969, p. 73)

Table 2.2: Estimated total urban population 1960-2000 for the world and its major areas

Area	Total urban population in millions:				
	1960	1970	1980	1990	2000
WORLD	990	1,330	1,780	2,370	3,090
Europe	245	280	310	345	375
North America	140	175	215	255	310
Soviet Union	105	140	190	250	300
Oceania	10	15	20	20	25
More Developed Areas	500	610	735	870	1,010
East Asia	180	245	325	420	520
South Asia	155	235	350	525	750
Latin America	105	160	245	360	510
Africa	50	80	125	195	300
Less Developed Areas	490	720	1,045	1,500	2,080

Source: United Nations (1969, p. 71)

particular, the low level of urbanisation pertaining to the less developed world regions prior to the Second World War is clearly discernible, as is its subsequent rapid increase. But once again it must be emphasised that the true magnitude of the current and future urbanisation process is only revealed if we examine the absolute numbers involved, as shown in Table 2.2. These projections of urbanisation are based on an estimated total world population of 6,112 millions by the year 2000 (United Nations, 1969). Of this, 51 per cent or some 3,090 millions will be found residing in urban settlements. Further, it is estimated that an 83 per cent urbanisation level will apply to the developed world in 2000, made up of some 1,010 million urban residents. However, whilst the overall level of urbanisation of the less developed world is estimated at 43 per cent, this represents a total urban-based population of 2,080 millions, double that residing in developed world cities. Indeed, South Asia alone in 2000 will house more urban dwellers than North America and Europe put together. It is sobering to reflect that the total urban population living in Third World cities is projected to double from 1,045 million in 1980 to 2,080 million only twenty years later in 2000 (Table 2.2). Such is the scope and challenge of the current Third World urbanisation process. These global statistics stagger by their sheer enormity and it must be remembered that they mean some 1,000 million extra urban dwellers in Third World cities between now and the end of the twentieth century. Hence, the need for homes, jobs, schools, hospitals and for sound planning and administration.

What will be the end product of this rapid contemporary process of world urbanisation? Doxiadis and Papaioannou (1974) have forecast the development of "Ecumenopolis" or a single World City by the middle of the twenty-first century (Figure 2.5). It is certainly the case that just as the early stages of world urbanisation witnessed new trends in urban form, so the physical configuration or layout of urbanised areas has been changing over the past twenty five years or so. In particular, as is shown in Figure 2.1a, during this period major urban areas have started to coalesce to form massive polynuclear urban systems. On the northeastern seaboard of the United States, for example, Gottmann (1957) recognised what he referred to as "Megalopolis" in the 1950s. This comprised a whole chain of metropolitan areas stretching the 600 miles between Boston and Washington and which housed 30 million people. More recently, Gottmann (1978) has recognised six such megalopolitan systems around the world, with populations in excess of 25 million. These include two in the less developed world: in Brazil based on Rio de Janeiro/Sao Paulo and in China, based on Shanghai. If a 10 million population threshold is taken, many other megalopolitan systems can be recognised, including those centred on Buenos Aires, Calcutta, Bombay, Cairo and Mexico City in the Third World. Doxiadis and Papaioannou's suggestion is that these complexes will eventually become linked in a chain-like form giving a series of interrelated megalopolitan systems (Doxiadis, 1967). This is shown in Figure 2.5 and it is envisaged that high density urban lineaments will eventually connect the urban cores within Latin America, Africa and

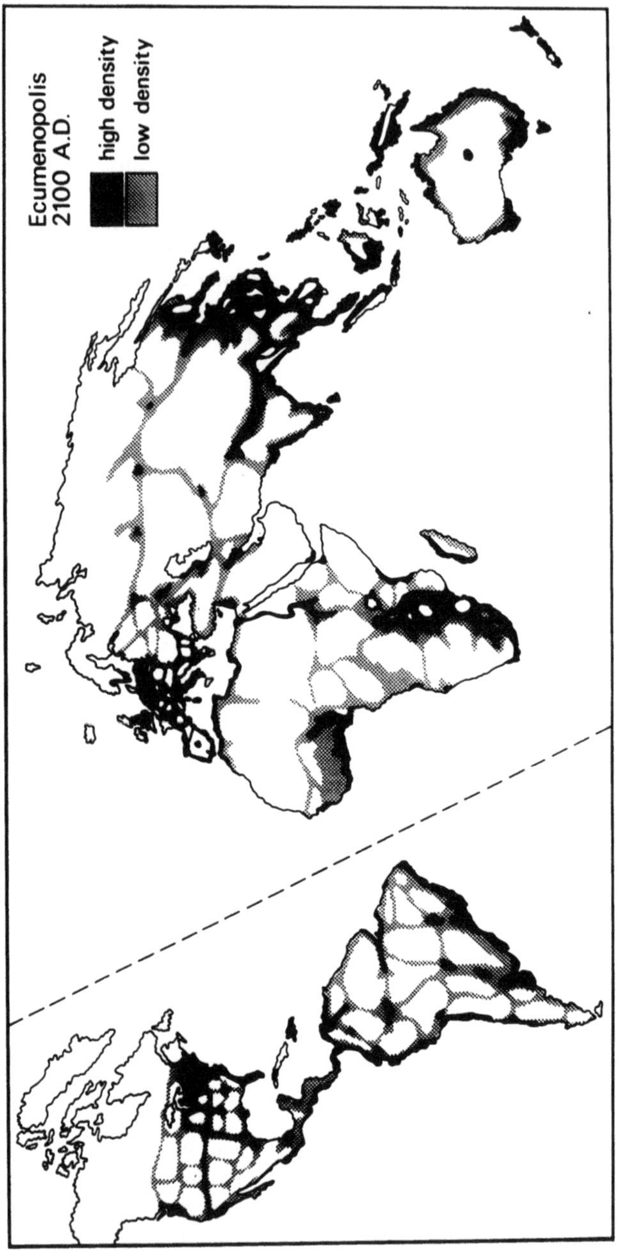

Figure 2.5: Ecumenopolis-the global prospect?
(Source: Doxiadis and Papaioannou ,1974)

Asia. The term 'ecumenopolis' or world city is used to indicate a functionally integrated urban whole and is obviously not intended to imply complete physical coverage of the world's land surface. In order to produce their forecast, Doxiadis and Papaioannou assumed world populations of 6,430 millions by 2000 and 9,600 millions by 2050, of which 71.5 per cent would be living in cities. They then looked at the habitability of different areas of the globe in 2100 according to climate, elevation, water supply and thereby ascertained areas for possible future urban development. These were then used to define a theoretical configuration of global urban centres and growth axes. The work subsequently projected the likely settlement pattern assuming world populations of 20,000 millions by 2100 and 50,000 millions by 2200. The authors sub-title their book "the inevitable city of the future" and although some may remain unconvinced of this inevitability, the historical course of world urbanisation and its present speed certainly make such predictions seem increasingly plausible.

CITIES, SURPLUSES, ELITES AND DEVELOPMENT: FINAL PERSPECTIVES

Cities and urban modes of living are here to stay and the ways in which individual cities and systems of settlements evolve over the next 70 years and beyond will be crucial variables in the process of global development. One important lesson that can be drawn from the review of urbanisation provided in the present chapter is that the character and form of urbanisation and urbanism cannot be divorced from the structure and development of societies as a whole. It follows, therefore, that structurally speaking, changes in societal organisation are needed to effect urban change.

We are witnessing the final stages of a process of rapid urbanisation that started some 5,500 years ago, but which has only become a truly global phenomenon during the past 150 years. Historically, it has been shown that the development of cities has been based on changes in social organisation, so that increasing occupational specialisation and social differentiation are direct correlates of the process of urbanisation. Thus, some writers argue most strongly that conflicts between social groups have had a direct bearing on the pattern and form of world urbanisation (Roberts, 1978; Gilbert and Gugler, 1982). Certainly, it is undoubtedly the case that cities have always served and been associated with elite groups, whether a religious and military group as in Mesopotamia, an expatriate elite as in the mercanitle-colonial city or the corporate interests of present-day multinational companies.

Associated with this is the fact that cities have been intimately connected with the generation of a surplus, indeed it can be argued that without a social surplus product cities cannot exist. One's view of the processes involved will be coloured by political predispositions. The issue has been treated, for example, from a Marxist viewpoint by Harvey (1973), who follows the traditional view that "Cities are formed through the geographic concentration of a social surplus product" (Harvey, 1973, p. 216). It can be argued that a surplus has two possible forms. In the first case it can be an amount of material product that is set aside to provide for improvements in human welfare. This may be regarded as the communist form of a surplus. A surplus is needed for advancing civilisation in a socialist society, but it does not necessarily need to have a class-based character, rather it can be used for socially defined purposes. Thus, Harvey notes that whilst a surplus is required in a socialist society, there are no a priori reasons why it should be spatially concentrated. However, Harvey goes on to argue that as investment may well be more efficiently deployed in concentrated form due to the operation of economies of scale and agglomeration, some type of urbanisation may well be acceptable. However, in so far as the surplus is distributed for the use of the population in general, it is in this regard that geographical concentration should be avoided. As an example of this, Harvey cites the instance of Cuba where a conscious attempt has been made to disperse medical facilities away from Havana. The second kind of surplus is described as an estranged or alienated version of the first. Simply, it is a quantity of material resources that is appropriated for one segment of society at the expense of another. Harvey argues that it is this form of surplus that has characterised Neolithic urbanism, rural-urban flows and contemporary urbanism under capitalism.

Under a capitalist formulation, it is posited that surplus value in the form of profits is invested in realising further profits. Hence, Harvey (1973) points to what he sees as the monumental architecture, conspicuous consumption and need-creation associated with the modern city. Some argue that under capitalism, urbanisation is legitimised by its contribution to gross national product (Gilbert and Gugler, 1982). Broadly, it is asserted that investment should be concentrated in the most profitable areas in order to maximise growth. Thus, advocates of the free-market system would argue that surpluses in the form of profits signal the demands of consumers to producers. It is a simple step to state that the city is a mechanism of economic efficiency and growth, and thus eventually, increasing prosperity for all. This thesis relates directly to the traditional idea that cities are intimately associated with economic development (see, for example, Gottmann, 1983). Certainly, the city has often been regarded as a centre of mixing, discovery and innovation, benefiting from scale and agglomeration economies. This is epitomised by Lampard's (1955, p. 92) statement that "The modern city is a mode of social organization which furthers efficiency in economic activity".

These two largely contrasting political views on the role of the city are in agreement in so far as they both recognise that cities are associated with the generation of wealth. Where they differ is in their assessment of the acceptability of the final outcome. This really boils down to an evaluation of the social and economic costs and benefits of urban agglomeration. This issue was clearly raised by Hoselitz (1955) who suggested that cities can be classified as either "generative" or "parasitic", according to whether or not they stimulate economic growth in their wider regions. Hoselitz suggested that most early colonial settlements were parasitic, that is they extracted surplus value from surrounding regions. But he maintained that they later generally became generative by widening out economic development over a more extensive area and an increasing proportion of the population. This is a critical issue and one that pervades any discussion of Third World urbanisation and planning. A further point that should be acknowledged is that the relationship between urbanisation and economic development is witnessing a radical transformation at the present time. Thus, as many analysts have observed, in almost paradoxical fashion cities in the contemporary developing world are more an expression of a lack of economic development, rather than a sign of it. Similarly, in Western societies, the demise of the inner city and counter-urbanisation trends in association with rapid developments in electronics and telecommunications make the future of the city as we know it look increasingly uncertain.

Whether cities and urbanisation are regarded as an evil, a necessary evil or the greatest of human achievements is a central, although highly contentious question. Indisputably, however, cities have always been closely associated with the generation of increasing inequalities. Facets of both global and national inequalities underlie patterns of present-day Third World urban development. Thus, as Gilbert and Gugler (1982, p. 11) point out, "In an unequal world, therefore, it is not surprising that cities should also be unequal". Accordingly, issues of inequality, both social and spatial represent the overall theme of the next two chapters, which look respectively at aspects of urbanisation and urban planning in less developed countries.

Chapter 3

Urbanisation in the Third World

Third World urbanisation cannot be understood in global isolation nor in an historical vacuum. Just as it is all too easy to omit that urbanisation and urbanism originated in what is currently referred to as the Third World, so it is tempting to succumb to the notion that contemporary Third World urbanisation represents an entirely distinct process. Whilst we should concur with Berry (1973) that there have been divergent paths in the urban experience of the twentieth century, equally, we must recognise that present-day urban development in poor countries is a direct outcome of the last five centuries of world development. To a considerable extent, therefore, the processes that led to mercantile trade, distant colonialism, imperialism and post war independence were also agents responsible in promoting contemporary urban forms in less developed countries. Thus, no account of contemporary Third World urbanisation can ignore the historical interdependence of the countries now comprising the so-called First, Second and Third Worlds.

The rapid urbanisation process that currently characterises the majority of Third World countries represents a major challenge to governments, politicians, planners and not least the communities, families and individuals involved. It was stressed in the previous chapter that contemporary Third World countries are urbanising at rates which are generally in excess of those that occurred in nineteenth century European countries. As Berry (1973) has summarised, world urban population quadrupled between 1920 and 1960. During the same time period, however, the total urban population of the developed countries increased by a factor of 2.75, but that of the Third World by a staggering 6.75. Meanwhile the large city population of the Third World increased much faster, by a factor of 9. Annual growth rates of urban population for individual Third World countries generally fall between 3 and 6 per cent, but in some instances reach as high as 10 per cent (for example, in 1970: Venezuela 5.6 per cent; Nigeria 6.0 per cent; Tanzania 8.6 per cent; Malawi 10.1 per cent). The growth of towns and cities in less developed nations is generally running ahead of industrialisation and the provision of jobs, infrastructure, welfare services and houses. This disparity is frequently referred to as 'urban inflation' or 'hyper-urbanisation'. Statistics included in Bairoch's (1975) survey of the economic development of the Third World in the twentieth century exemplify this observation. In 1970, in the non-communist less developed countries taken as a whole, 21 per cent of the population was urban, whilst some 10 per cent of the active population was engaged in manufacturing, so that urbanisation exceeded manufacturing employment by 110 per cent. By comparison, for the countries of Europe in 1930, 32 per cent of the population was urban and 22 per cent employed in manufacturing, giving an urbanisation surfeit of only 45 per cent. However, we must heed Robert's (1978) warning that it is misleading to view Third World urban problems as simply being the outcome of urban development without industrialisation. Rather, he argues that these problems are an integral part of immediate and past processes of economic growth in less developed countries.

Whilst in the past it may well have been fitting to have argued that rural-to-urban migration was a consequence of urbanisation and not its cause (Friedmann and Wulff, 1976), this seems far less valid today for the less developed nations. In these countries, age-selective migration is a major factor in urbanisation along with the demographic multiplier associated with the in-migration of youthful age groups. The present chapter considers the causes and consequences of rapid urbanisation in the Third World. The account follows Chapter 2 in arguing that urbanisation is associated with the generation and spatial concentration of a social surplus product. Hence, uneven spatial development and inequalities appear to have been basic to urbanisation under capitalism throughout history. In turn, the perceived attractions of different regions of the national space relate closely to socio-economic inequalities, and migrations and demographic variables also closely mirror them. Similarly polarised socio-economic conditions come to characterise the internal structures of Third World cities. Accordingly, the present chapter first views the development and structure of urban systems in less developed countries. Subsequently, in the second half of the chapter, the focus is on socio-economic conditions existing within contemporary Third World towns and cities.

THE HISTORICAL EVOLUTION OF THIRD WORLD URBAN SETTLEMENT SYSTEMS: THE MERCANTILE MODEL

The need for an historical approach to the study of urban settlement systems and the disappointingly ahistorical nature of much past work in this field was stressed in Chapter 2. Broadly speaking, the term "urban system" is employed to denote the total set of towns and cities that comprises the urban settlement fabric of a given area. In the 1960s and 1970s, it became fashionable to bring the full force of general systems theory to bear on the study of urban settlement patterns (see, for example, Bourne, 1975). Such an approach was not without its advantages, for it stressed that urban settlement systems are adaptive, continually shifting their structures and external linkages. By definition, they are open systems that should be studied in relation to their wider contextual environments. Viewed in this manner, the need for a distinctly social-historical approach to the study of urban settlement systems in Third World countries becomes apparent.

A comprehensive historical framework of the evolution of national urban settlement systems has not as yet been developed, but a tentative outline has recently been provided by Johnston (1980). This framework is based on the five stage model of societal development outlined in Chapter 2, that is the progression from reciprocity to capitalism via the systems of rank redistribution, money exchange and mercantilism. As we have already noted, reciprocal societies are preurban in all respects and urbanisation only commences under a system of rank redistribution. The incipient stages of urbanisation will be characterised by a simple urban pattern and the only impetus to change comes with the need for increased trade. Johnston suggests that this was generally first met by small scale internal colonial expansion, forming perhaps a simple hierarchy of control settlements along the lines envisaged in classical central place theory (Christaller, 1933).

Historically, however, it was with the development of Mercantile societies from the fifteenth century onwards that urban systems started to evolve along more complex lines. Initially, the emergence of merchants was associated with the development of periodic or ring market systems. The main development, however, came with colonialism, for continued economic growth required greater land resources. Frequently this was initially met by local colonial expansion via trading expeditions. By the seventeenth and eighteenth centuries, however, this need was increasingly being fulfilled by distant colonialism, that is a trans-oceanic version of local colonialism. The implications of these historical developments have been well summarised by Vance (1970, p. 148):

> The vigorous mercantile entrepreneur of the seventeenth and eighteenth century had to turn outward from Europe because the long history of parochial trade and the confining honeycomb of Christaller cells that had grown up with feudalism left little scope there for his activity. With overseas development, for the first time the merchant faced an unorganized land wherein the designs he established furnished the geography of wholesale-trade location. By contrast, in a central-place situation (such as that affecting much of Europe and the Orient), to introduce wholesale trade meant to conform to a settlement pattern that was premercantile.

The development of Mercantilism has already been summarised in Chapter 2. During this phase, ports naturally came to dominate the evolving urban systems of both the colony and also the colonial power. In the colony, once established, ports acted as gateways to the area. Subsequently, evolutionary changes occurred that first saw increasing spatial concentration at certain nodes and later, lateral inter-connection of the coastal gateways and the establishment of new inland regions of expansion. The settlement pattern of the homeland also underwent considerable change, for social surplus products flowed into the capital city and principal ports, thus serving to strengthen them considerably.

These historical facets of trade articulation led Vance (1970) to suggest what amounted to an entirely new model of urban settlement evolution. This is referred to as the Mercantile model, and it is reproduced in Figure 3.1. In the figure, the model is summarised in five stages. In each of these the colony is shown on the left of the diagram, the colonial power or mother country on the right.

1. Initial search phase of Mercantilism - involving the search for economic information on the part of the prospective colonising power.

2. Testing of productivity and harvest of natural storage - sees the periodic harvesting of staples such as fish, furs and timber. However, no permanent settlement is established in the colony.

3. Planting of settlers who produce staples and consume manufactures of the home country - the settlement system of the colony is established at a point of attachment. The developing symbiotic relationship between the colony and colonial power is witnessed by a sharp reduction in the effective distance separating them. The major port in the homeland becomes pre-eminent.

4. Introduction of internal trade and manufacture in the colony - penetration occurs inland from the major gateways in the colony, based on staple production. There is rapid growth of manufacturing in the homeland to supply both the overseas and home markets. Ports continue to increase in significance.

5. Final stage - sees the establishment of a mercantile settlement pattern with central place infilling in the colony; and the emergence of a central place-type settlement system with a mercantile overlay in the homeland.

The mercantile model stresses the historical-evolutionary viewpoint in examining the development of national urban systems. The framework offers what Vance sees as an alternative and more realistic picture of settlement structure, based on the fact that in the seventeenth and eighteenth centuries, mercantile entrepreneurs turned outward from Europe. Hence, the source of change is external to the developing areas, so that an exogenic or truly open and adaptive settlement system is envisaged. Vance argues that in contrast, the theory of central places is endogenic, for it models only parochial or local demand, thereby rendering what is essentially a closed settlement system. Thus, there is more

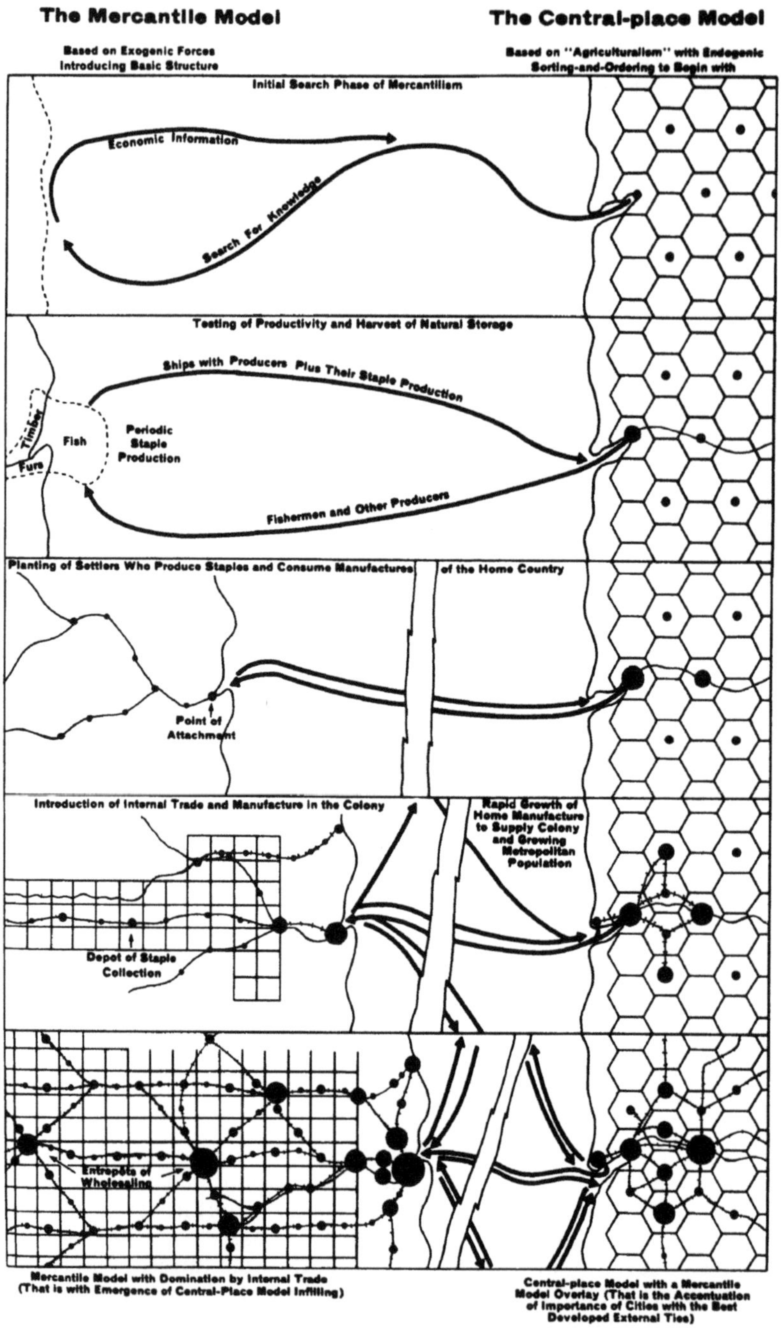

Figure 3.1: Vance's mercantile model of urban systems development (Source: Vance, 1970)

than a grain of truth in the assertion that whilst central place theory applies to the Feudal period, it cannot cope with the realities of our mercantile-capitalist past (Vance, 1970).

The hallmark of the Mercantile settlement system is its remarkable linearity. This spatial pattern takes two forms. The first is the alignment of settlements along coasts, particularly in colonies. Secondly, linear settlement patterns also occur along the routes of trade which developed between the coastal ports of attachment and the staple producing interiors. Interestingly, these two alignments are also recognised in the four-stage historical model of transport expansion in less developed countries advanced by Taaffe, Morrill and Gould (1963), based on the settlement and transport histories of Brazil, Malaya, East Africa, Nigeria and Ghana.

The virtues of the Mercantile model are many. Principally, it stresses that the development of settlement systems in most Third World countries amounts to a form of dependent urbanisation. Certainly we are reminded that the high degree of urban primacy and the littoral orientation of settlement fabrics in Africa, South America, the Caribbean and Asia are all the direct product of colonialism, not accidental happenings or aberrant cases. Ports and other urban settlements became the focus of economic activity and of the social surplus product which accrued. A similar but somewhat less overriding spatial concentration also applied to the colonial power. Hence, a pattern of spatially unequal or polarised growth emerged strongly several hundred years ago with the strengthening of this symbiotic relationship between colony and colonial power.

URBANISATION AND POLARISED DEVELOPMENT IN THIRD WORLD COUNTRIES

The Vance settlement system provides us with a model of how the initial stages of mercantile-capitalism led to increasingly unequal development over space. A similar polarisation is envisaged in John Friedmann's (1966) centre-periphery model, but only during the early phases of economic progress. Thus, from a theoretical point of view, Friedmann's central argument was that "where economic growth is sustained over long periods, its incidence works toward a progressive integration of the space economy" (Friedmann, 1966, p. 35). This process is made clear in a four stage ideal-typical sequence of development, which is shown in Figure 3.2 and summarised below:

1. Independent local centres with no hierarchy - this represents the preindustrial stage and is associated with a series of isolated, self-sufficient local economies. This stage is regarded as being relatively stable (Figure 3.2a).

2. A single strong centre - it is posited that as a direct result of some form of 'disruption', normally externally induced,

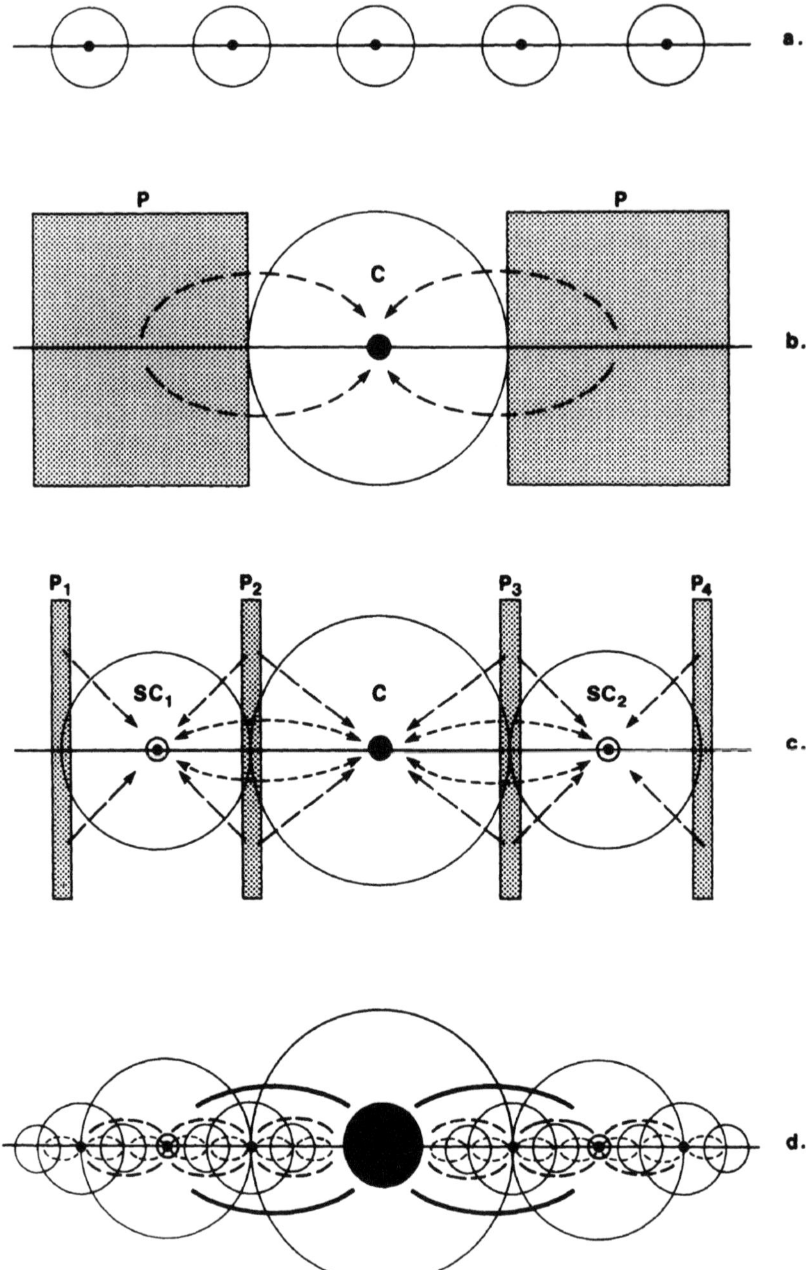

Figure 3.2: Friedmann's core-periphery model of development (Source: Friedmann, 1966)

the former stability breaks down. Growth occurs rapidly in one main region and urban primacy is the spatial outcome. The centre (C) feeds on the rest of the nation, so that the extensive periphery (P) is drained. Advantage tends to accrue to a small elite of urban consumers located at the centre. However, Friedmann regarded this stage as inherently unstable (Figure 3.2b).

3. A single national centre with strong peripheral subcentres - as a result of such instability, it is argued that with time, the simple centre-periphery pattern is progressively transformed to a multinuclear one. Sub-cores develop (SC_1, SC_2) leaving a series of intermetropolitan peripheries (P_1 to P_4), as shown in Figure 3.2c.

4. A functionally interdependent system of cities - this fourth and final stage, described by Friedmann as "organised complexity" is one where progressive national integration continues, eventually witnessing the total absorption of the intermetropolitan peripheries. A smooth progression of cities by size is envisaged as resulting (Figure 3.2d).

The first two stages of the centre-periphery model obviously relate directly to the history of many less developed societies, and they parallel closely Vance's model in envisaging increasingly unequal growth in the early stages of development. Similarly, both models see the initial breakdown of the stable preindustrial system as being a result of the exogenic forces. Thus, Friedmann commented that the centre-periphery relationship is essentially a colonial one, his work having been based on the history of regional development in Venezuela.

The principal idea behind the centre-periphery framework is that early on, factors of production will be displaced from the periphery to the centre, where marginal productivities are higher. Thus, at an early stage of development, nothing succeeds like success. However, the crucial change is the transition between the second and third stages, that is where the system tends toward equalisation and equilibrium. Friedmann's model is one which argues that in theory, economic development will ultimately lead to the convergence of regional incomes and welfare. But at the same time, from an empirical perspective, Friedmann observed that in reality there was evidence of persistent disequalisation:

> There is, however, a major difficulty with the equilibrium model: historical evidence does not support it. Disequilibrium is built into transitional societies from the start (Friedmann, 1966, p. 14).

It is precisely this issue that in different guises has dominated economic development theory over the past forty years or so. The centre-periphery or dualistic relationship had, in fact, been discussed in a wide variety of contexts well before Friedmann focussed upon it. Thus, he cited and drew specifically on the works of Meier and Baldwin (1957), Prebish (1950), Perloff and Wingo (1961), Schultz (1953), Perroux (1950), Myrdal (1957) and Hirschman (1958), all of whom had examined the relationship, albeit at different scales.

The general economic development theories of Myrdal and Hirschman, concerning the degree to which free market forces accentuate or reduce regional inequalities are of direct relevance here. Myrdal (1957), a Swedish economist, took a basically pessimistic view, maintaining that capitalist development is inevitably marked by deepening regional and personal income and welfare inequalities. He followed the arguments of the vicious circle of poverty in presenting his theory of "cumulative causation". Thereby, it was posited that once differential regional growth occurs, thereafter internal and external economies will perpetuate the pattern. This is the outcome of the "backwash" effect whereby population migrations, trade and capital movements all come to focus on the key growth points of the economy. Increasing demand, associated multiplier effects and the existence of social facilities also serve to enhance the core region. Whilst "spread" effects will undoubtedly occur, principally via the increased market for the agricultural products and raw materials of the periphery, Myrdal basically concluded that under free-market forces these would nowhere match the effects of backwash. Myrdal's thesis leads to the advocacy of strong state policy to counteract what is seen as the normal tendency of the capitalist system to foster increasing regional inequalities.

Hirschman (1958), whose work was published a year after that of Myrdal, advanced a noticeably more optimistic view (Hansen, 1981). Specifically, he argued that polarisation should be regarded as an inevitable characteristic of the early stages of economic development. This represents the direct advocacy of a basically unbalanced economic growth strategy, whereby investment is concentrated in a few key sectors of the economy. It is envisaged that the growth of these sectors will create demand for the outputs of the other sectors of the economy, so that a "chain of disequilibria" will lead to growth. The corollary of sectorally unbalanced growth is geographically uneven development and, in fact, Hirschman cited Perroux's (1955) idea of the natural "growth pole". The forces of concentration were referred to as "polarisation", a notably less derogatory term than Myrdal's "backwash". It is clear that very strong psychological effects must also underline the forces of polarisation. Crucially, however, Hirschman argued that eventually, development in the core will lead to the "trickling down" of growth to backward regions. These trickle down (or spread) effects were seen by Hirschman as an inevitable and spontaneous process. Thus, the policy implications of Hirschman's thesis stand in almost complete contrast to those of Myrdal. They suggest that governments should not intervene to reduce inequalities, for at some appropriate juncture in the future, the search for profits will promote the spin off of growth-induced industries to backward regions.

This process whereby spatial polarisation trends give way to spatial dispersion out from the core to the backward regions has subsequently come to be referred to as "polarisation reversal" (Richardson, 1977, 1980). Clearly, the turning point can be related to the third and fourth stages of the core-periphery model of Friedmann (Richardson, 1980).

The true significance of these ideas concerning polarised development extends beyond their use as a basis for understanding the historical processes of urban-industrial change. Significantly, they came to represent an explicit framework for regional development policy (Friedmann and Weaver, 1979). Thus, the doctrine of unequal growth gained both positive and normative currency in the first post-war decade and the path to growth was actively pursued via urban-based industrial growth during the 1950s. The policies of non-intervention, enhancing natural growth centres or creating new induced sub-cores became the norms. As Friedmann and Weaver (1979, p. 93) observe, the "argument boiled down to this: inequality was efficient for growth, equality was inefficient", so that, "Given these assumptions about economic growth, the expansion of manufacturing was regarded as the major propulsive force".

The pursuit of unequal development as a matter of policy came in particular to effect the newly independent, formerly colonial territories. It was perhaps inevitable that in seeking to develop they should come to equate development with industrialisation. This was hardly surprising given that the conventional wisdoms of development economics stressed this very connection. For many Third World countries, post-war decolonisation gave political independence and promoted the desire for the economic independence to accompany it. In the words of Friedmann and Weaver (1979, p. 91), such countries:

> ... took it for granted that western industrialized countries were already developed, and that the cure for 'underdevelopment' was, accordingly, to become as much as possible like them. This seemed to suggest that the royal road to 'catching up' was through an accelerated process of urbanization.

The trend toward industrialisation in Third World nations had been particularly associated with the policy of import substitution. This represents an obvious means of increasing self-sufficiency, for less developed countries have always traditionally imported most of their manufactured goods in return for their exports of primary products such as sugar, coffee, rubber, textiles and cotton. Key industrial sectors for development were those which were relatively simple and where a substantial home market already existed, for example, food, drink, tobacco, clothing and textile production. By now, most Third World countries have followed this path toward import-substitution industrialisation, although as Dickenson et al (1983) observe, few have progressed much beyond it. The development of basic heavy industries such as steel, chemicals and petrochemicals along the lines of the USSR model, has not been possible for most Third World countries. Such a policy, which following Rostow's (1960) unilineal model of development might seem attractive, requires a level of population and effective demand not normally present in most Third World countries. Further, the competition from developed countries along with capital and infrastructural shortages and problems of lumpy investment, technological transfer and capital rather than labour intensity also militate against such heavy industrial development. An exception, however, is provided by India which has achieved a high level of industrial self-sufficiency since 1945 (Johnson, 1983). Many less developed countries are presently encouraging light industrialisation by means of the making available of fiscal incentives to foreign companies. These so called 'enclave industries' consist of the branch-plants of multinational companies, which often produce goods entirely for the overseas market. Overall, quite high rates of industrialisation have been recorded for many less developed countries. For example, from 1938 to 1950, the rate of growth in manufacturing was around 3.5 per cent per annum, and from 1950 to 1970, 6.6 per cent per annum for the less developed countries taken as a whole (Dickenson, et al, 1983).

Industrialisation has occurred, but the vital question is whether or not it has been accompanied by spread effects or trickling down. Here, however, there is much empirical evidence showing that unfortunately, Third World countries are following the prescriptions of Myrdal, rather than those of Hirschman. Thus, many would argue that inter-regional disparities are showing little or no sign of reduction with development in the Third World. In Venezuela, for example, in the 1960s, rural incomes were on average only 40 per cent of those in urban areas (Gilbert and Gugler, 1982). Similarly, in Barbados, wages in the agricultural sector average only 42 per cent of those in the non-agricultural sector (Hope and Ruefli, 1981). Williamson's (1965) oft cited paper suggested that the pattern of regional income inequality follows an inverted U-shaped path over time; that is, inequality increases during the youthful stage of development, but decreases thereafter. Thus, Williamson argued that Myrdal's analysis was excessively dismal and he supported the view of Hirschman that regional convergence would ultimately occur in a spontaneous and unplanned manner. It has been envisaged that similar bell-shaped curves also characterise changes in personal income inequality (Kuznets, 1955), geographical concentration, demographic change and indeed development stages over time (Alonso, 1980). Significantly, a number of writers have broadly agreed with Williamson's views (Mera, 1973, 1975, 1978; Alonso, 1968, 1971). But for many present-day Third World countries, the actual record seems to be one of marked and consistent inequalities, as shown by the comparative data included in Table 3.1. Thus, Gilbert and Goodman (1976) argue that for every Third World country that shows regional convergence, there is another that displays divergence. Similar views have recently been expressed by Stohr and Taylor (1981) and Friedmann and Weaver

Table 3.1: Regional inequalities for a selection of developed and less developed countries

Country	Gross regional product per capita, US dollars, 1976		Ratio of richest to poorest region
	Richest region	Poorest region	
Developed Areas:			
Belgium	4,380	2,616	1.67
France	5,918	2,833	2.09
Germany	7,022	2,683	2.62
United Kingdom	3,667	2,566	1.43
Netherlands	4,032	2,578	1.56
Italy	3,384	1,538	2.20
Japan	5,555	1,900	2.92
Less Developed Areas:			
India	217	97	2.24
Korea	582	270	2.16
Thailand	1,358	215	6.34
Malaysia	730	202	3.62
Iran	3,132	313	10.07
Colombia	1,342	199	6.75
Mexico	1,067	198	5.39
Brazil	1,102	109	10.14
Venezuela	1,354	237	5.72
Argentina	3,706	397	9.33

Source: Renaud (1981, p.118)

(1979). The latter authors argue that regional development planners have hoped that the turning point was not far off, perhaps only requiring greater labour mobility, agricultural transformation and the transnational movement of capital to facilitate the change from inter-regional divergence to convergence. However, a major point of concern is that perceptions of inequality may well have more effect on the perpetuation of regional inequalities than actual conditions (von Böventer, 1975; Clark, J. A., 1982). Thus, regional perceptions may become immutable stereotypes, a possibility which has strong links with Myrdal's theory of cumulative causation, and which is further investigated later in the present text.

Enough Third World countries appear to be characterised by spatial polarisation and backwash effects to suggest that disequilibrium cannot be regarded as merely transitional. The issue really boils down to the apparent conflicts existing between economic efficiency and social equity in the short run. Accordingly, the balance achieved between these two goals stands as a crucial political and governmental issue in Third World development planning (see also Parr, 1974). Proponents of laissez-faire capitalism would argue that regional and social inequalities are the necessary and acceptable price of economic development. The socialist view, on the other hand, would stress the need for social equity, even if its promotion entailed a reduction in the overall level of economic growth. This is an important bone of contention in development planning practice and we shall return to it in the next chapter.

QUESTIONS OF CITY SIZE, MODERNISATION AND SPATIAL DIFFUSION

Thus, the historical trends of urban concentration described by Vance's model have generally been bolstered by post-war development planning in less developed countries. In short, the advantages of capital cities and major ports have been reinforced in most non-socialist developing nations since 1945. It is customary, in fact, to assert that the condition of urban primacy is typical of less developed countries. Urban primacy, formally recognised by Jefferson (1939), is a condition where the largest city in a country is superordinate in both size and national influence. This contrasts with the rank-size or lognormal city size distribution which is commonly regarded as being more typical of advanced industrialised countries. The rank-size rule which is normally attributed to Auerbach (1913) and Zipf (1949), envisages a situation where cities follow a smooth progression by size. It is notable that in both the mercantile and core-periphery models, the early phase of economic development is characterised by increasing primacy. Similarly, in Friedmann's model, development is witnessed by the transition from a primate city size distribution to a smooth, or lognormal one (see Figures 3.1 and 3.2).

However, once again all is not as simple as might at first appear. Although it is indeed true that many underdeveloped countries exhibit high levels of urban primacy, at the national scale, relative economic development must be set aside as a single explanatory variable. Thus, there are a number of highly developed countries which show strong urban primacy, for example, Denmark, Austria and France. Equally, it is well known that several less developed countries show low levels of primacy, at least as measured in relation to the upper echelons of their urban size distributions. Brazil, India and Kenya stand as particularly good examples of this.

As Richardson (1973a) notes, explaining these facets of city size distribution has fascinated urban and regional analysts for decades. The most important early paper was published by Berry (1961) and specifically sought to examine any possible connection existing between city size distributions and economic development. Berry looked at a sample of 38 countries at different levels of economic development, 13 of which showed distinct lognormality, 15 primacy and 10 distributions classed as intermediate between these two types. This research demonstrated first, that there was no clear statistical relationship between a country's city size distribution and its level of urbanisation. Secondly, Berry concluded that different city size distributions were in no way related to the relative economic development of countries. Thereby, Berry was the first to suggest that at the national level, no single variable influences city size distributions. Rather, he proposed a temporal model which suggested that a whole complex of forces influence relative city size patterns. Thus, it was argued that if a few strong forces operate, primacy is frequently the outcome. Fewer forces will affect the urban structure of a country the smaller it is, the shorter its history of urbanisation, the simpler its economic and political life, and the lower its overall degree of development. In contrast, where the reverse of these conditions pertains, it was envisaged that a country will develop a whole range of specialised cities performing a variety of different functions, that is, a rank-size distribution.

Linsky (1965) took up this research theme, examining the relationship existing between levels of primacy and six other variables. It was found that the strongest relationship was a negative one between urban primacy and the size of countries (-0.37). The only other significant correlation was a positive one between primacy and rates of population growth (+0.33). Although not assuming statistical significance, urban primacy was further shown to be positively associated with the proportion of the total workforce engaged in agriculture, the degree of export orientation and the length of colonial history of countries. Linsky's work was important in showing that whilst primacy may be regarded as particularly characteristic of small countries which have low *per capita* incomes, a high dependence on exports, a colonial history, an agricultural economy and a fast rate of population growth, it is certainly not precluded elsewhere. Almost identical results were reached in a study carried out at about the same time by Mehta (1964).

A similar line of reasoning was taken by Vapnarsky (1969) in his work based on the historical evidence of city size and economic development in nineteenth and twentieth century Argentina. Here it was suggested that the rank-size and primate patterns are not mutually exclusive. Rather, Vapnarsky argued that each is produced by a different set of circumstances. Primacy is seen as being positively related to the degree of closure of the economy, that is its dependence on overseas trade. With increasing closure, urban primacy is thought to reduce, other things being equal. This accords well with the later stages of the core-periphery model. Rank-size or lognormality on the other hand, is regarded as being affected by the level of interdependence existing in a country, that is the extent of interregional linkage. Thus, as internal interdependence increases, so lognormality is progressively approached. The cross classification of these two key variables gives four principal types of city size distribution, as shown in Figure 3.3. Following these from the one shown at the top left hand corner in a clockwise direction provides an interesting parallel with the city size distributions envisaged in the core-periphery developmental sequence. High closure with low interdependence is typical of very underdeveloped areas, and is associated with no clear city size distribution pattern. If interdependence remains low, but closure diminishes as a function of increasing trade, strong urban primacy is likely to result. If the state of low closure is now increasingly accompanied by internal interdependence, the result is primacy at the upper level of the urban system, juxtaposed by lognormality further down. Finally, if closure now increases, this plus the existing internal interdependence leads to the classical pattern of rank-size distribution.

These research papers suggest that other things being equal, primacy is frequently associated with a history of foreign dependence and low levels of internal interdependence. These are both characteristic features of former colonial countries. But clearly, a range of other factors are related to urban primacy, especially the size of a country. This latter variable has been stressed by Johnston (1971) in the case of the development path of New Zealand. Here, increasing closure and interdependence have in a seemingly paradoxical manner been associated with increasing primacy. Johnston explains this by virtue of New Zealand's small size, which has meant that import-substitution industrialisation has given rise to single plant industries, located in Auckland. Further, Johnston notes that during the period of European settlement, strong regional primacy became the norm, reflecting the orientation of local economies to Britain. Thus, given a doctrine of unequal growth, primacy may well be a logical response, especially in small countries and/or in open developing ones. In an important paper, El-Shakhs (1972) has, in fact, shown that if less developed countries are considered on their own as a group, a positive relationship exists between levels of urban primacy and relative levels of economic development. This would appear to relate to the bell-shaped curve of geographical concentration over time (Alonso, 1980) and to the point of polarisation reversal.

Clearly, there is a whole complex of forces interacting to

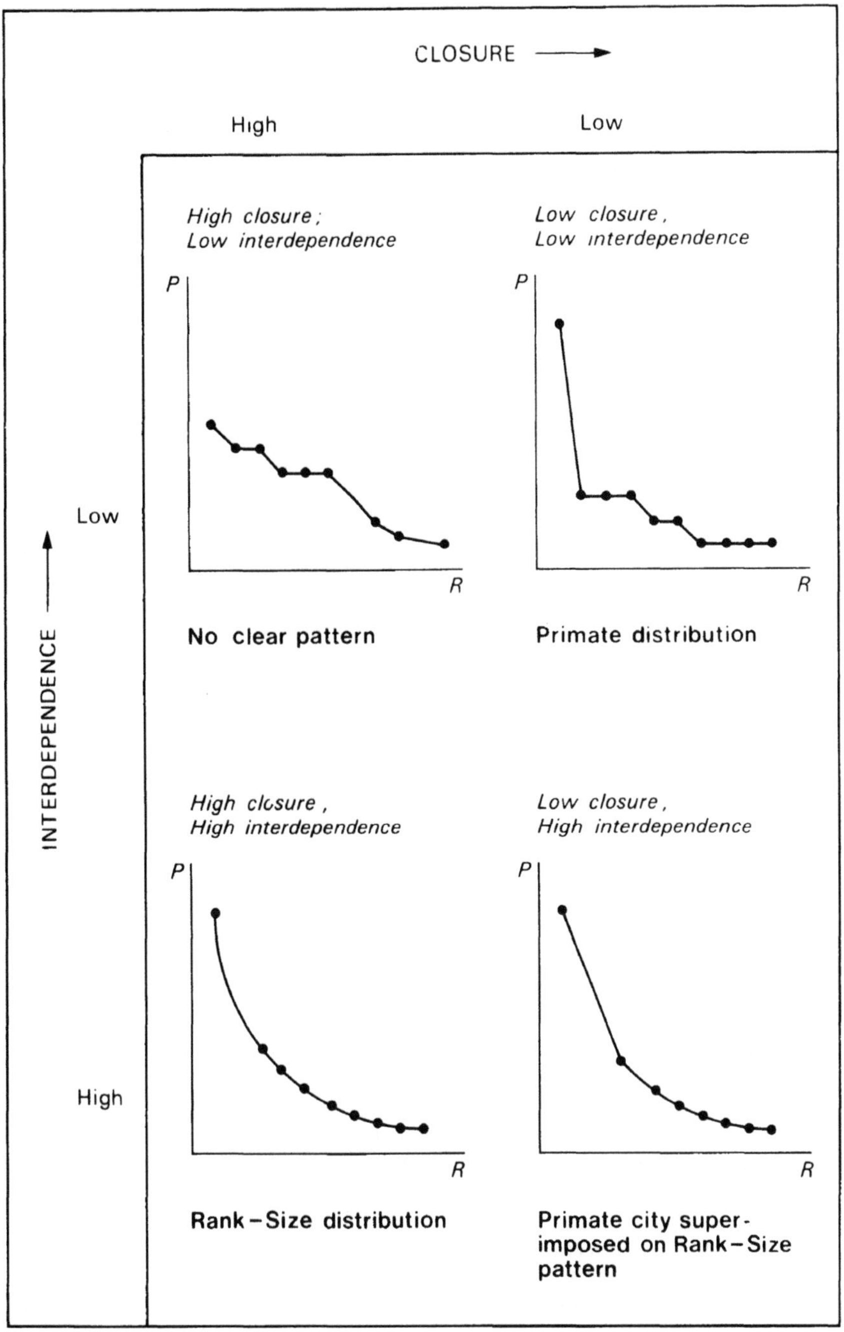

Figure 3.3: Vapnarsky's model of city size distributions

mould the relative size characteristics of cities at the national level. However, Johnston (1980) notes a further crucial point. It is that primacy is normally examined at the level of the nation-state. Further, he argues that these political divisions may have been of little or no real consequence at the time of rapid urban systems development. In particular, as the New Zealand case clearly implies, colonisation from Europe through a series of gateways is likely to have led to strong regional primacy, rather than national primacy. This is, of course, particularly relevant in the case of large countries. For example, India shows strong regional primacy in terms of Calcutta, Bombay and Madras in descending size order. The same is also true of China and Brazil. Thus, strong urban primacy may exist and be maintained at a regional level, but when aggregated, may sum to a distribution akin to lognormality. Interestingly, this argument was advanced from a purely theoretical standpoint in the early 1960s by Berry and Barnum (1962). Significantly, therefore, even in countries where the national level of primacy is low, production, incomes and welfare levels may all still be highly uneven at the regional scale. Thus, primacy must not be regarded as being synonymous with polarisation, but merely one form of it. For example, only 15 per cent of the total population of Brazil lives in Sao Paulo (Renaud, 1981), but as Table 3.1 clearly shows, Brazil's richest region records a per capita regional product over ten times greater than that of its poorest region. Similarly, a primacy level of only 16 per cent applies to Malaysia, but its regional inequality ratio is 3.62 (see Table 3.1). Thus, although primacy does frequently occur in developing countries, it is not always apparent at the national level. Rather, the historical process of polarised growth via a number of cores has led to strong regional primacy and inter-regional disparities in most countries.

Aspects of city size distribution have also come to be linked with cognate ideas concerning modernisation surfaces and spatial diffusion processes. The modernisation thesis was developed largely in the field of political science, but was picked up by a group of geographers in the late 1960s (see, for example, Soja, 1968, 1974; Gould, 1970; Riddell, 1970). In this work, sets of indices held to reflect modernisation were subjected to multivariate statistical analyses, and the resulting components mapped. The "modernisation surface" of Tanzania derived by Gould (1970) employing this approach has already been shown in Figure 1.3. The modernisation approach did serve to emphasise that core urban systems and the transport corridors running between them are the focus of dynamic change in many developing countries (Brookfield, 1975). On the other hand, the approach was entirely descriptive and in the words of Friedmann and Weaver (1979, p. 120) only succeeded "in mapping the penetration of neo-colonial capitalism".

The true contribution of such work was that it stressed that modernisation is a temporal-spatial development process, and that if and when spread effects occur they do so as a form of spatial diffusion. This parallels the findings of empirical studies of the growth of settlements within urban systems, which suggest that city growth is itself a special case of the hierarchical diffusion of entrepreneurial innovations. These studies have been almost exclusively carried out in developed countries. Thus, for example, Robson (1973) examined the growth of towns in England and Wales from 1801 to 1911, and concluded that key urban inventions such as gaslighting, building societies and telephone exchanges diffused down through the urban hierarchy from large to small places, rather than spreading in particular regions. Some similarities were revealed in a study of the growth of the American urban system from 1790 to 1960 completed by Borchert (1967). As early as 1911, Schumpeter in his general economic theory had argued that the essence of development is a volume of innovations. Opportunities tend to occur in waves which surge after an initial innovation. Thus, Schumpeter argued that development tends to be 'jerky' and to occur in 'swarms'. The crucial role played by spatial innovation diffusion in the process of urban systems development was, however, stressed in a number of papers, starting with that of Hudson (1969). In this, the basic ideas of Hagerstrand (1953) were applied to the central place system. Thus, innovations can travel through the settlement system by a process of contagious spread, where there is a neighbourhood or regional effect of clustered growth. Alternatively, diffusion can occur downwards through the settlement hierarchy, the point of introduction being the primate or capital city. Pedersen (1970) argued strongly for a strict hierarchical process of innovation diffusion, an assertion which seemed to be borne out by Robson's empirical work. Pedersen was also instrumental in making a clear distinction between domestic and entrepreneurial innovations, the latter being the instruments of urban growth. In another important paper, Berry (1972) also argued strongly in favour of a hierarchical diffusion process of growth-inducing innovations, this principally being the outcome of the sequential market searching procedures of firms and imitation effects. But rather disappointingly, Berry's analysis was based on the diffusion of a domestic as opposed to an entrepreneurial innovation, namely that of television receivers.

In reality it is most likely that a combined hierarchic cum neighbourhood diffusion process occurs (see Haggett, Cliff and Frey, 1977, p. 241). The point of introduction of a key innovation is likely to be the capital city, especially in a less developed country. Although there is likely to be a neighbourhood spread in the immediate vicinity of the core region, the next major development is then the spatial leapfrogging of the innovation to the national subcores. Again although there will be some local spread, the innovation continues to percolate down the urban hierarchy. Such a process would, of course, lead to a highly polarised map of economic change and modernity in the short to medium run periods. There is, in fact, some evidence of the operation of processes of this sort. For example, again arguing from American experience, Pred (1973, 1977) suggests that the growth of the Mercantile city was based on circular or cumulative causation, linked to multiplier effects. Further, Pred has argued that the growth of large cities was based on their interdependence, so that large city stability occurred. However, Pred maintains that key adoption sequences were not always hierarchic. This is of especial relevance, as it can be argued that today, urban systems growth is likely to be increasingly based on the locational decisions of multinational firms and government organisations.

These ideas have been picked up by Lasuen (1973) specifically in relation to Third World countries. Lasuen considers that large cities are the principal adopters of innovations, so that natural growth poles tend to become evermore hierarchical in nature. The author also argued that the spatial spread of innovations is generally slower in developing countries, due to the frequent existence of single plant industries and generally poor infrastructure. Thus, Lasuen stresses what he sees as the difficulties involved in overcoming spatial inequalities in developing countries, noting that basically, such countries have two policy choices. Firstly, the major urban centres can continue to adopt innovations before previous ones have spread down through the settlement system. On the other hand, the adoption of further innovations at the top of the urban hierarchy can be delayed until the filtering down of earlier innovations has occurred. The former policy will result in increasing economic dualism, but in all probability, a higher rate of national economic growth. On the other hand, the latter option will lead to increasing regional equality, but like as not, a lower overall rate of national economic growth. As Lasuen points out, most developing countries have adopted policies close to the former alternative of unrestrained innovation adoption, stressing economic growth rather than social equity.

We have come full circle, back to the doctrine of urban-industrial concentration and unbalanced growth as an explicit policy option in Third World nations. We shall return to this important issue when looking at urban systems planning in Chapter 4.

EXAMPLES OF THE DEVELOPMENT AND STRUCTURE OF THIRD WORLD SETTLEMENT SYSTEMS IN THE CARIBBEAN

To conclude this lengthy review of ideas concerning the growth and structure of urban settlement systems in less developed countries, we turn to the Caribbean region for a number of illustrative examples.

Urbanisation and urban growth are major contemporary forces in the Caribbean (Broom, 1953; Stevens, 1957; Clarke, 1974, 1975; Cross, 1979; Hope, 1983). By now, around 52.2 per cent of the total population of the region is to be found living in urban areas, having increased from a level of 38.2 per cent in 1960. Urban populations are commonly increasing at rates between 3 and 6 per cent per annum, and rates twice as high as these are not uncommon (Clarke, 1974). For example, between 1960 and 1980, the growth rate of urban population in Jamaica was 7.3 per cent and in both Trinidad and Guyana, 4.2 per cent per annum (Hope, 1983). The growth of individual towns can be just as startling, so that Castries, the capital of St Lucia increased in population from 16,566 in 1946 to 32,334 in 1960, that is at an annual rate of increase of around 6.79 per cent (Dyson,1967).

This trend of rapid urbanisation may be recent, but as stressed by Clarke (1974), the region has a very long history of urban development, going back to the fifteenth and sixteenth centuries and the process of European colonialism. Clarke also argues that a major consequence of colonialism was that Caribbean towns were customarily ports, administrative and retail centres, but never manufacturing centres. But, as Cross (1979, p. 9) has recently observed, "Colonial expansion, on the other hand, tended to capture rural economies in a web whose centre was the mercantilist city oriented to Western economic needs".

As many authors have noted, this orientation is clearly reflected in the strikingly coastal agglomeration of population that is found in most Caribbean territories, and particularly in their high levels of urban primacy. Levels of overall urbanisation and primacy for individual Caribbean territories are shown mapped for the latest available date, generally in the mid-1970s, in Figure 3.4. High levels of urbanisation apply to countries such as Cuba (60.3 per cent), Puerto Rico (58.1 per cent), the Bahamas (57.9 per cent), Trinidad (50.0 per cent), Martinique (50.0 per cent), Guadeloupe (48.0 per cent) and Barbados (45.0 per cent). The diagram also shows that high levels of urban primacy are common in the region, and in a number of countries, as much as 30-40 per cent of the total population is to be found living in the largest settlement. Figure 3.4 gives the distinct impression that urban primacy is most prevalent in the more developed Caribbean countries such as Barbados (38.8 per cent) and Trinidad (30.9 per cent), and also in small countries like the Netherlands Antilles (59.4 per cent) and St Lucia (39.8 per cent). Overall levels of economic development for Caribbean territories are summarised in Figure 3.5, which shows Gross Domestic Product *per capita* in US dollars c. 1980, and the percentage contribution of agriculture to G.D.P. This map serves to demonstrate the marked variations in levels of development which characterise the Caribbean, ranging from middle-income developing countries such as Puerto Rico (4,081 US dollars), Trinidad (3,429) and Martinique (3,349), to very poor countries, for example, Haiti (278), Dominica (394), Grenada (410) and St Lucia (588).

The interrelationships that exist between these variables summarising levels of economic development and aspects of urbanisation are revealing. As might be anticipated, there is a strong positive relationship between the levels of urbanisation of Caribbean countries and their levels of G.D.P. *per capita* ($r = +0.74$). Further, urban primacy is positively correlated with both levels of urbanisation (+0.43) and G.D.P. (+0.31). These three variables are thus closely interrelated, so that within the Caribbean, generally speaking, the more developed and urbanised a country, the higher its level of urban primacy. This finding, of course, accords well with the ideas of El-Shakhs (1972), reviewed previously. Thus, at one extreme, countries such as Trinidad and Tobago, Bahamas and Barbados show relatively high levels of overall economic development, strong urbanisation and urban primacy, whilst the reverse is true of St Vincent, Grenada, Haiti. The possible influence of the size of countries on city size profiles is indicated too, for overall, national population size is nega-

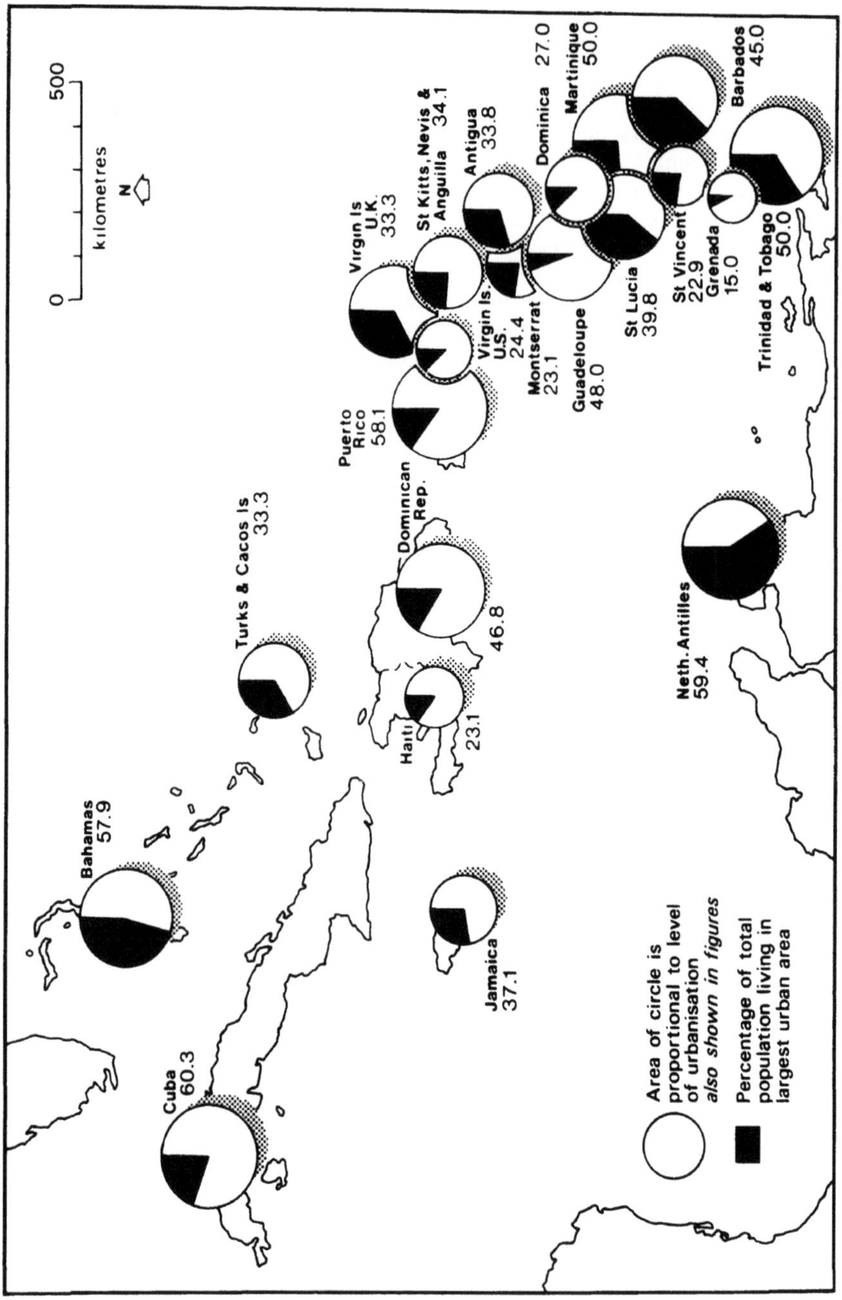

Figure 3.4: Levels of urbanisation and primacy in the Caribbean

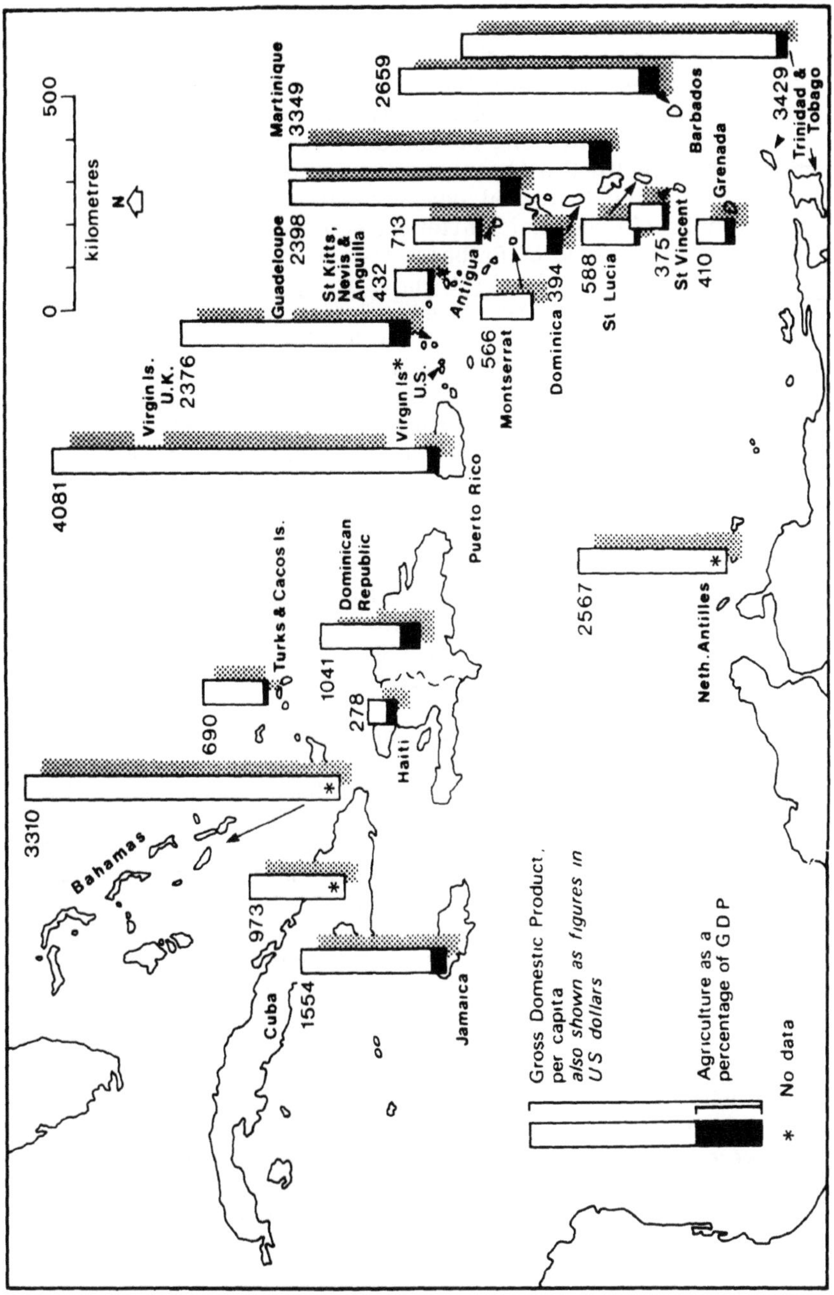

Figure 3.5: Levels of economic development in the Caribbean

tively related to levels of primacy r = -0.27), in the manner envisaged by Linsky (1965), and others.

The West Indies were discovered in error by Columbus in 1492 (Blume, 1974) and as Lowenthal (1972) observes, colonialism is the whole history of the Caribbean. The main period of colonisation came in the sixteenth and seventeenth centuries. The example of one Commonwealth Caribbean country, namely Barbados, will afford insights as to the historical development of the settlement system and internal socio-economic patterns (see also Potter, 1983d).

Although the island of Barbados was discovered by the Portuguese in 1536, it was not settled by them as it lay to the west of longitude 46°37'W, a line which separated colonisation by the Spanish and Portuguese under the Treaty of Tordesillas (1494). It was almost ninety years later, in 1625, that the first English ship landed and claimed the island for King James. Two years later, in 1627, the first settlers, some 80 in number arrived and started laying the foundations of the present-day settlement of Holetown, St James, then known as Jamestown. This original settlement was located midway along the sheltered leeward or western coast of the island (see Figures 3.8 and 3.9 for locations). The settlers started to clear the land and planted food crops, cotton and tobacco (Hoyos, 1979). By 1628, a rival group of English settlers had established themselves in the present-day area of Bridgetown. By this time there were as many as 2,000 settlers in all. Subsequently, the settlements of Oistins and Speightstown were established, all of these being located on the western and southern coasts.

The early settlement history of Barbados is obviously highly reminiscent of Vance's model, and the linear-coastal patterning of settlements attests to the operation of basically mercantile forces. The striking point is the durability of this highly skewed settlement pattern, for 350 years later, it still persists. Thus, the early pattern that emerged, which was geared to the mother country, focused on what remain today as its four main urban population clusters of Bridgetown, Speightstown, Oistins and Holetown. By the nineteenth century, Speightstown functioned as the principal port, being referred to as 'Little Bristol', reflecting its West Country trade links. In the early years, the Bridgetown region was seen as less favourable due to its low-lying, swampy and mosquito ridden nature. Later, however, its natural harbour proved to be increasingly attractive.

The settlers at first grew cotton and tobacco for export, but they were soon unable to compete with that produced in Virginia, North America. In the 1640s, colonists learned from Dutch traders, the methods by which the Portuguese in Brazil were extracting sugar from sugar cane. By 1645, sugar had become the major crop in Barbados, and with it the Plantation System developed. The trend toward large scale operations reflected their economic efficiency and the need for machinery to process the cane immediately it was cut. As a consequence, large numbers of unskilled labourers were also required, but the supply from Britain had dwindled rapidly. By 1643 slaves were being brought from West Africa and a total population of some 37,200 had been reached. Sugar was always a rich man's crop, and as noted by Lowenthal (1972), control and often possession of the sugar properties was soon vested in the mercantile houses of Amsterdam, Paris and London. Many proprietors were absentees, whilst resident planters generally spent much money abroad, on the education of their children, for example. Thus, a vast social surplus product flowed out from the West Indies to Europe.

By the nineteenth century, Bridgetown had developed as the principal seaport and settlement. In 1884, it housed 28 per cent of the island's total population. Its primacy increased steadily through the late nineteenth and early twentieth century. Thus, by 1900, it housed 30 per cent of the population and by 1970, some 42 per cent. This trend toward increasing geographical concentration with early development is further attested by the spatial patterns of population change on the island between 1871 and the present (Figure 3.6). Throughout this period, growth focused on the southern and western coastal parishes, whilst population decline or below national average rates of increase were persistent in the eastern and southern ones.

As demonstrated elsewhere in greater detail (Potter, 1983d), these demographic trends of concentration were paralleled by processes of increasing socio-economic differentiation, even on an island as small as Barbados. Thus, although only 430 square kilometres in areal extent, no matter which index is taken, striking disparities are revealed between the western and southern coastal areas and the rest of the island. This point is well-demonstrated by the data mapped in Figures 3.7A to 3.7F. Figure 3.7A, for instance, picks out the eastern parishes as being predominantly rural and agricultural, whilst a mirror image is displayed if the proportion of the workforce engaged in manufacturing is mapped for 1970. The metropolitan parish of Bridgetown-St Michael stands out in this respect (Figure 3.7B). Levels of car ownership (Figure 3.7C), the proportion of the population having attended university (Figure 3.7D), the number of professional workers (Figure 3.7E) and high-income earners (Figure 3.7F) all pick out the strong spatial inequalities which characterise Barbados.

These figures attest to the fact that the pattern of coastal-oriented development which started in the seventeenth century exists just as clearly today in respect of aggregate socio-economic disparities. In fact, it can be argued strongly that these patterns have been intensified by post-war economic developments in both manufacturing and tourism. With respect to the former, approximately 15,000 jobs are now provided by the industrial sector, which made an 11.3 per cent contribution to G.D.P. in 1979, higher than that of sugar at 5.4 per cent. Since 1955, when the Barbados Development Board (now the Industrial Development Corporation) was set up, a programme of industrialisation has proceeded (Potter, 1981), starting firstly with the policy of import substitution. Following Puerto Rico's successful "Operation Bootstrap", Barbados launched its "Operation Beehive". A series of fiscal incentives were offered to lure enclave industries, those manufacturing entirely for export

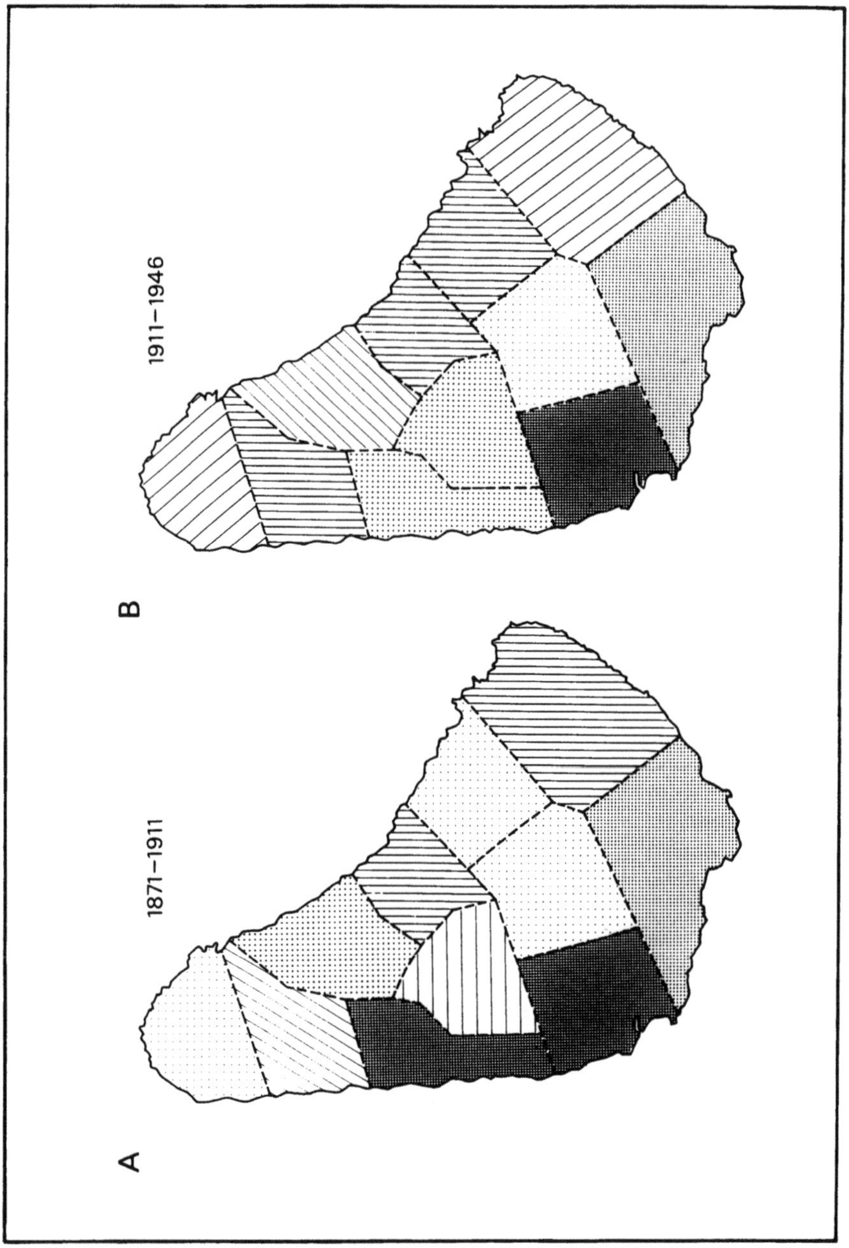

Figure 3.6: Spatial population change in Barbados, 1871-1980

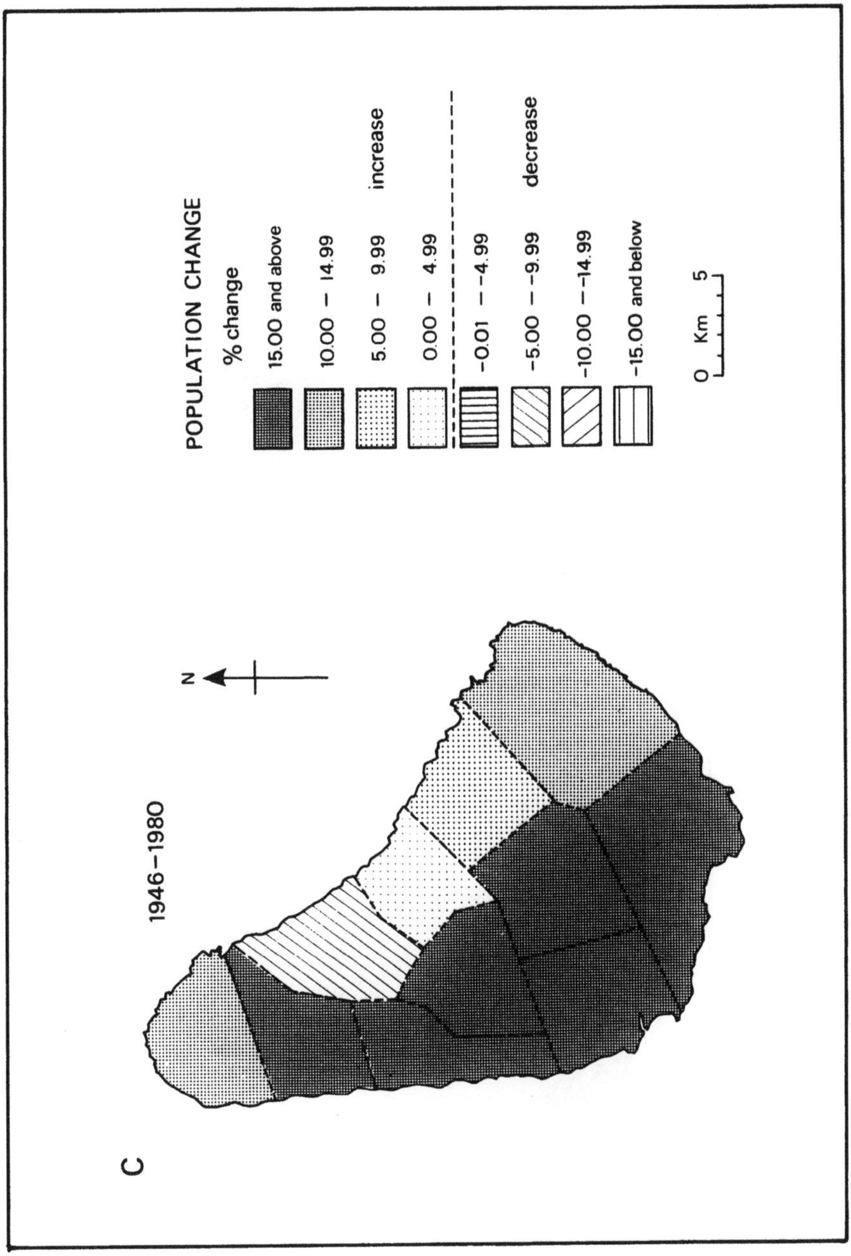

Figure 3.6: Cont.

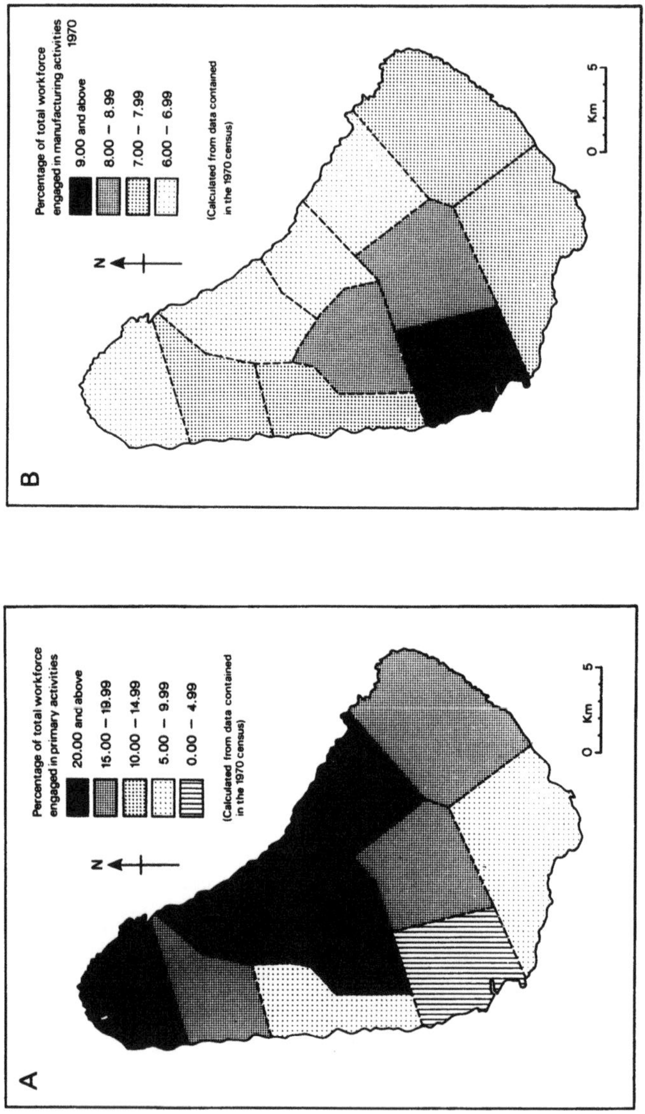

Figure 3.7: Socio-economic patterns in Barbados

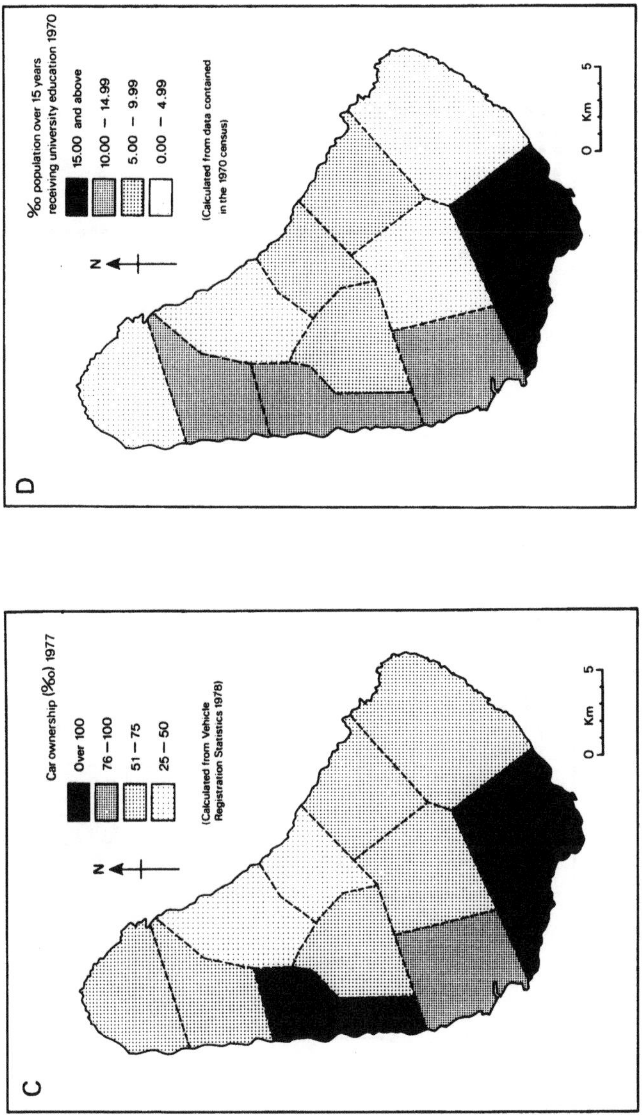

Figure 3.7: Cont.

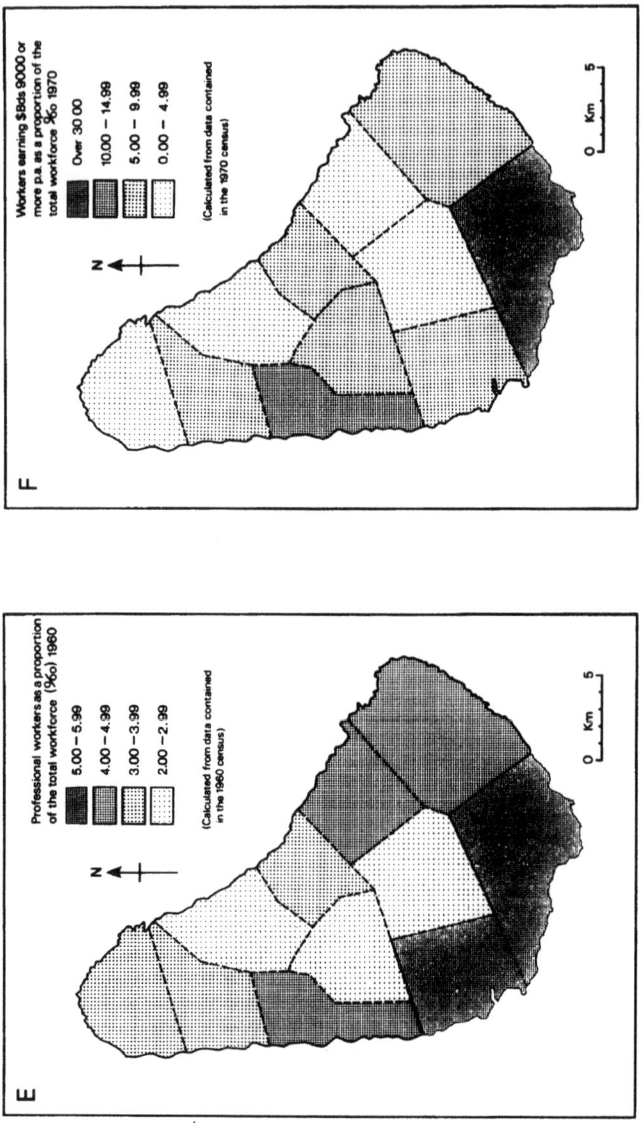

Figure 3.7: Cont.

to home markets. A notable feature of the industrialisation trend has been the development of nine industrial estates by the Industrial Development Corporation (Figure 3.8). All of the early industrial parks were located in the metropolitan parish of St Michael, whilst the late 1960s saw the construction of three slightly further afield, but all notably still in the neighbouring parish of Christ Church. It can be argued that recent industrial development has tended to reinforce spatial polarisation in Barbados. Similar arguments can be advanced with respect to tourism, the other major post war growth industry, which in 1979 contributed 11.8 per cent to G.D.P. (Potter, 1983a). As Figure 3.9 demonstrates, virtually all of the tourist hotels, apartments and infrastructure are to be found located in the same favoured areas of the south and west coastal zones, particularly those of the parishes of Christ Church to the south of the capital, and St James to its north. Although total tourist expenditure was estimated as exceeding $Bds 360 millions in 1979, much of this is lost to the island. A substantial proportion is leaked away on food imports for tourists, whilst another factor is the foreign ownership of hotels. Thus, 44 per cent of all tourist accommodations are foreign owned, but more significantly, 74 per cent of first class hotels. A whole series of social problems can be traced to tourism too (see Karch and Dann, 1981; Potter, 1983a, for example).

The case of Barbados demonstrates the temporal stability of polarised growth in many Third World nations. The linear pattern of spatial development, which stems from the seventeenth century process of mercantile colonialism remains entrenched. In fact, growing disparities or divergence seems to be the experience, even in this small country. Whilst the relative population share of the Bridgetown urban area declined for the first time between 1970 and 1980, showing some sign of polarisation reversal, increasing concentration is still occurring in terms of the wider south-western coastal zone (Potter, 1983b). It is difficult not to see elements of the neo-Marxist interpretation of Third World urbanisation in this process. Historically, urban areas appear to have acted as points of extraction of surplus value from the labours of agricultural and other workers. The result is a core and periphery arrangement that applies to most aspects of life. Thus, for example, Jones (1977, 1981) shows how the adoption of birth control closely corresponds with this spatial pattern of metropolitan dominance in Barbados. Historically, much of the surplus has been transferred abroad, either through foreign ownership and control, or unequal patterns of trade, and this remains characteristic. As Lowenthal (1972) has argued, over one hundred years after emancipation, the colour-class system still prevails and West Indians have achieved the appearance of freedom without its substance. Thus, new forms of dependency or neo-colonialism characterise the region, relating to tourism, manufacturing, the role of multinational corporations and technological developments. It is hard to resist the conclusion that without genuine intervention on the part of governments, both spatial and social inequalities will continue to increase in Barbados, as in most other Third World nations.

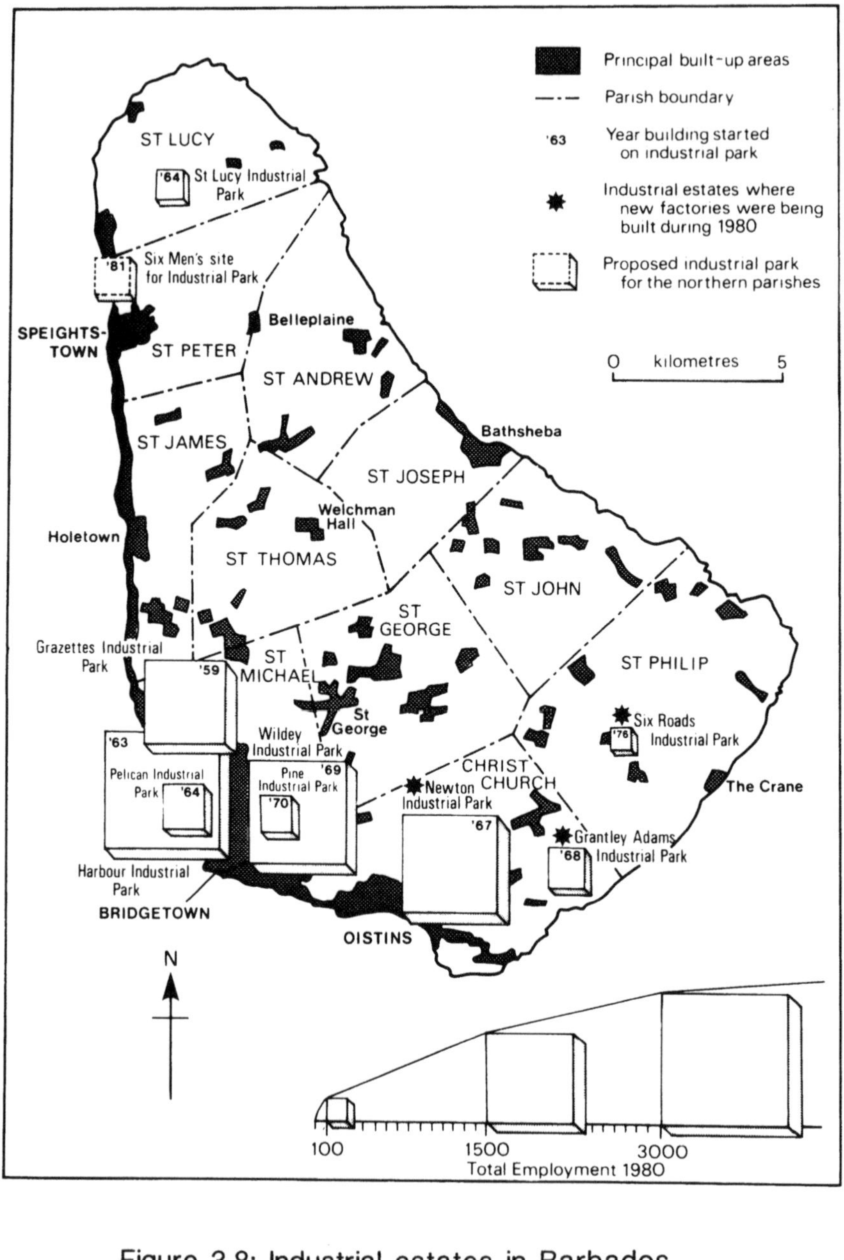

Figure 3.8: Industrial estates in Barbados

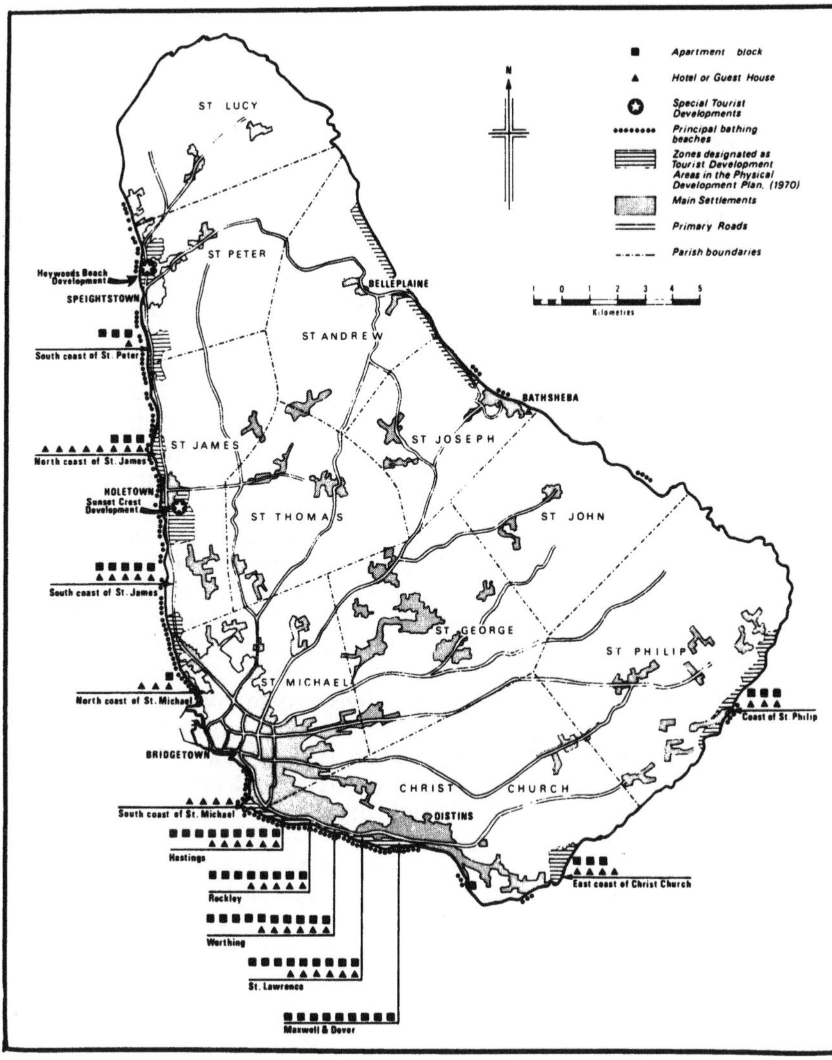

Figure 3.9: Tourist zones and infrastructure in Barbados

RURAL TO URBAN MIGRATION

Rural to urban migration in Third World countries can be seen as a direct perceptual-behavioural response to the marked regional and personal inequalities in income and welfare which exist. Thus, as Jones (1980, p. 98) observes, "migrants act not upon objective characteristics, but upon their perceptions of them". This central point was emphasised in Chapter 1, especially via Figure 1.1.

The behavioural foundations of migration at all spatial scales were stressed by Wolpert (1965, 1966), who employed the notion of 'place utilities', that is the positive and negative satisfactions that people possess with regard to different places and regions. Similarly, Lee (1966) in an important theoretical paper suggested a useful framework, seeing migration as the outcome of a disparity in the balance existing between sets of perceived attracting and repelling forces at the present location (origin), and a possible future one (destination). Thus, if the balance between the pluses (or attracting forces) and the minuses (or repelling ones) lies in favour of the destination, migration may occur (Figure 3.10a). The information stream available to individual decision-makers is therefore extremely important, especially its spatial extent and overall accuracy. Lee's ideas give rise to the by now well-known concepts of 'push' and 'pull' factors of migration. In the Third World setting, the pull factors relate to the rural-urban gap as epitomised by the better incomes, employment prospects, public amenities such as education, training, medical care, piped water, housing and sanitary facilities existing in the towns. On the other hand, push factors reflect the massive rural poverty and unemployment which prevail. Although some might be tempted to argue that conditions are so poor in Third World cities that migration is irrational, this view misunderstands the scale and severity of rural poverty. Thus, it is by now well-established that urbanward migration occurs for structural-economic reasons and not merely due to the lure of bright lights:

> Wage differentials everywhere explain the move to the towns, even in Jamaica where unemployment is far higher in Kingston than in the rural areas, though it is rarely perceived to be so by prospective migrants (Dickenson _et al_, 1983, p. 176).

It is often the perceived possibility of a better life, either now or more often in the near future that encourages cityward migration. This reflects rural-urban real income and social welfare differentials, even if the immediate prospects of securing a job and a decent home are slim. As already noted, in the majority of Third World cities, tertiary economic activities predominate rather than manufacturing activities. Thus, for example, in Caribbean cities, unemployment frequently runs at between 20 and 30 per cent and the jobs which do exist are mainly to be found in domestic service and the tertiary refuge or informal sectors of the economy.

The impact of cityward migration on urban growth in Third World countries is obviously considerable, but nevertheless is sometimes

a. **Perceived attracting and repelling factors in migration**

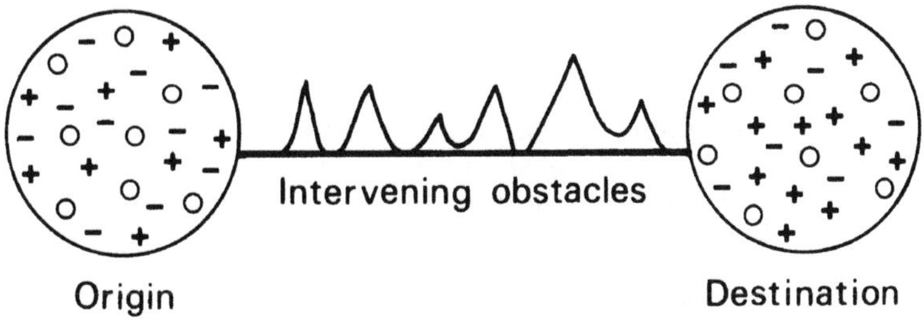

b. **The mobility transition**

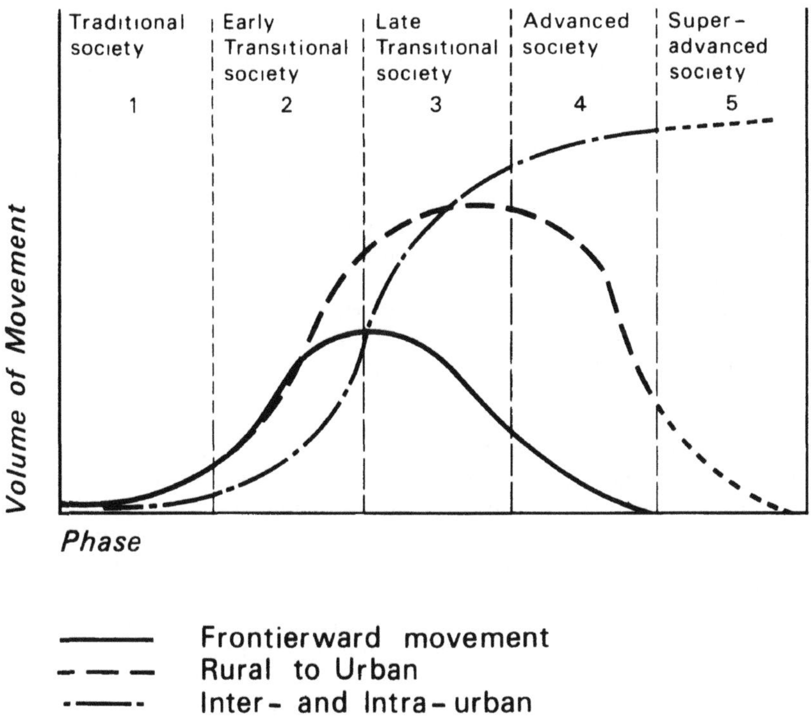

Figure 3.10: Basic aspects of spatial migration

over-stressed. In fact, debate has always raged as to whether urban areas grow by natural increase or migration (see Keyfitz, 1980, for instance). Davis (1965) and others stress that European cities during the industrial revolution grew mainly by in-migration. But as we have already noted in Chapter 2, as a result of mortality reduction in Third World countries, cities of the less-developed world are growing as much by natural increase as by rural to urban migration. Generally, natural increase and migration contribute about equally to total urban growth in developing countries. For example, in Bombay from 1952 to 1961, migrants accounted for 52 per cent of urban population increase, and the equivalent figure for Nairobi (1960-69) was 50 per cent, and for Bogota (1965-66) 33 per cent. The fact that migrants are for the greater part young, means that urban reproductive capacity will be high, so that high rates of natural increase are inextricably linked to rapid urban in-migration.

What is true, however, is that rural to urban migration represents the predominant type of population migration in transitional or developing societies. This is clearly exemplified in the mobility transition suggested by Zelinsky (1971). As shown in Figure 3.10b, it can be envisaged that the volume of cityward migration also follows a bell-shaped curve with development stages. Little or no permanent migratory movement occurs in traditional societies. Frontierward and rural to urban migration come to characterise the second stage, that of the early transitional society. By the late transitional period, however, cityward movement is predominant. Toward the end of the late transitional period, inter-urban and intra-urban movements become the most numerous and this pattern becomes the established norm in advanced and super-advanced societies (see also Simmons, 1968; Short, 1978). The developmental trends shown in Figure 3.10b are of course the demographic corollary of core growth and increasing spatial dualism.

Having looked at the why side of rural to urban migration, we must now turn to how. The popular image of cityward migration is the direct movement of the rural poor and unemployed from the countryside to large cities. Direct migration, where rural to urban transfer takes less than six months is indeed often true of West Africa, East and South East Asia, reflecting the perceived attractions of large cities and the lack of intervening opportunities. In many other cases, however, the movement between the countryside and city is far more gradual, perhaps spanning several years and being composed of a number of distinct steps. Thus, a rural dweller may move to a small town, and then later up the settlement hierarchy to a large town and thence on to the primate city. This is known as step-wise migration, and there is much evidence of its operation in both Latin America and the Middle East. It is important to note that step-wise migration can itself occur in a number of different ways, as clarified by Conway (1980). As shown in Figure 3.11, spatial step-wise migration can be distinguished from hierarchical step-migration. In the former, the migrant makes a series of short-distance moves, always moving closer toward the major city, which is the ultimate destination (Figure 3.11a). In the hierarchical step-migration process, there is an almost aspatial movement of people upward through the country's urban hierarchy (Figure 3.11b). More realistically, perhaps, the idea of a series of spatial step-wise migrations can be juxtaposed with that of a hierarchical step-wise progression. Such a hierarchical-cum-spatial step-wise migration process, shown in Figure 3.11c, has been supported by a number of researchers (Riddell and Harvey, 1972; Thomas, 1972; Kosinski, 1976; Conway, 1980). A third distinct type of cityward migration is "fill-in" migration, as discussed by Gilbert and Sollis (1979). Here, as an individual moves from a small to a large city, a gap is left in both employment and residential terms, which subsequently will be filled by somebody from a rural area.

Much research remains to be done on rural to urban migration paths and processes in Third World countries. At the present time, a major conclusion must be that such phenomena are far from uniform, reflecting different socio-economic environmental conditions, as well as different social and economic groups of migrants. This diversity is expressed in Third World urban residential and employment patterns, a point that will be exemplified subsequently in the present chapter.

THE GROWTH AND STRUCTURE OF THIRD WORLD CITIES

Just as massive poverty, manifest inequalities and rapid population growth in Third World countries may be regarded as correlates of the rural-urban gap and of cityward migration, so these factors are closely mirrored in the current growth and form of cities in the less developed realm.

Two distinctive characteristics of present day Third World cities are their rapid recent growth and the pronounced internal divisions and social inequalities that characterise them. Thus, Mexico City, currently with a population of 10.9 million is growing at an annual rate of 4.4 per cent and it is estimated that it will reach 31.5 million by the year 2000. Similarly, Cairo is increasing at 3.6 per cent per annum and its population will grow from 6.9 millions to 16.9 millions from 1975 to 2000. With regard to inequalities, clearly decipherable traditional and modern sectors exist in industry, retailing and housing, in fact, in all realms of city life.

Frequently, Third World cities are seen as being synonymous with the concept of the pre-industrial city. Thus, Dickenson _et al_ (1983, p. 193) state that "In origin Third World cities and towns - both pre-colonial and colonial - are quintessentially pre-industrial". the idea of the pre-industrial city as a distinct urban form was first stressed by Sjoberg (1960). Typical attributes of such cities were seen as their walled structure, narrow streets, jumbled morphology and multi-functional economic character. In geographical terms, Sjoberg identified three broad generalisations relating to the spatial structure of pre-industrial cities. The first, was that the elite social groups typically tend to reside near to the urban centre, in close proximity to the prestigious, administrative and symbolic buildings,

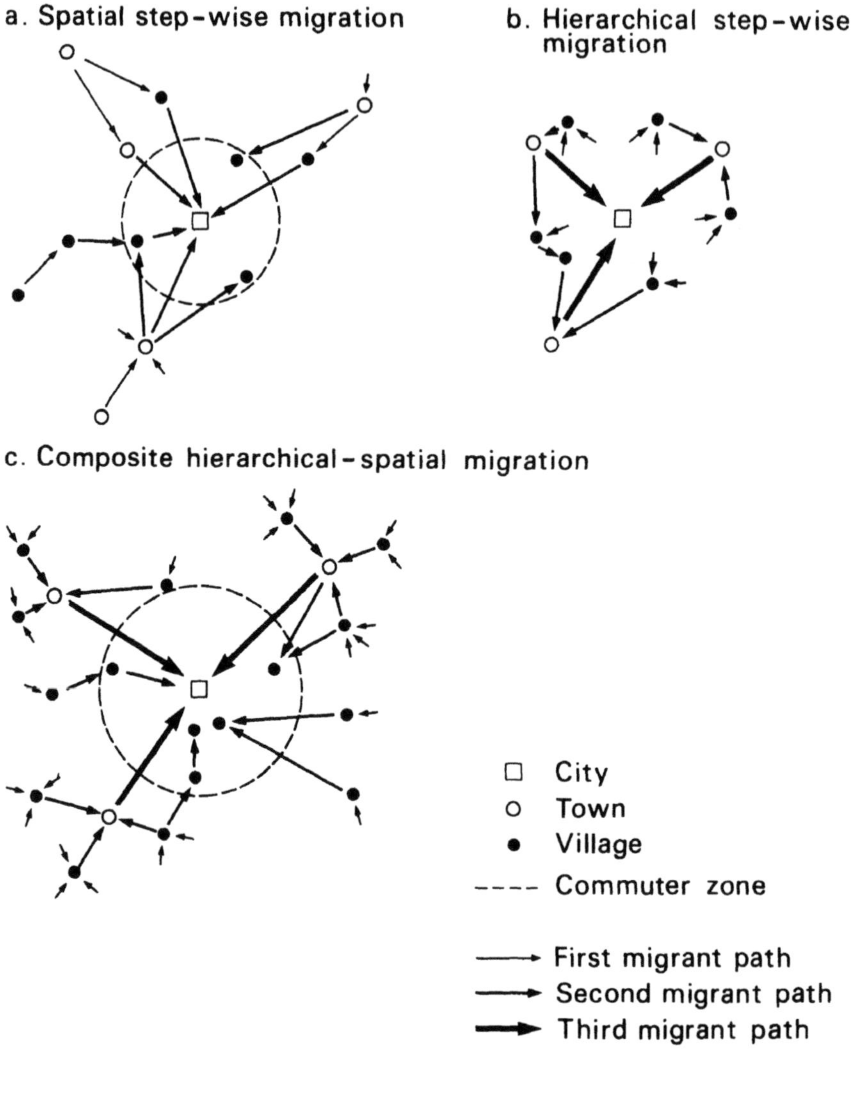

Figure 3.11: Spatial and hierarchical step-wise migration (Source: Conway, 1980)

whilst the poor live towards the periphery. Secondly, such broad social class areas are usually further broken down into distinct quarters or precincts. Thirdly, Sjoberg maintained that pre-industrial urban areas tend to lack clear functional differentiation of their land-use patterns, with most plots being put to multiple uses, such as work and residence. There has been general confirmation of the salience of these features, especially the declining social gradient with movement outward (Langton, 1975, 1978; Radford, 1979). This, of course, is the reverse of the social patterning that is found in most industrial and post-industrial cities (Burgess, 1925).

A number of authors have seriously questioned whether a singular description such as this can be applied to cities occurring in widely different geographical, cultural and social settings, and covering a 5000 year time span. McGee (1971) has argued that the biggest shortcoming of Sjoberg's two-fold division of cities is that it ignores the colonial city, which developed as the consequence of the interaction between two civilisations. Horvath (1969, 1972) has made a similar point in classifying the colonial city as a distinct urban form, resulting from the dominance of one group by permanent colonising settlers. It is argued that these imposed colonial characteristics have often endured to the present-day, making them particularly significant.

Although the pre-industrial or inverted Burgess social pattern is still observable in some Third World cities, notably those of Asia and Africa (Dwyer, 1975; Mabogunje, 1968), research has shown that it is fast becoming less typical, especially in Latin America. Thus, in a study of Bogota, Quito, Lima and Santiago, Amato (1970) has shown how the elite groups have tended to desert the centre of the city and now generally occupy one or more sectors of the outer urban fringe. This sort of spatial patterning is of course highly reminiscent of the Hoyt (1939) scheme of urban growth and structure in developed countries. Similar conclusions have been reached by other writers, for example, Morris (1978) for Caracas, Venezuela, and Conway (1981) for Port of Spain, Trinidad. This feature has also been assigned as of prime significance in a model of contemporary Latin American city structure recently put forward by Griffin and Ford (1980). In this, the elite residential area occupies a distinct sectoral zone, surrounding a commercial and industrial spine which radiates from the C.B.D. But interestingly, in the other areas of the model city, residential quality is still seen as decreasing as one moves out in concentric zones, due to the location of large peripheral squatter settlements. Although other authors have argued against such a simple assignment of squatters to the periphery in any model of Latin American city structure (Morris, 1978), pressures for land often do mean that the largest are peripheral. Certainly, such residential zones are effectively becoming the ubiquitous hallmark of rapid urbanisation in Third World countries. This point is examined in detail in the following section.

LOW INCOME HOUSING IN THIRD WORLD CITIES

Perhaps the most conspicuous and emotive signs of Third World urban poverty and social inequality are to be found in the field of housing. One way or another, the urban majority live in accommodation which viewed from a Western standpoint is shabby and lacking in the most rudimentary of facilities. Thus:

> Despite man's unprecedented progress in industry, education, and the sciences, the simple refuge affording privacy and protection against the elements is still beyond the reach of most members of the human race (Abrams, 1964, p.1).

The problem is not just a reflection of the existence of squatter settlements or shanty towns in Third World cities, but rather reflects an inability or unwillingness to tackle the overall housing situation. As Abrams (1964) stresses in his well-known book, Housing in the Modern World, there are at least three distinct types of poor urban dweller in developing cities. First, there are the homeless and the street sleepers. Secondly, a large group are to be found living in slums and tenements, frequently paying relatively high rents. Thirdly, Abrams recognises the squatters and occupants of the shanty towns. The term "low-income housing" is used in the present account to cover all types of housing for the poor in Third World cities.

The Third World housing problem has really emerged since the early 1940s (Dwyer, 1974). The scale of the problem is easily exemplified. In Calcutta, it was estimated that in the early 1960s, some 600,000 people slept on the streets, whilst in Bombay, one in every 66 people were homeless and a further 77,000 lived under stairways, on landings and the like (Abrams, 1964). In many parts of the British West Indies, people live in mobile shacks or 'chattel' houses reflecting the fact that although they own their self-built houses, they rent the land on which they are located. In Cairo, the severe housing shortage is reflected in several quite distinctive and novel ways. The old city, or madina has become a vast area of tenement slums. Perhaps more surprising are the 'tomb cities' or 'cities of the dead', which are to be found located on the eastern edge of the city. Here, since the second World War, the tomb houses built for caretakers or for relatives visiting graves are now occupied by the poor as permanent homes. Another response is living on the rooftops of apartments, for as long as structures placed on the roofs are not constructed of permanent materials, they are not illegal. It is estimated that in the region of half a million people live in such roof top dwellings in Cairo (see Brunn and Williams, 1983; El-Shakhs, 1971; Abu-Lughod, 1971).

The most ubiquitous sign of rapid Third World urbanisation is, however, the "squatter settlement" or "shanty town". Such dwelling types also go under a wide variety of other descriptions, among them "spontaneous settlements", "uncontrolled settlements", "makeshift", "irregular", "unplanned", "illegal", "self-help", "marginal" and "peripheral settlements". This diverse nomenclature is added to by an almost bewildering array of local names such as the "barriadas" and "pueblos jovenes" of Peru, the "favelas" of Brazil, the "colonias proletarias" of Mexico, the "gecekondu" of Turkey, "bustees" of India, "bidonvilles" of Algeria and Morocco and "gourbivilles" of Tunisia. These developments all reflect an excess of housing demand over supply and a situation where the erection of a dwelling is frequently primarily in the hands of the household itself. Such responses have occurred in developed countries, for example, the United States, Germany and France in times of economic hardship (Harms, 1982). Indeed, as Abrams (1964) reminds us, before the development of the capitalist mode of production, self-help represented the predominant form of housing provision. In the contemporary Third World city it reflects the failure of the private and public housing sectors to provide adequate low-income housing.

The variety of terms employed points to an important characteristic of such settlements, namely their extreme diversity with regard to formulation, physical fabric and inhabitants. The terms "squatter settlement" or "illegal settlement" are frequently used, but are now increasingly regarded as being disparaging (Drakakis-Smith, 1981) and also potentially misleading. Thus, squatter settlements are those where individuals have settled without legal title to land, or alternatively without planning permission. Such housing is frequently located on government or church-owned, rather than private land. But illegality is not always characteristic. Some low-income houses are owned, the plots having been sub-divided and sold. Similarly, some homes and/or the land on which they are sited are rented. Such rentyards are common in many Caribbean countries and also in Mexico City (Clarke and Ward, 1976; Ward, 1976). Another common characteristic and description is that they are "makeshift settlements" or "shanties", built from whatever materials are available. Thus, basic shelters made from cardboard cartons and newspapers, or from packing cases and fish barrels have been described in the Moonlight City area of West Kingston, Jamaica by Clarke (1975). More typically, perhaps, recycled scraps of wood and corrugated iron may be employed for both outside walls and roofs, as exemplified by Plates 3.1 and 3.2, which show dwellings in Eastern Port of Spain, Trinidad and central Caracas, Venezuela respectively. Other materials that are frequently employed in construction include flattened tin cans, straw matting and sacking. Such settlements may also be makeshift in the sense that they have none of the basic urban services such as water, electricity or sewerage when they are initially developed. But even in this regard caution must be exercised, for with time such services may well be acquired, and likewise, brick built houses may come to predominate in a formerly makeshift area. Certainly, the basic but frequently overlooked distinction between squatter settlements characterised by their illegality on the one hand, and shanties identified by virtue of their poor physical fabric on the other, must be clearly appreciated. Yet other settlements are characterised by their "unplanned" or "irregular nature", or by their origin in rapid mass "invasions" of land. Such haphazard or speedy development is epitomised by the description

Plate 3.1: A basic shelter in the John John area of East Port of Spain, Trinidad

Plate 3.2: Makeshift dwellings at right Caracas, Venezuela

"spontaneous". All of these terms are highly appropriate in certain cases, but potentially misleading in others. Thus, whilst many low-income settlements are unplanned in a professional sense, many are the result of much careful forethought, especially those involving organised land invasions. The latter are typical of some Latin American squatter settlements (see, for example, Gilbert, 1981). However, in Africa, Asia, the Middle East and the Caribbean, the development of low-income housing is typically a much more gradual process based on slow infiltration and individual initiative. In this sense, such developments are characterised by the very antithesis of spontaneity. Similar concern can be expressed concerning the lack of universal applicability of descriptions such as "peripheral" or "marginal settlements", whether used in a strictly geographical or an economic context. Finally, although the terms "self-help" and "autoconstruction" are useful in signifying that the building of such dwellings is not normally undertaken by professionals, it would be highly erroneous to give the impression that such houses are built entirely by their present or previous occupants.

In short, the very character of these settlements necessitates a catholic definition, and the adoption of a single description merely for the sake of convenience. The term "low-income housing" is preferred by the present author, so that the range of housing responses of the poor can be seen together. But even this runs the risk of implying a misleading economic homogeneity among inhabitants. Recently, the term spontaneous settlement has come into widespread use as a generic term. However, the salient point is that these settlements are highly diverse, varying with respect to basic attributes such as materials, tenure, legality, speed of occupation, size, location, the origin and nature of residents, their perceptions and aspirations.

There are by now a number of excellent texts on spontaneous settlement and housing in the Third World, especially those by Dwyer (1975), Lloyd (1979, 1980) and Drakakis-Smith (1981). Similarly there are many papers by geographers concerning low-income housing in different areas of the world, such as Latin America (Ward, 1976, 1978; Gilbert and Ward, 1981, 1982a, 1982b; Stadel, 1975; Buksmann and Rowley, 1978), the Caribbean (Clarke, 1966, 1974; Eyre, 1972), Asia (Jackson, 1974; Ulack, 1978; Mountjoy, 1982; Aitken, 1981) and Africa (Wells, 1969; Winters, 1982), as well as more general accounts (see, for example, Mountjoy, 1976, 1978). The growth of numbers in spontaneous settlements is often running at between 12 and 15 per cent per annum, that is frequently at rates which are far ahead of the growth of total urban populations (Turner, 1967; Dwyer, 1975). Statistics in Dwyer (1975), for example, show that there are 800,000 squatters in Rio de Janeiro, and 767,000 in Manila. Frequently, spontaneous settlements account for at least 20-30 per cent of the total urban population, as in Kingston (25 per cent), Calcutta (33 per cent), Manila (35 per cent), and Rio de Janeiro (30 per cent). But in a number of instances, this percentage is far higher, as for Bogota (60 per cent), Casablanca (70 per cent) and Addis Ababa (90 per cent).

Quite simply, the Third World urban poor cannot afford to purchase

Figure 3.12: Typical sites for spontaneous settlements (Source: Dwyer, 1975)

houses that are properly surveyed, built and serviced. Where rental property is available, either in private tenements or in government housing schemes, rents are exorbitantly high (see, for instance, Clarke, 1966, p. 176, in relation to the Jones Town area of West Kingston). Thus, squatters construct peri-urban structures on land which has not previously been used for building purposes. Typical sites for spontaneous settlements are illustrated in Figure 3.12, all of them constituting what may best be described as marginal land. In the old walled section of Manila, spontaneous settlements have infilled small vacant lots (Figure 3.12a). Another typical site is on steep hillsides, as shown in Figure 3.12b for a part of Caracas. The same is also true of cities like Hong Kong and Rio de Janeiro. Land which is swampy or subject to flooding offers further opportunities, as shown by the example of Singapore (Figure 3.12c). Similarly, land adjacent to railways is also often occupied by squatters, as in the case of Kuala Lumpur (Figure 3.12d).

Dwyer (1975) has presented an extremely useful model of the location of spontaneous settlements, despite warning that "As yet there is no generally accepted theory of spontaneous settlement location" (p. 32), due to a scarcity of comparative research. The first point that can be noted is that spontaneous settlements are frequently located throughout contemporary Third World cities. However, the largest are almost always peripheral. Secondly, the city centre with its urban job opportunities, both formal and informal, can be regarded as the principal attracting force for spontaneous settlements (Figure 3.13). As the urban area expands, Dwyer notes that the ideal major squatter location may well come to be moved out centrifugally by the normal processes of invasion and succession. Other prime locative factors for spontaneous settlements are good water supplies and tolerable relief. However, with growth, spontaneous settlements may be displaced outward toward difficult topographical sites and those which lack water, as shown in Figure 3.13. Similarly, basically agricultural spontaneous settlements on the urban fringe may eventually be replaced by urban-oriented spontaneous settlements.

Dwyer's locational model is also useful in stressing that there are diverse paths of migratory movements of low-income populations into spontaneous settlements. The commonly held view that spontaneous settlements are populated by migrants who have moved directly to them from small towns and the countryside is indeed frequently true of many African cities. However, for most Asian, Latin American and Caribbean cities, the evidence suggests that spontaneous settlements are not settling areas for new arrivals from the rural areas. More typically, new urban arrivals go directly to the city centre in order to be close to job opportunities (outer right-hand loop of Figure 3.13). Thus, inner city tenement slums often act as the reception areas for new migrants, who only later move to spontaneous settlements, often after establishing themselves in the urban environment. Increasingly, this distinction is seen as quite basic and suggests yet another dimension of diversity in that squatters are frequently not the poorest of the poor, a point that

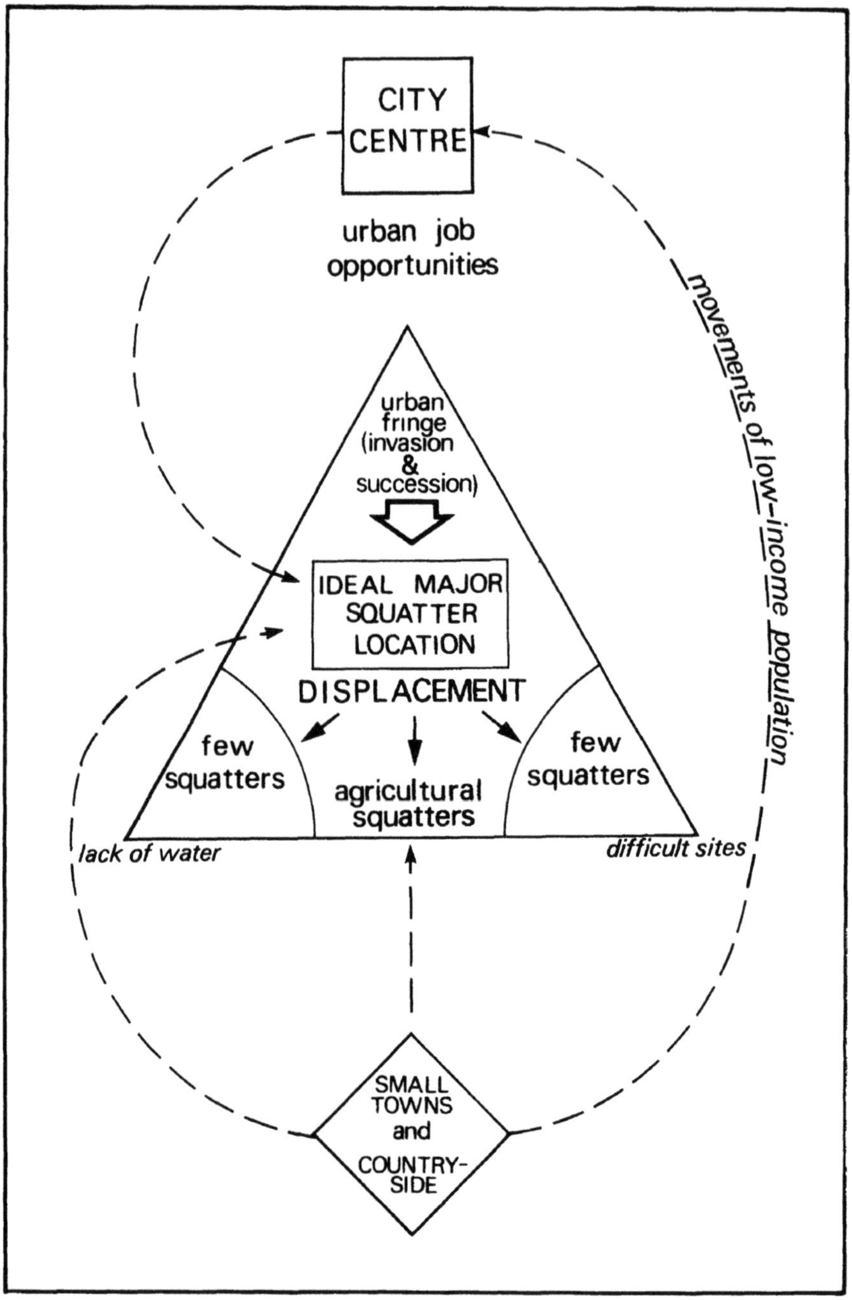

Figure 3.13: Dwyer's locational model of spontaneous settlements (Source: Dwyer, 1975)

will receive detailed attention shortly.

Reference to a specific case study should help to illustrate some of these points regarding the structure and population dynamics of spontaneous settlements. In 1972, Eyre published an excellent study of the shanty towns of Montego Bay, the second largest settlement in Jamaica, which had a population of approximately 50,000 at that time. Eyre noted that strong spatial variations in income and wealth characterised the city. Further, it was suggested that Montego Bay could be split into six broad urban zones: the central business district, the tourist sector, the free-port commercial area, an outwardly expanding arc of middle- and upper-class residences, and finally, a group of shanty towns. In all, ten such shanty towns were clearly identifiable, mainly located some 4-6 km from the urban core, and between them housing some 20,100 persons, that is 40 per cent of the total population. The structure of Montego Bay and the migration paths underlying it are clearly shown in Figure 3.14.

Eyre identified five main flows of population (shown from A to E in Figure 3.14), and the relative importance of these substantiates the earlier discussion and illustrates the salience of Dwyer's locational typology. Thus, the predominant flow (A) is of rural migrants into the low-income inner city slums. Most of those involved in this movement are reported to be single and poor and looking for their first job. The second most significant flow (B) is shown to be that from the city centre slums to the shanty towns. These individuals are seen as wanting to escape from the high rents and other negative aspects of inner city life (e.g. crowding and crime) and are often those with a family and perhaps with some accumulated capital too. The other flows involve far fewer people. They include the movement of the upwardly mobile from inner city slum areas to the upper and middle class zones (flow C) and an even smaller flow from shanty towns to the middle class ring (flow D). Finally, the flow of rural migrants directly to the shanty towns (flow E) is shown to be of only very minor significance. Eyre (1972, p. 403) summarises as follows:

> ... the majority of shanty town dwellers are urbanites of long standing, not rootless rural migrants who have drifted cityward ... The average household head had lived eleven years in Montego Bay, including his residence in the shanty town. More than three-quarters of the population of all ten shanty towns had been born within the city limits as defined. The rural migrant stream is directed primarily toward the inner city, where rows of slum tenements specifically cater for such transients. A central location is more convenient for job seeking, since a peri-urban location incurs travel expenses.

THE ROLES OF SELF-HELP AND COMMUNITY INVOLVEMENT

These comments concerning low-income housing areas in Third World

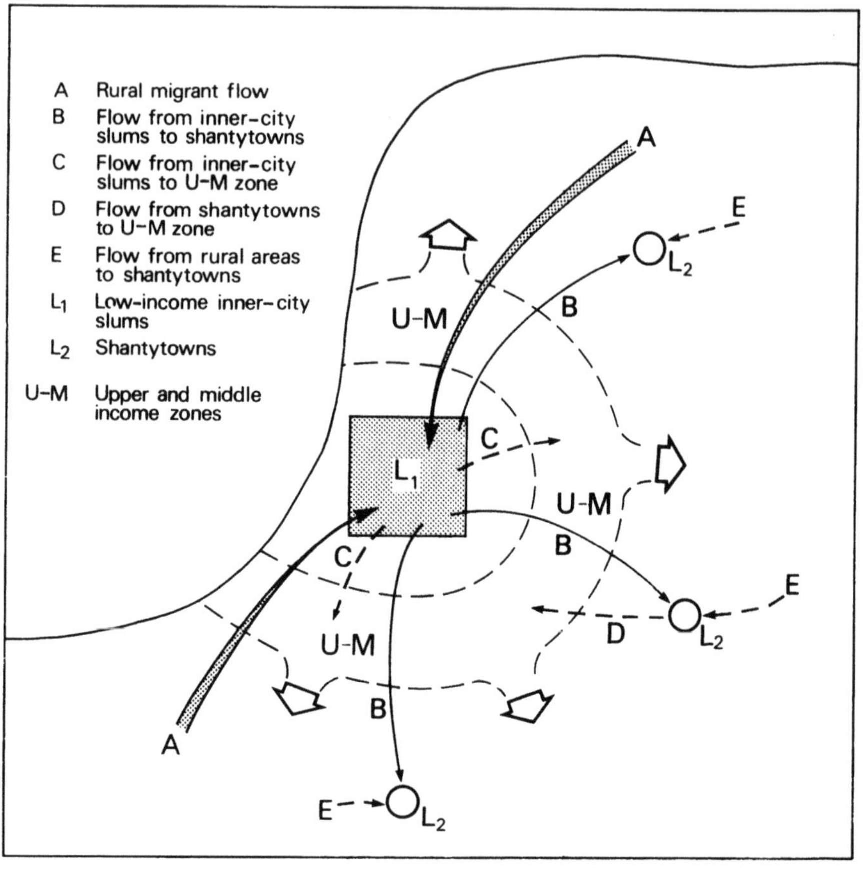

Figure 3.14: Squatter settlements and migrations in Montego Bay, Jamaica (Source: Eyre, 1972)

cities bring us to an important issue, that is the practical and ideological implications of such settlements. Are they good or bad? Will they improve or deteriorate over time? Do they merely reflect poverty or do they constitute a dynamic force and desire for change and improvement? There are two parts to all of these questions. First, if left to themselves, would such settlements improve and become a beneficial part of the urban fabric? Secondly, therefore, how have and how should governments, both local and national, react to spontaneous settlements?

Turning to the first question, it seems that academic views on spontaneous settlements have changed over time in response to popular attitudes at large (see Lloyd, 1979, pp. 53-7, for instance, on this point). Attitudes have, in fact, changed quite markedly since the 1950s, and can by and large still be characterised as ambivalent, perhaps not too surprisingly when the pressing nature of the urban housing problem is fully appreciated. Many of the early views were essentially negative and pejorative, and in the 1960s, came to be dominated by Oscar Lewis's (1959, 1966) concept of the culture of poverty. Lewis, an American anthropologist, had worked in Mexico, Puerto Rico and India. He argued that the poor are locked into an inescapable cycle of poverty and thereby come to form a culturally separate group within society. As Drakakis-Smith (1981) observes, recent views of squatters as socially and economically marginal groups, participating only in the informal sectors of the economy carry some connotations of the culture of poverty. But such ideas have been strongly attacked of late, especially on the grounds of lack of rigour (Lloyd, 1979; Drakakis-Smith, 1981). Further, it is suggested by some that such views are convenient for the wealthy and powerful, insofar as they suggest that "poverty is the poor's own fault" (Gilbert and Gugler, 1982, p. 84). It is views such as these that are reflected in popular descriptions of squatter/spontaneous settlements as "urban cancers", "festering sores", and "urban fungi" and the like.

But Stokes (1962) in a paper presenting an overall theory of slums, made a clear distinction between what he saw as successful and unsuccessful poor communities, referring to these as 'slums of hope' and 'slums of despair' respectively. Although Stokes's division represents a partial refutation of Lewis's concept of the culture of poverty, it probably did result in a certain degree of tunnel vision and inflexibility of thought. However, the distinction between settlements which are inhabited by the upwardly mobile and achievement-oriented on the one hand, and by the downwardly mobile on the other is an interesting one. Certainly, such a dichotomy is useful in stressing the extreme socio-economic diversity which often characterises spontaneous settlements, as is perhaps exemplified in Plates 3.3 and 3.4 taken in squatter areas of Castries, St Lucia and Caracas, Venezuela respectively.

Such writing heralded a more positive view of spontaneous-type settlements when viewed both in terms of the individual family and the community at large. In particular, "self-help" housing, where homes are built, added to and improved over time was increasingly

Plate 3.3: Settlements of hope or despair? Residents of Castries, St. Lucia

Plate 3.4: Settlements of hope or despair? Barrio residents of Caracas, Venezuela

being seen as a potentially beneficial force. It was argued that the poor are not indolent, dishonest and disorganised, but rather quite the reverse (Abrams, 1964; Turner, 1963). However, a major change in attitudes was to be precipitated by the experiences of two academics who were working in Peru: William Mangin, an American anthropologist, and John Turner, a British architect (see, for example, Mangin and Turner, 1968). Both Turner and Mangin were to advocate strongly self-help housing as a positive and efficacious process in Third World housing provision. Arguably the single most important paper was written by Mangin (1967), the title of which conveys the essence of the argument presented: Latin American squatter settlements: a problem and a solution. In this, Mangin described most of the predominant views on squatter settlements as myths: they were not disorganised, a drain on the urban economy, populated by criminals and radicals, nor were they made up of a single homogeneous social group. Rather, Mangin argued that most squatters were in employment, were socially stable and had been residing in the city for some considerable time. Illegally occupying land gave them the opportunity to avoid paying high rents, and at the same time, to build their own homes at their own pace. Turner, in like manner, worked for over eight years in Peru, and for a considerable proportion of that time was involved with self-builders in barriadas. His partly autobiographical essay (Turner, 1982, pp. 99-103) is particularly instructive in this regard. His overall attitude is clearly summarised in one concise quotation:

> Like the people themselves, we saw their settlements not as slums but as building sites. We shared their hopes and found the pity and despair of the occasional visits from elitist professionals and politicians quite comic and wholly absurd (Turner, 1982, p. 101).

Turner argues that all that had to be done in order to assist the self-builders was to "approve simple sketch plans, and distribute cash in appropriate stages" (p.102) and further commented that the economies of self-help were founded upon "the capacity and freedom of individuals and small groups to make their own decisions, more than on their capacity to do manual work" (p. 102), so that as a consequence, "never before did so many do so much with so little" (p. 102). However, perhaps the most positive message put forward by Turner was that such settlements improve progressively over time. Thus, houses that were originally constructed from straw matting later acquired walls, services and paved streets. In the terminology of Stokes, they were clearly slums of hope, characterised by *in situ* housing improvements and the upward social mobility of their populations (see also Turner, 1969, 1976).

The clearest indication of Turner's basic ideas concerning housing improvement are contained in his typology of low-income housing groups in Third World cities, shown in Figure 3.15 (see Turner, 1968). Put in simple terms, Turner argued that in making residential choices

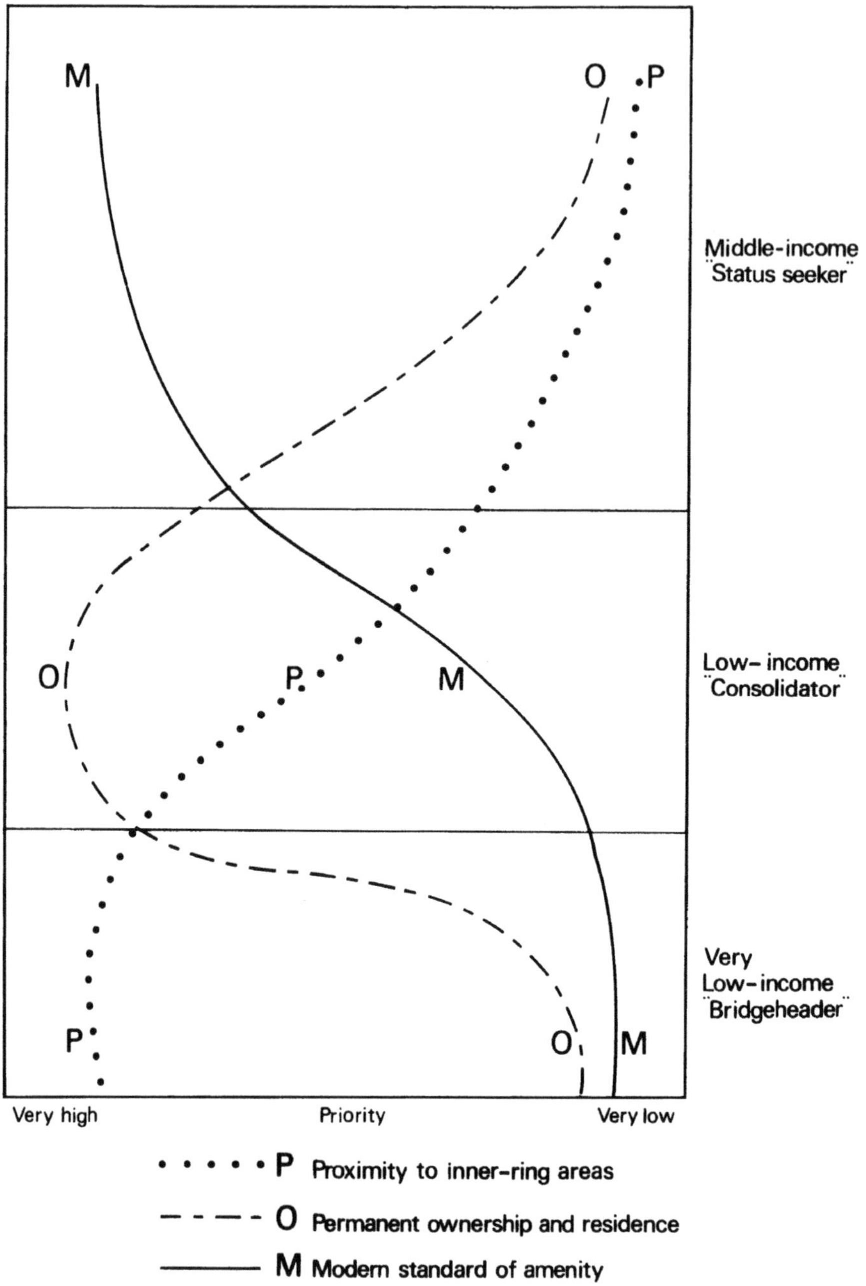

Figure 3.15: Turner's model of low-income housing groups

all individuals are influenced by three major groups of factors. First, there is the need to live in close proximity to the inner-urban ring, so as to be near available job opportunities (see curve PP in Figure 3.15). Secondly, many individual households have the desire for permanent ownership and/or security of tenure (OO). Thirdly, the quest for modern standards of amenity (MM) affects many. But the relative importance of these three desires varies for different sets of low-income residents, giving at least three major sub-groups: "bridgeheaders", "consolidators" and "status seekers" (Figure 3.15). These three groups reflect our recurrent theme of the socio-economic diversity existing among low-income housing groups.

The very low-income "bridgeheader" is normally a recent rural migrant who is seeking to establish himself. His priority will be to find a job and thus to reside close to the urban core, whereas his need for permanency, ownership and high standards of amenity will be slight (Figure 3.15). In contrast, the low-income "consolidator" is one who has already gained a firm foothold in the urban area. It is more than likely that his income will be rising and that a high priority will be placed on attaining better standards of amenity and ownership (Figure 3.15). Turner's model thereby parallels the division of settlements, and to some considerable degree their inhabitants, into the categories of 'hope' (consolidation) and 'despair' (bridgeheading). The model also allows for a third group of fully upwardly-mobile individuals, who are described as middle-income status-seekers, for whom modern standards of amenity become the ultimate goal, whilst proximity to inner-ring areas and permanent ownership are factors of little or no consequence (Figure 3.15).

The major policy implication of Turner's work is that governments are best advised to help the poor to help themselves by facilitating spontaneous self-help, and by sponsoring aided self-help schemes such as squatter upgrading, core housing and site and services schemes. These local intra-urban planning responses are fully considered in the next chapter. Whilst Turner's views have certainly helped to change public and governmental attitudes to spontaneous settlements, they have been attacked by some for implying that the poor can exercise choice between the variables examined (Gilbert and Gugler, 1982), although in fairness, the Turner model does deal with the latent aspirations and cognitions of groups, rather than their revealed behaviour. But perhaps far more fundamental are the criticisms of Marxist writers who maintain that Turner's thesis, and indeed the policy aspects of self-help housing aid capitalist development and rationalise mass poverty (see, for example, Burgess, 1977, 1978, 1981, 1982; Conway, 1982; Harms, 1982). In particular, Burgess (1982) has dubbed Turner's advocacy of self-help housing "a curious form of radicalism". We shall return to these arguments shortly.

There is, however, considerable evidence suggesting that consolidation of spontaneous housing does occur if the circumstances are right. Plate 3.5, for example, shows a quite typical scene in the urban area of Caracas, Venezuela where houses are actively being improved, with some brick-block houses acquiring new storeys, whilst a few somewhat more humble wooden shacks are still to be

Plate 3.5: The consolidation process at work: Caracas, Venezuela

Plate 3.6: Low-income housing in East Port of Spain, Trinidad showing signs of consolidation

seen between them. Similarly, some of the dwellings which started out as simple timber abodes on the hills extending eastwards out of Port of Spain, Trinidad have undergone progressive improvement and may now be hard to distinguish from what may be regarded as more 'regularly' built homes (Plate 3.6). Two conditions, however, are generally seen as being essential prerequisites to spontaneous housing consolidation; firstly, a sufficiently high income amongst residents and secondly, and possibly more importantly, security of tenure. Without these, neither the means nor the desire for substantial housing improvement and upgrading will exist in the first place.

In this connection, it is salient that much of the attitude changing work described above was carried out in the Latin American context, where despite the existence of wide income differences, economic conditions have generally been more healthy than elsewhere in the less developed realm. Thus, incomes on average have probably been high enough to give the consolidation process a chance to work. In Africa, Asia and much of the Caribbean, on the other hand, conditions by and large are much poorer. Thus, there is less chance that self-help imperatives will be able to alleviate housing problems. This is an important cautionary note, which is certainly borne out when contemplating the housing problems of say, Calcutta. It also serves to remind us that the Third World cannot not be treated as a homogeneous entity, an argument stressed in Chapter 1 in relation to the futility of trying to unearth universal solutions to so-called 'Third World problems'.

Turner's ideas are just as noteworthy and perhaps as contentious too, in suggesting that the public know what is best for them. Clearly, Turner's overall thesis is anti-elitist planning and architectural practice. Similarly, ideas concerning the acceptability of individual and community self-help dovetail with those advocating appropriate or intermediate technology (Dunn, 1978). Self-help housing, involving the original building of homes, their improvement and the lobbying of local politicians and administrators for services can all be seen as forms of citizen participation in the planning process. Whilst it is all too easy to talk about public participation in planning, it is far more difficult to establish and implement it in a practical manner, and there are special problems which may be encountered in the Third World setting (Conyers, 1982). This theme is treated at length in Chapter 5. A principal difficulty for decision-makers and researchers is the identification of the needs, attitudes and perceptions of members of the public. Lloyd (1979, p. 182), for example, describes as "vacuous in the extreme the types of questions that are typically asked in surveys of Third World urban residents". However, this need not necessarily be the case, and quite clear policy-related recommendations can be derived by such methods, as shown in Dann's (1984) recent sample survey of the quality of life in Barbados. Another good example is afforded by the work of Andrews and Phillips (1970), who by means of sample surveys, examined the attitudes of inhabitants of selected barriadas in Lima to the provision of 26 basic services. Here it was found that dissatisfaction was most acute in respect of the availability of property titles, the location of medical services and the provision of utilities such as street paving, lighting and water.

As already noted, although the promotion of self-help appears to have democratic appeal, some have argued strongly that it has other less laudable connotations as well. The principal advantages of self-help have been well-summarised by Ward (1982) and include the provision of alternative housing, labour rather than capital intensity, overtones of intermediate technology, institutional acceptance and powerful backing by organisations such as the World Bank and the 1976 Habitat International Conference. However, a growing number of analysts maintain that the acceptance of an explicitly self-help based housing policy can be used as the front for a policy of neglect by governments. Those of the hard left of the political spectrum thereby argue that such policies can be employed merely as a diversionary exercise, acting as a conspiracy to maintain the status quo of monopoly capitalism. This line of attack has been put extremely forcibly by Burgess (1982), as previously noted. Even Abrams (1964) observes that the urban poor in less developed countries are doubly exploited: once at work and once at home. It is certainly tenable to suggest that this overall thesis casts some interesting light on governmental reactions to illegal squatter settlements. Sometimes this amounts to a policy of what is best described as benign neglect, where the issue is effectively ignored. On other occasions, especially in the 1960s, eradication was often the principal response, as, for example, in Manila where some 2,877 shacks were demolished in a two week period (Dwyer, 1975). But the eradication of even sub-standard dwelling units amounts to a bizarre form of housing policy unless the occupants come to acquire more satisfactory shelter as a result. It is tempting to suggest that the state's posture of turning a blind eye in some circumstances, reflects the fact that this strategy may well afford it power over the groups which illegally occupy land. Further, the provision of key services and utilities is not unknown immediately prior to an election. Similarly, the question of political patronage crops up again, when, as is common, the occupants of squatter settlements state that they have been encouraged to occupy government-owned land by the political party in opposition. Such an action might well gain electoral support at the same time as creating a problem for the existing government.

It seems almost inescapable that the positive aspects of low-income housing should come to be used as a principal plank of housing policy in Third World cities. But as with all palliatives, there is the real chance that its very success may serve to prevent examination of the fundamental causes of the problem. These are undoubtedly structural and relate to income inequality, land ownership patterns and mass poverty. The very heterogeneity of spontaneous and squatter settlements, in terms of their physical fabric, location, migration paths, incomes and the aspirations of their residents, clearly attests to the scale and pressing nature of the urbanisation process in the Third World.

HOUSING IN THIRD WORLD CITIES: EXAMPLES

In this final section, a series of four examples drawn from Latin America and the Caribbean are provided in an effort to illustrate some of the general points made in the second half of this chapter, concerning the structure and growth of Third World cities.

Caracas, Venezuela

Caracas, the capital and primate city of Venezuela exemplifies the scale and severity of many of the problems that face Third World cities today. Various views of the urban scene of Caracas have already been shown in the plates included in this chapter, and Plates 4.1 and 4.2 in the next chapter give further impressions of the size, location and general features of the city. Caracas is located in a narrow west to east oriented valley, some 25 km long by 5 km wide, which runs parallel with the Caribbean Sea (Figure 3.16). After the development of the oil industry in the 1920s, Caracas grew very rapidly, having become the magnet for migrants. It grew from 0.6 millions in 1950 to around 1.3 millions by 1961, and by the latest census in 1981, had reached the 2.07 million mark. Whilst from 1961 to 1971 the population of the Federal District had increased by 48 per cent, this overall level of increase fell to 11.4 per cent between 1971 and 1981, the first time in recorded censuses that the area had shown an overall rate of growth substantially below the national average (Eastwood, 1983). However, major socio-economic issues remain to be dealt with. The distinctive feature of urban growth in Caracas has been the rancho, or makeshift dwelling, which accounted for well over 25 per cent of all households even by 1950. Whilst most ranchos have electricity, few have water supply and adequate sewerage facilities. A characteristic location is on hillsides and ravines (see Figure 3.12b and Plate 3.2), with the earth face of the hillside forming one wall of the dwelling. The main rancho areas are on the hillsides to the north, west and south of the city centre (Figure 3.16). A very large peripheral barrio area is to be found in the east at Petare. The intervening eastern area of Caracas is given over predominantly to middle and upper class residences (Penfold, 1970; Morris, 1978; Franklin, E.H.M., 1979). Groups of ranchos are referred to as barrios, and Jones (1964) notes that they are by no means made up of the universally poverty-stricken (see Plate 3.4), many occupants having quite well-paid jobs such as clerks in government offices, so that "they are not expressions of despair, but of hope and activity and courage" (Jones, 1964, p. 424). Jones also notes how the consolidation process frequently works in Caracas, with shacks eventually becoming brick-built structures (see Plate 3.5). A major response to squatter settlements came in the 1954-58 period, when many barrios were bulldozed and 97 high-rise, generally 15 storey, apartments were built. Their location is shown in Figure 3.16 and they are clearly visible in Plates 3.2 and 4.2. Such measures were far from successful, however, the apartments were relatively expensive to rent and more squatters

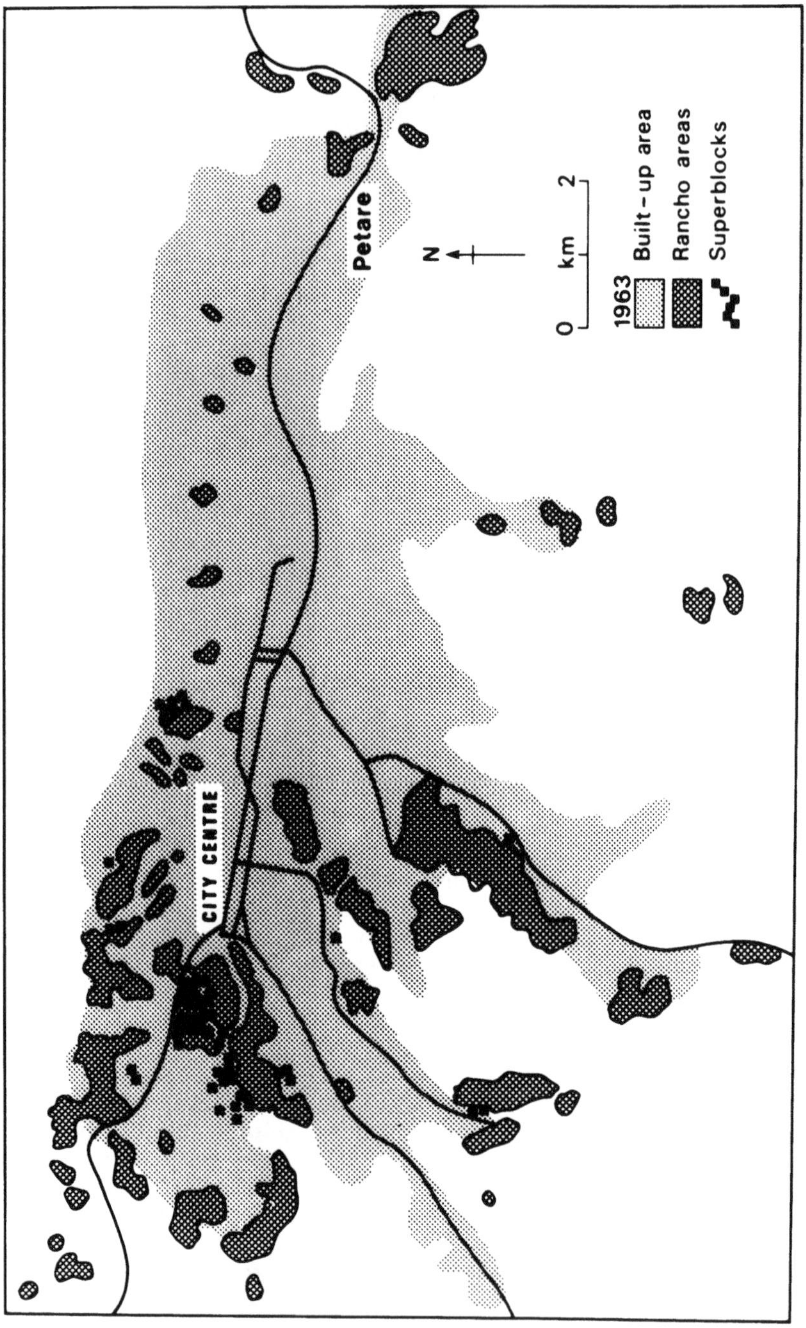

Figure 3.16: The structure of Caracas, Venezuela (Source: Jones, 1964)

were soon attracted, many of them to the open spaces existing around the blocks (Dwyer, 1975). We shall return to this example in the next chapter.

Port of Spain, Trinidad

The urban region of Port of Spain, Trinidad, is by contrast far smaller, only housing a population of 250,000. But this represents some 22 per cent of the total population of Trinidad and Tobago, making it the second largest city in the Commonwealth Caribbean. In fact, the wider capital region has a population of 506,400, or 44 per cent of the entire national population, but more significantly, 53 per cent of its employment opportunities. Regional incomes are highly unequal, with a pronounced urban bias, which is matched by sharp personal income inequalities, with the top 5 per cent accounting for 25 per cent of total incomes (Potter, 1983c). The population of Trinidad is very markedly concentrated on the western coastal zone. Port of Spain is situated in the north west on the Gulf of Paria. Its site is constricted to the north by mountains, and in the south by the Laventille Swamp (Figure 3.17). Its early growth was areal, followed by encroachment of a number of valleys such as those of Maraval and Diego Martin. In fact, the north and north-eastern valleys became high status sectoral wedges (Conway, 1976, 1981), as is indicated by Figure 3.17. In contrast, the hills which extend eastward out from the central business district of Port of Spain became the site for low-income groups. Today this area is populated by about 38,000 people and the town planners have divided it into nine distinct environmental areas, plus a zone used as a pilot area in a social survey. These areas are clearly denoted in Figure 3.17. A planning survey judged that 54 per cent of all dwellings in the Pilot Area, centred on John John, were structurally poor. In environmental area 1, based on Eastern Quarry and Prizgar Lands, 43 per cent of the houses were structurally poor, 86 per cent depended on pit latrines and 82 per cent only had access to water via a standpipe. Some of the types of houses present in these areas are shown in Plates 3.1 and 3.6, both of these photographs having been taken in the Pilot Area at John John. Various planning surveys and reports have been prepared for this area (for example, Urban Redevelopment Council, 1972; Town and Country Planning Division, Ministry of Finance and Planning, 1973). The 1972 report stated that environmental area 1 was "an area of spontaneous unplanned settlement where squatters form 23 per cent of the total population and in-migration from the other small islands of Grenada and St Vincent is great" (p.3). The report also noted that these were areas of rapid population growth and that as any area is cleared, so "squatters converge on it and scores of shanties go up" (p. 16), particularly on Crown Lands. Conway (1981) picked up these general assertions made by the planners and found them to be generally erroneous. Firstly, he argued that the scale of the problem appears to have been greatly exaggerated by the planners. Thus, Conway notes that while it is true that some 25-30 per cent of the total

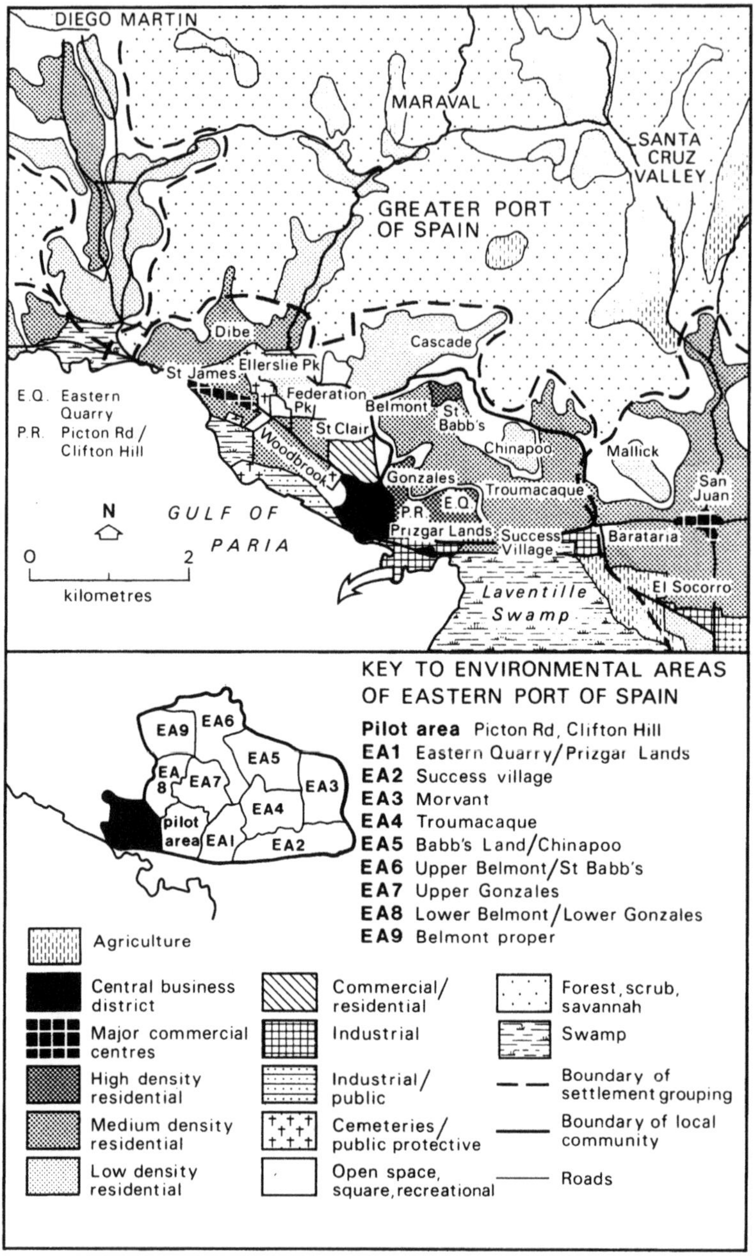

Figure 3.17: The structure of Port of Spain, Trinidad

population of the Pilot Area and EA1 are illegal squatters, they represent only a very small fraction, some 5 per cent, of the total population of the entire eastern periphery of Port of Spain. However, those squatter areas that do exist are clearly visible from the Eastern Main Road, as shown by the view in Plate 3.6. Secondly, Conway demonstrates that of those residing in these areas who were born outside Trinidad, most had lived in other areas of the city for some considerable time before relocating to it. Clearly then, the area is not a reception point for recent migrants, and indeed most in-migrants who were residing in the area were shown to have arrived in the 1940s and 50s. Finally, it is argued that most areas are not burdened down with squatters as is implied by the planning reports. In fact, the most common form of tenure, the 'chattel rental' where the land is paid for on a month by month basis without legal title, accounts for between 40 and 50 per cent of all households. In a further paper, Conway (1983) has shown that direct internal migration to the eastern innercity sections of Port of Spain does occur, although many "bridgeheaders" move in a step-wise manner to the eastern suburban commuter zone first. The Port of Spain example illustrates the need for careful analysis and interpretation when viewing the growth and characteristics of low-income settlements for planning purposes, if misleading conclusions are to be avoided.

Bridgetown, Barbados

The housing situation in Bridgetown, the capital city of Barbados, also serves to exemplify the importance of the rental sector in Third World towns and cities. There is little or no squatting in Barbados, reflecting the fact that land is almost exclusively in private ownership. But the quality of housing is a different matter. The 1970 Census showed that 75.91 per cent of all houses on the island were constructed of timber and only 9.65 per cent of concrete. Timber dwellings can be ideally suited to tropical conditions, but unless the wood is treated and painted regularly, the average life expectancy may be as low as 10 to 15 years, due principally to termite attack. On the other hand, Dann (1984) observes that in 1970, 70 per cent of all dwellings were over 20 years old. Although facilities are often sound, 70 per cent of houses have pit latrine toilets, and in a sample survey carried out by Dann (1984), 24.6 per cent of households were without running water and 11.2 per cent lacked electricity. Most homes are single storey bungalows and there are no high rise buildings. Hence, although net residential densities are low, houses are often located very close together. There are other paradoxical facts. Approximately 85 per cent of Barbadian homes are owner-occupied, and there is great pride in home ownership (Dann, 1984). But, on the other hand, few Barbadians own the land on which their homes are built. This reflects the long-enduring system of land tenure, which is virtually without counterpart elsewhere in the world. Its roots lay in the sugar plantation system where workers were assigned to a small housing site or 'spot'. The worker would then place his chattel house on the site. This would normally be a two-roomed timber cabin, some 10 by 18 feet, set on a rock pile base, and with an iron sheet or shingles roof (see Plate 4.4, next chapter). If employment changed, the house could be moved to another spot. This type of dwelling became the typical low-income house in Barbados and it possessed the advantages of self-build and the possibility of gradual extension as savings allowed. In fact, over time, essentially the same housing pattern was transferred to the urbanising scene (Abrams, 1963). For the majority of such houses, spot rents were paid on a week to week or month to month basis, without long-term security of tenure. But rental agreements were customarily quite stable up to the early 1960s, when a sharp increase in land values occurred due to tourism and other developments. This situation accelerated moves and evictions to an alarming degree and long-standing home owners were shown clearly to be at the mercy of site owners (Abrams, 1963, 1964). Efforts to reduce tenants' insecurity of tenure are now being made, and in October 1980, the Tenantries Freehold Purchase Act was passed with the aim of enabling many to purchase their house lots at reasonable prices. With regard to wider housing policies in Barbados, planners distinguish between chattel houses and slums. Surveys carried out around 1970 identified four categories of priority slum clearance in central Bridgetown, as shown in Figure 3.18. These amounted to a total of 250 acres, and the Emmerton shanty area to the east of the port was designated as the single top priority slum clearance area. The planners have generally followed Abram's (1963, 1964) argument in stating that slum clearance should be held at a minimum where there is a housing shortage. In order to provide public housing, especially for low income families, the Barbados Housing Authority was established in 1955. By 1981, it had constructed some 4,000 units (Dann, 1984) but demand greatly exceeds the supply; for example, there were 10,000 applications for such dwellings between 1956 and 1965. However, rent arrears soon became a major problem affecting such housing, and by 1963, 90 per cent of renters and borrowers from the Authority were said to be in default (Abrams, 1964). Within the Bridgetown urban area, housing has been provided on four main peripheral housing estates (Figure 3.18). However, although it is envisaged that a proportion of the new schemes could be based on self-help principles or site and service schemes, the general planning presumption is that the wooden chattel house will remain as the most efficacious way of providing low-income housing for some time to come in the future.

Castries, St Lucia

By its very nature, the urbanisation process tends to focus attention on very large settlements. Thus, the fourth and final example considered here stresses that the problems of urbanisation and housing are by no means confined to large Third World cities. As was noted earlier in the chapter, Castries, the capital of St Lucia, has a population of around 40,000, or approximately 40 per cent of the national total. Its growth has been very rapid in the post-war period, reflecting its economic dominance and a strong pattern of rural to urban drift.

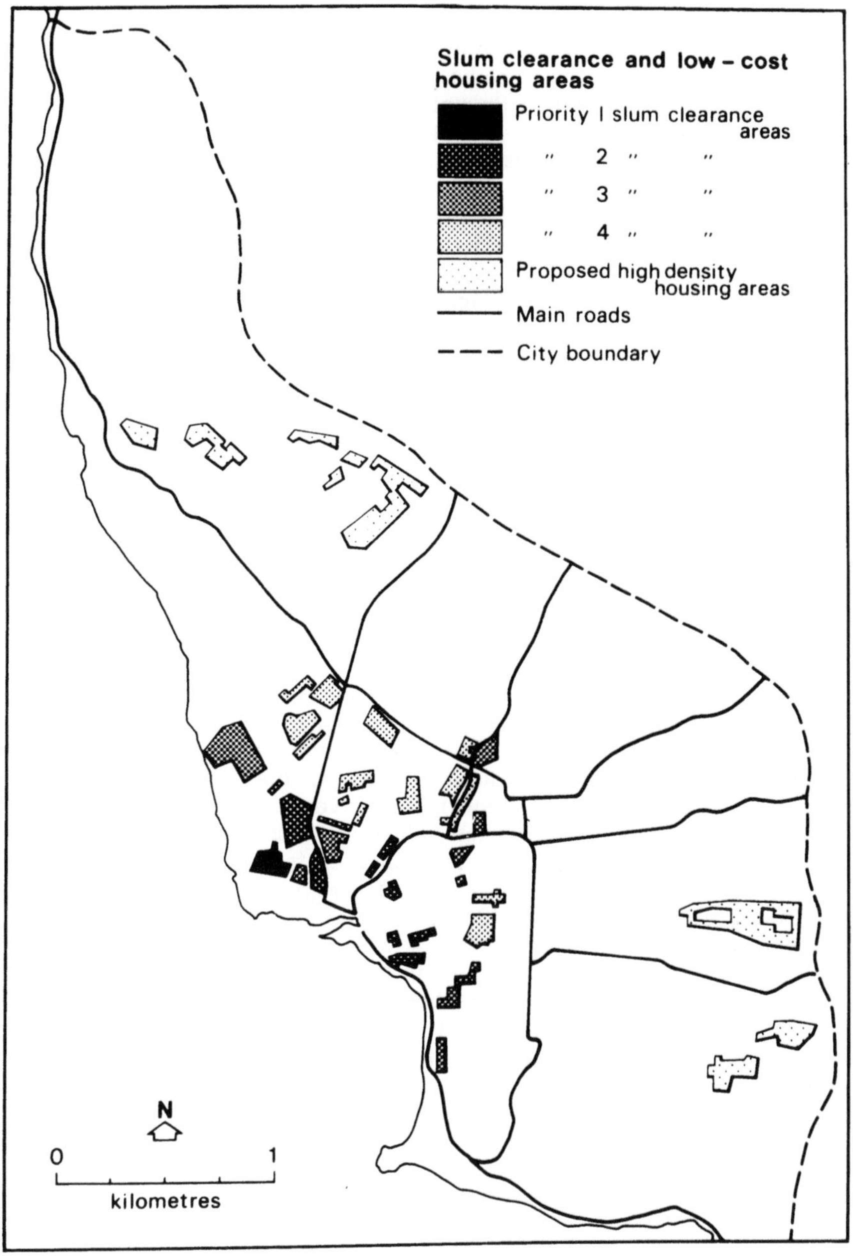

Figure 3.18: Slums and low-cost housing areas in Bridgetown, Barbados

Thus, Castries' population was only 16,566 in 1946, but increased thereafter at an average rate of 6.79 per cent per annum. The site of Castries, on a protected inlet with highland all around, owes much to defensibility, and the town has a long history of military association (Figure 3.19). As noted by Paquette (1965), the town's physical growth has been blocked by the mountainous slopes existing on both its southern and eastern sides. Consequently, it has tended to spread in three clear directions: inland along the Castries River, along the narrow coastal strip to the south of the harbour and toward the northern beaches (Figure 3.19). The surrounding ring of mountains is not entirely devoid of settlement however. To the south, for example, on the flanks of Morne Fortune, there was a scattering of upper class dwellings in the 1950s, and this has grown as a relatively high status sector since then. The town has also grown north and northeastward, giving three high-income residential areas, those of La Vigie, La Clery and Vide Bouteille (Figure 3.19). However, low-income wooden houses are far more typical of most areas of the town, especially the lower slopes of Morne Fortune overlooking the harbour, the mountainous slopes east of the town, and the Castries river valley in the southeast, toward Marchand and Bagatelle. There are also two distinct areas of spontaneous-type settlement, located respectively to the north of the harbour at Conway and to its south at Four a Chaud (Figure 3.19). These two areas are somewhat different in character. The community at Conway, for example, is a longstanding one and the settlement consists of an unplanned scattering of wooden dwellings and associated workshops. Although Paquette (1965, p. 139) described the Conway area as a "sickening sector, that is fortunately due to be reclaimed soon as an open space", there is considerable evidence to suggest that the inhabitants of the area are doing their utmost to thrive in a situation of severe social and economic deprivation. Further, despite plans to redevelop the area as an extension to the commercial core, as yet no such changes have occurred. The area of irregular settlement at Four à Chaud, on the southern side of the harbour is, however, of more recent origin. In fact, part of the site has only been reclaimed during the past few years. The area again consists of an irregular grouping of wooden dwellings and workshops such as furniture makers and the like, and despite much poverty, there is a genuine air of improvement and progress about the settlement. A residential scene in the area is shown in Plate 3.3 included in this chapter, and a wider areal view is shown in Plate 4.3 in the next.

CONCLUSIONS

This chapter has sought to provide an outline of the nature and consequences of the rapid urbanisation process that is now characterising much of the Third World. At both the inter-urban or national urban systems level, and the intra-urban or local urban systems level, the urbanisation process appears to be characterised by marked and widening social and economic inequalities. Further, it has been

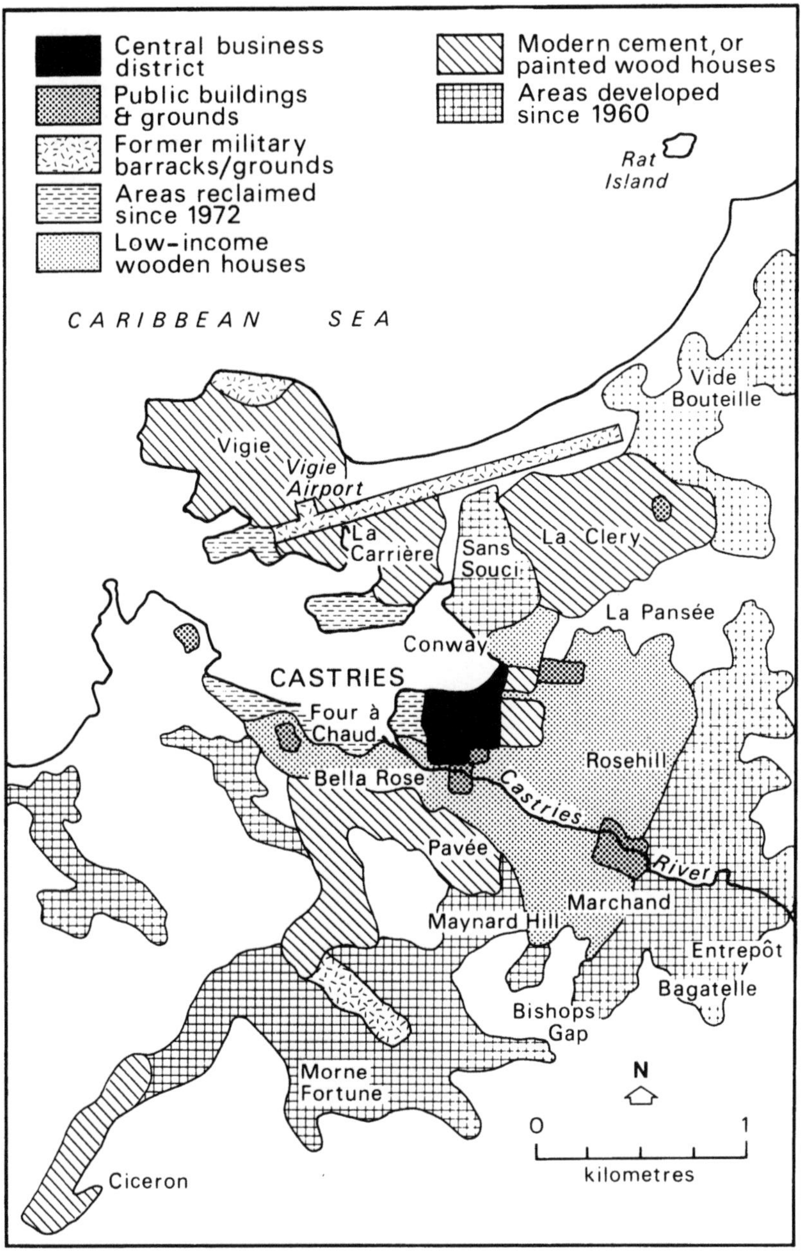

Figure 3.19: The structure of Castries, St Lucia

shown that the spatial perceptions of individuals closely reflect these physical patterns of inequality and heterogeneity. This applies equally with regard to both the perceptions of rural to urban migrants and those of self-builders within towns and cities. Hence, there is a basic need for all those concerned with the urbanisation process to pay careful attention to the aspirations and views of migrants and urban dwellers, a theme which is explored in detail later in this volume. However, basic issues at the present are whether these disparities will increase or decrease over time if left to themselves, or whether the state should intervene, and if so, in what manner. These questions are addressed in the next chapter which looks at urban planning in Third World countries at the national, regional and local urban scales.

Chapter 4

Urban Planning in the Third World

There is little doubt that if a panel of experts, or indeed laymen was convened to consider the problems of Third World urbanisation and to recommend remedial action, they would agree unanimously that first and foremost, sensible and responsive urban planning is required. That a consensus on the need for sound planning might exist is possibly misleading, however. Simply, it can lead to the implicit view that planning exists as a distinct entity, with clearly defined and uncontroversial goals and methods that result in some prespecified outcome. Nothing could be further from the truth. Planning is in reality a general blanket term for an extremely catholic set of activities. Put at its most crude level, urban planning means different things to different people, and more significantly, to different states and governments. If there is agreement about the ends of urban planning, there is commonly dissent concerning its means. More typically perhaps, there is often little if any consensus even as to the desired end product of planning.

The present chapter reviews the major aims, approaches and problems of urban planning in Third World countries. The argument presented follows from Chapters 2 and 3 in suggesting that two major sets of problems face urban planners in the Third World. First, the highly uneven pattern of regional development and secondly, the acutely skewed distribution of incomes, employment, welfare, housing and other benefits within cities. In general terms, there is a frequently expressed suspicion that absolute poverty is increasing in much of the Third World, and that there is also increasing income divergence, regionally, personally and at the intra-urban level (Taylor and Williams, 1982). For example, the top 10 per cent of urban households in the Philippines account for over one-third of the total urban income and there is an almost non-existent middle class. Similar income shares for the top 10 per cent of households are recorded for India (35.2 per cent), Venezuela (35.7 per cent), Malaysia (39.6 per cent), Brazil (50.6 per cent), as shown by the World Bank (1980). Taylor and Williams (1982) stress how this personal income disparity in Third World cities occurs alongside the urban-rural disparity in wealth, so that a "fourlayer pyramidal structure occurs" (p.9). Thus, generally, a small wealthy urban elite exists at the apex, above secondly, government servants and those employed in modern industries, and thirdly, the vast numbers of the urban poor. Finally, the mass of rural poor exist at the broad base of the welfare pyramid. These inequalities represent the basis of the planning problem. Accordingly, the present chapter falls into two broad divisions, with the first looking at planning at the level of the urban system. Here the issues involve questions of city size and efficiency, notions of optimal city size and national urbanisation strategies. The second part examines planning at the level of the individual urban area, where the focus is frequently on the provision of homes, jobs and services. But before focusing on these specific issues in detail, wider matters concerning the general nature of urban planning in the Third World are considered in the next section.

THE NATURE AND PURPOSE OF URBAN PLANNING

The notions of cumulative causation and progressive regional and personal income divergence suggest that without some form of planned intervention, economic development will favour certain geographical localities and particular social groups. Whether such concentration is seen as desirable or not will depend largely on political views. In an environmental context, planning may be defined as the deliberate achievement of objectives by means of assembling actions into an orderly sequence (Hall, 1974). In somewhat broader terms, McLoughlin (1973) has suggested that it essentially involves foreseeing and guiding environmental change. An important implication of planning is that it represents the direct intervention of the state over the market mechanism. A central argument here is that the market would otherwise allocate resources in an undesirable manner. This opens up the spectrum of planning from the socialist state where virtually all decisions are taken centrally, to that of laissez-faire or free-market capitalism. In all cases, the suggestion can be made that in theory at least, planning seeks to reconcile competing claims, and to promote equity and efficiency. Many of these attributes of planning are well-expressed by Taylor and Williams (1982, p. 23), when they state that:

> ... planning is also a mechanism to provide an environment for living which all may desire but which would not be obtained through the fragmented decisions of individuals. It is a means to organise the public goods of society.

Many individuals, households, firms and other decision-making units are of course planning on a day-to-day basis, but they do so from a largely personal and highly idiosyncratic standpoint. It follows that another avowed aim of planning is to arbitrate between possibly conflicting interests, and to decide upon consensus objectives. This function is implicit in the phrase 'which all may desire' in the above quotation. However, we must not be naive, for such an aim begs a series of questions: what is desirable and what is undesirable, and for whom, and in what circumstances? In other words, planning cannot be seen as a set of value-free and entirely objective procedures and associated techniques by means of which an optimal solution to societal problems is to be achieved. We shall return to this point shortly, for in a divided society, and indeed given a divided space, one person's problem is likely to be another's advantage. Thus, determining consensus goals is much more difficult than it might at first appear, and is a basic stumbling block in all forms of planning. An elitist version of planning would assign the key decision-making role to the professional planner. Conversely, advocates of participatory planning suggest that the planner should seek to follow the express wishes of the people. A closely related point is that planning is therefore an activity with inherently political connotations. This is so for two principal reasons. First of all, where there are conflicting socio-economic or environmental interests, planning decisions will inescapably tend to favour one at the probable expense of the others. Secondly, planners basically serve their political masters, so that we must be aware that it is a myth to regard planners as entirely impartial and benign public servants. There is a related point regarding the style and ideology of planning, and that is whether, particularly at the regional scale, development should be attempted from 'above' or 'below' (Stohr and Taylor, 1981). 'Development from above' describes efforts to spread growth by functional integration at the national and international scales (Hansen, 1981), normally by following programmes of unequal growth and involving strong urban bias. In contrast, 'development from below' maintains that the way ahead is for the re-establishment of a territorial basis to planning, which thereby emphasises greater regional self-sufficiency and selective regional economic closure (see Stohr and Todtling, 1978; Friedmann and Weaver, 1979; Stohr and Taylor, 1981). As emphasised by Stohr (1981), development from below does not merely entail altering the level at which decisions are taken, but rather the very criteria of development. Thus, the approach is related to the movement away from a monolithic economic concept of development toward a broader socially-oriented one. It is also noted by Stohr that the doctrine of development from below carries a spatial implication in the form of development from the economic periphery inward, rather than the other way around. The above and below, inward and outward and elitist and participationist contrasts in views of urban and regional planning will be recurrent themes in this and the next chapter.

Finally, planning covers a large number of activities varying in both scale and overall focus. With respect to scale, planning varies from that carried out at the national level where normally, broad economic targets are set in economic development planning (Tinbergen, 1967), via the regional scale where strategic spatial-locational issues become predominant, down to the sub-regional and local urban (project) scales. Similarly, the systematic focus of planning varies and it is customary to distinguish between comprehensive and sectoral planning. Three principal types of sectoral planning are normally identified: economic, social and physical development. In the infancy of planning, the emphasis was generally placed on physical development or land use planning at the local scale (Franklin, G.H., 1979; Taylor and Williams, 1982), and on economic planning at the national level. Of late, the social perspective is increasingly being stressed at all scales, especially in the Third World context (Conyers, 1982). Further, Franklin, G.H. (1979) has made the plea for a greater emphasis to be placed on physical planning at the national scale in developing countries.

THE QUESTION OF CITY SIZE AND EFFICIENCY

A major and controversial issue in urban planning at the regional-national level is the relationship between urban size and economic efficiency. The unquantifiable relation between the actual and perceived opportunities of cities of various sizes is at issue here. As was shown in the previous chapter, the perceived greater variety of opportunities existing in large cities tends to further increase their size. Hence, large cities often grow faster than their smaller counterparts and take a disproportionate share of overall economic activity. Planning may then seek to go with this seemingly natural trend, thereby enhancing capital-intensive, unbalanced and unequal growth. In fact, it is precisely this type of primarily economic argument that is used to examine the issue of urban size. On the credit side, it is argued that large city regions allow the concentration which is required to facilitate internal and external economies of scale, such as the size of the market and the existence of other firms of the same type in the area respectively. These add to urbanisation economies, that is the advantages deriving from the overall level of economic activity existing in a city, to give the catch-all concept of agglomeration economies. On the other hand, however, there is the notion that diseconomies of scale are likely to result after a certain, but unspecified urban size level has been reached. This will be due to factors such as congestion, administrative inefficiencies, financial problems, the deterioration of social conditions, pollution and sometimes,it is suggested, human degradation (Wirth, 1938).

This argument concerning the economies and diseconomies of city size gives rise to the question of whether or not there is an optimal city size, where the difference between the benefits and costs are maximised in favour of the former. This has been the focus of much attention and debate, especially by economists. It would obviously be a paramount planning consideration if a concrete optimal city size could be ascertained. However, precious little is known about the hard facts of city size. Evidence undoubtedly supports the hypothesis that _per capita_ incomes increase with city size (see, for instance, Hoch, 1972), but whether this reflects higher productivity _per se_ or greater accumulated investment is hard to say. Despite these problems, however, there have been efforts to produce theories concerning how costs and benefits vary with city size. An early example is afforded by the work of Isard (1956) who examined hypothetical urbanisation economies and diseconomies for different urban activities, but was forced to conclude that "there are many logical objections to this procedure" (p. 186). However, the most developed statement has come from Richardson (1972, 1973b, 1978). Here, the author starts by looking at theoretical average cost and average benefit curves with increasing city size. It is hypothesised that the average cost curve of a city will take on the familiar U-shape, with the minimum cost point existing at a relatively low population level. The average benefit curve on the other hand is assumed to take an S-shape, initially rising much faster than city-size, but later flattening out and perhaps even declining. These two curves, depicting average benefit (AB) and average cost (AC) with increasing city size are shown in Figure 4.1. The marginal cost curve (MC) takes the required statistical form, crossing the average cost curve at its lowest point, and the marginal benefit curve (MB) loops below the average benefit curve at its highest point (Figure 4.1).

If it is assumed that these curves are realistic, a whole series of critical city sizes are revealed, although not all of them may be described as optima. Point D on Figure 4.1 indicates the minimum viable city size population P1. At this size level, average benefit equals average cost, with benefits rising and costs reducing. Further along the curves, at point E, the least-cost city size is reached at the lowest point on the average cost curve. This is not an optimal situation, however, for the benefits of city size are entirely disregarded. At a population of P2, net benefits per capita are maximised, for here the difference between average benefits and average costs is greatest. At this point, the slopes of the average benefit curve (at F) and the average cost curve (at G) are equal. Whilst this size level may be beneficial for the city's existing occupants, it may not necessarily be good for society taken as a whole. Average benefits are maximised at point H on the curves, but this is of course as one-sided as the minimum cost city size. However, the point I, associated with a population level of P3 is far more significant. This is the social planning optimum city size. At this point, marginal benefits are equal to marginal costs, and the total benefits generated by the city are maximised. As Richardson notes, however, there is a major problem with regarding this as an optimum city size level. Potential in-migrants are likely to base their decision to move to the city on the gap that still exists between average benefits and average costs at this size level. Thus, migrants may continue to flood into the city, until point J is reached, where average cost is again equal to average benefit, and where the market equilibrium is reached. Thus, there may well be a discrepancy between the optimum at I and the actual or market solution at J.

The many criticisms of such a theoretical approach have been well-summarised by Richardson (1978). Firstly, there is no evidence stronger than pure guesswork to support the shapes of the assumed curves. Further, doubt must be cast on whether we can measure the costs and benefits involved, especially given the interdependenc of activities and also, the impossibility of including the costs o

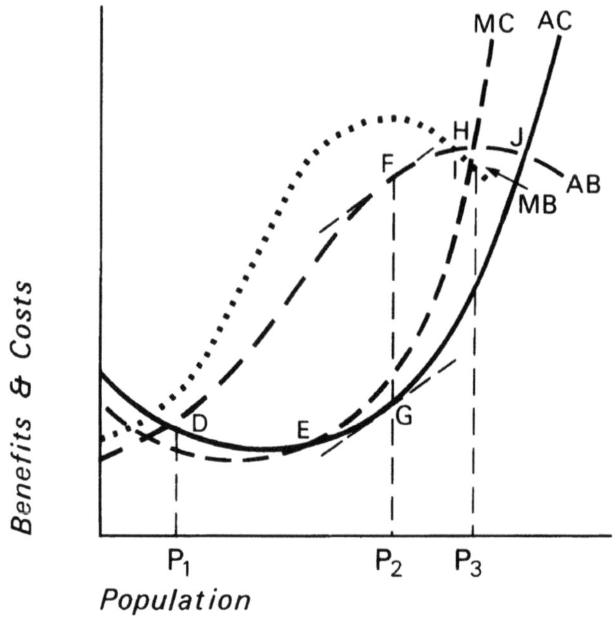

Figure 4.1: The theory of optimal city size (Source: Richardson, 1978)

externalities such as crime and public goods benefits. More fundamentally, however, the concept of a unique optimal city size loses all meaning in the context of a national urban system, especially a hierarchy, within which centres specialise. In such conditions, efficient city size will vary between levels. Therefore, the need is to promote efficiency within all city size ranges. Richardson (1972) concludes that optimal city size is neither a useful problem nor a suitable basis for the formulation of national urban planning policies. In fact, elsewhere he calls it the "optimal city-size game, because it is fun even though it gets nowhere" (Richardson, 1978, p. 322). Others have commented that pronouncements of optimal city size are too rigid and are doomed to failure and that "the goal of an optimal city remains illusory" (Bourne, 1975, p. 25).

Although the search for a unique optimum has thus been decried, a number of scholars have recently criticised the growth of large cities, especially those in developing countries. From a primarily economic standpoint, Gilbert (1976) has questioned the argument that higher productivity in big cities is caused by economies of scale, and suggests instead that it is achieved at the expense of productivity elsewhere. He then suggests that "If infra-structure of the same quality were provided in medium-sized centres, then productivity in these cities would rise" (Gilbert, 1976, p. 29). However, the arguments of the anti-large city lobby conflict with those of a growing band of regional economists who argue forcefully that large metropolitan centres produce more benefits than costs, and that any attempt therefore to retard the growth of the largest cities is likely to reduce national rates of economic growth (see, for example, the discussions between Richardson, 1976; and Gilbert, 1977; also Alonso, 1971; Hoch, 1972; Richardson, 1973b). Such arguments are extremely difficult to evaluate for most data on this topic relate to cities in the developed world and it is often impossible to acquire data on _per capita_ product for developing countries. However, an interesting case study has recently been provided by Gwynne (1978) based on data derived from the Chilean Institute of Statistics' 1970 survey of retail prices in urban areas. Here it was demonstrated that the average index of retail prices by increasing city size takes the form of an inverted U-shaped distribution. Thus, prices tended to be low for small towns, those in the 20,000-60,000 population bracket, presumably as a result of their relatively simple but highly efficient marketing structures. Similarly, the average price level was just as low in the primate capital Santiago. It is argued that the capital city has benefited most from modern large-scale developments in marketing and retailing. On the other hand, prices were high in cities of the intermediate size order of between 60,000 and 250,000. Gwynne maintains that these are too large for efficiency in traditional marketing systems, but too small to benefit from the economies of scale afforded by large marketing organisations.

Even the ostensibly simple and objective issue of optimal city size in reality boils down to political viewpoints concerning the efficacy of unequal development. But there is a more salient point, and that is that arguments concerning large cities should not be based on economic reasoning alone. Quite simply the issue depends on the goals that are set, and these are essentially socio-political. This takes us back to the thorny argument raised in the previous chapter, concerning the balance between social equity and economic efficiency. A related point is that the issues of urban size and urban primacy are quite often confused. In some small countries, although the capital city is highly primate, by international standards its absolute size is limited. This is frequently the case in small open-economies, as for example in the Caribbean region. Thus, in such contexts, the economic reasons for urban size control may be non-existent, especially as it was shown in Chapter 3 that urban primacy may well be efficient in a small country. But the same line of argument can be applied and there may be very strong social reasons for urban size control (see, Potter and Hunte, 1979 and Potter, 1983b). Hence, although Richardson (1978, p. 326) concludes that "Tampering with city size is too much like utopian social engineering", perhaps it is precisely this type of social perspective that is frequently required.

NATIONAL URBAN DEVELOPMENT STRATEGIES: A REVIEW

Despite these conflicting views on large cities, attempts to decentralise people, jobs and social infrastructure away from primate cities and other large metropolitan areas can be described as the single most frequent urban planning objective in both developed and developing countries alike. As suggested above, these policies may often reflect social, political and strategic motives rather than economic ones. However, there is frequently the suspicion that the size of urban areas is being treated as if it is the only problem, rather than a manifestation of wider societal problems.

We shall shortly return to the continuing debate concerning spatial centralisation and decentralisation when we examine the effectiveness of various urban policy options. First, we look in broad terms at the sorts of policies we are talking about. In a recent paper, Richardson (1981) although having previously argued from a strongly pro-large city standpoint, stressed the importance of national urban development strategies (or NUDS) in developing countries. In this paper, 11 prototype strategies of NUDS representing "attempts to change the inter-urban distribution of population in the pursuit of some longterm policy objectives" (p. 267) are recognised. These have been given a spatial expression by the present author in Figure 4.2. Given an existing large urban area, then there are three policies which can be regarded as enhancing concentrated urbanisation. A hypothetical primate city is shown at the centre of a circular national territory in Figure 4.2i. The primate city region is shown by a dotted circle, and the other important urban areas beyond it by five black circles. The first policy of concentrated urbanisation is the <u>laissez-faire</u> one of doing nothing and letting the market take its course. If, however, problems of congestion and diseconomies are recognised

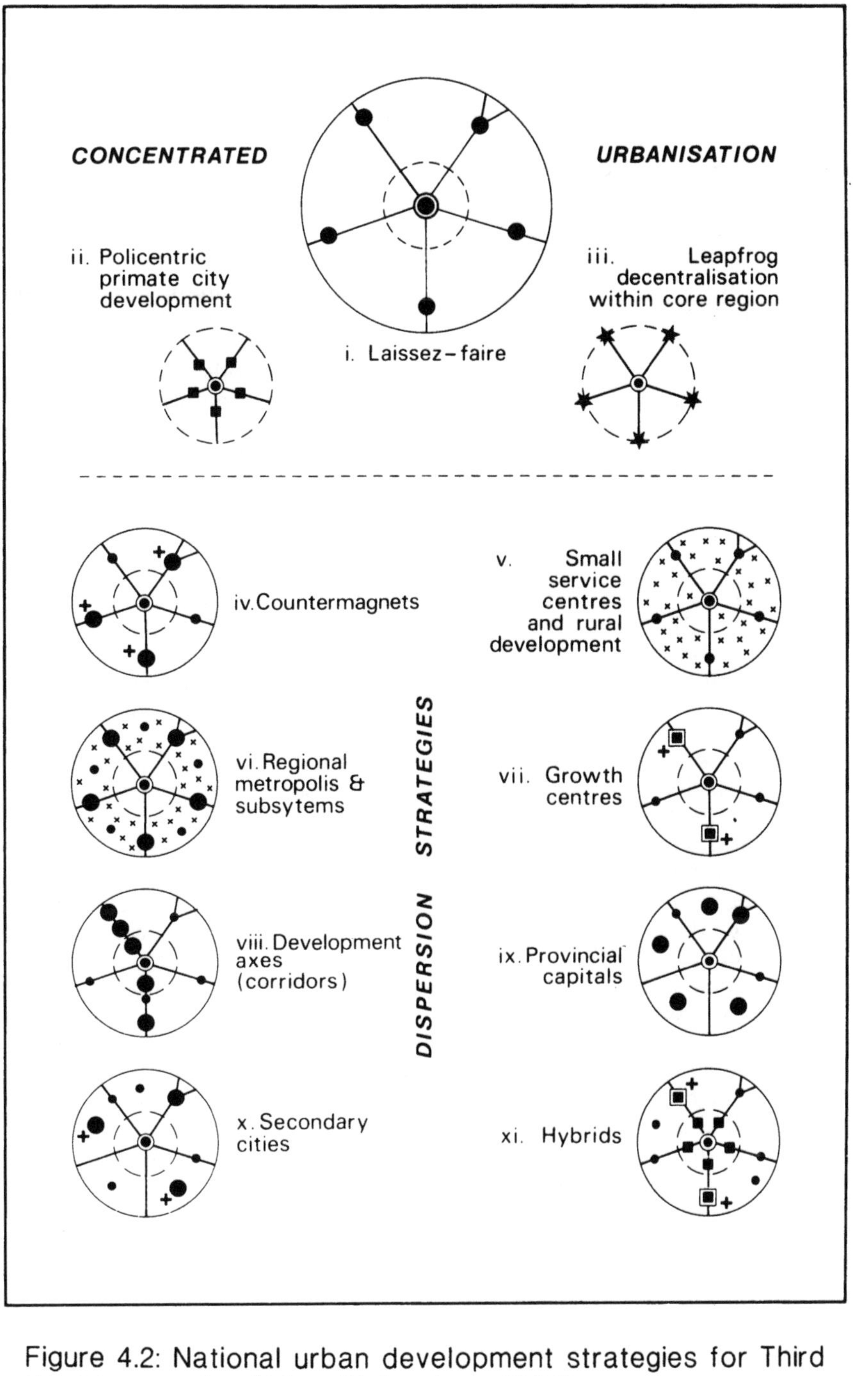

Figure 4.2: National urban development strategies for Third World countries (After Richardson, 1981)

in the primate city, efforts may be made to decentralise, but merely within the core region. Some may regard this as an intra-urban response to what is essentially an inter-regional problem. Thus, a policentric pattern of development may be promoted in the primate city (ii), or a form of leapfrog decentralisation to the edge of the core region (iii) may be envisaged (Figure 4.2).

Strategies iv to xi in Figure 4.2, on the other hand, all represent forms of dispersion at the inter-regional scale, and involve spatial deconcentration from the primate core area. In strategy iv, counter-magnets are formed by strengthening a few major distant cities. At the other end of the scale, a dispersed policy might be favoured, involving the promotion of small service centres and rural development throughout the periphery (strategy v). A compromise is suggested in strategy vi, described as a regional metropolis and subsystem, whereby a regional hierarchy is developed. The promotion of planned growth centres (strategy vii) can be seen as a further policy, the effectiveness of which we shall debate shortly. By means of strategy viii, corridors of development can be channelled along the principal inter-city transport axes. Other forms of decentralisation can be effected by the promotion of provincial state and departmental capitals (ix) and the creation of a limited number of secondary cities (strategy x). These strategies are not mutually exclusive and indeed, several of them are very similar indeed. Further, they can be combined together to give hybrids. For example, in strategy xi depicted in Figure 4.2, policentric development of the primate city has been combined with the development of growth centres and secondary cities.

Richardson is at pains to stress that the reasons for the implementation of such policies may be socially rather than economically based, so that the "key goals of a NUDS are the same as societal goals in general" (Richardson, 1981, p. 270). Further, it is stressed that NUDS have to be highly country-specific if blunders are to be avoided, in particular, it is stressed that the size of country is a crucial dimension. Richardson finishes by still arguing that the efficiency merits for slowing down urban primacy by NUDS are dubious; but it must be stressed that he is still arguing from a basically economic, growth-oriented, point of view.

A theme which runs through all of the strategies is the underdevelopment of the middle tiers of the settlement hierarchy. This line of argument was strongly followed in the Third World context in a well-known book published by Johnson (1970), an economic historian. In this book on spatial organisation in developing countries, and in his earlier work set in India (Johnson, 1965), he was taken with the general paucity of towns in developing regions. For example, Johnson (1970) showed that the countries of Europe had approximately ten times as many central places (towns over 2,500 people) per village as Middle Eastern ones. It was posited that the gap in the national settlement system must serve to lower agricultural productivity in Third World countries. This led Johnson to argue strongly for town-building programmes as a major plank of economic development policy in less developed areas. Johnson's work was clearly influenced by the classical central place theories of Christaller (1933) and Losch (1940). As an example, the planning of the national settlement system of Israel after 1948 followed some elements of this type of approach. Thus, in 1948, 43 per cent of the total population lived in Tel Aviv. In the next 20 years, some 34 new development towns were created, accounting for 21.3 per cent of the national population. As a result, by 1970, Tel Aviv's share of the national population had fallen to 33 per cent, and a tendency toward rank-normality was increasingly discernible. In assessing Johnson's overall thesis, some authors have argued that in reality the causation is the reverse of that which he suggests, and that the lack of large urban centres reflects the low level of agricultural production (Gilbert and Gugler, 1982). Certainly, it is hard to see how changes such as those suggested by Johnson would achieve the desired end in the absence of more radical reforms. However, central place theory is still quite frequently employed in relation to development studies and national urban settlement strategies (see, for example, Brookfield, 1975). Although as noted in Chapter 3, central place theory is often associated with a rigidity and ahistoricism (Vance, 1970), its normative aspects should perhaps not be overlooked for planning purposes. But this is merely to say that if broader reasons and policies suggest the efficacy of decentralisation, then central place frameworks might assist in the task of spatial planning. In this regard, an interesting development is the general hierarchical framework of central places of Parr (1978, 1980, 1981). The systems developed by Parr do not depend on fixed numerical hierarchies, so that different k values can operate at different levels of the settlement system. Secondly, the models have a dynamic character which allows structural change to be accommodated within them. For a simple introduction to such systems see Parr (1978) and Potter (1982, pp. 49-52). Such flexibility is well-illustrated by the example systems shown in Figure 4.3 and 4.4. A traditional hierarchy according to the transport principle is shown in Figure 4.3A, so that $k = 4$ at all hierarchical levels. If a new level is now formed within the system by downgrading some of the second level centres, a variable k hierarchy results ($k = 2$, $k = 2$, $k = 4$). Further, the market areas of the settlements performing functions at the pre-existing second order level become rectangular (Figure 4.3B). Two further examples of dynamic change in central place systems which involve flexible market area geometries, non-standard and even fractional k values are shown in Figure 4.4A-D. It can be ventured that such models offer a far more flexible and realistic normative framework for urban settlement planners, especially as work on central place systems in developing countries still continues, both in relation to general theory and the analysis of actual conditions (see Funnell, 1973, and Singh, 1979, for instance).

The thesis of central places, and in particular the missing middle size orders, is more interesting if viewed in rounded policy-related terms. An American planner, Rondinelli (1982, 1983a, 1983b) has noted that since the 1970s, an increasing emphasis has been placed on rural development as a corrective to the earlier urban bias in

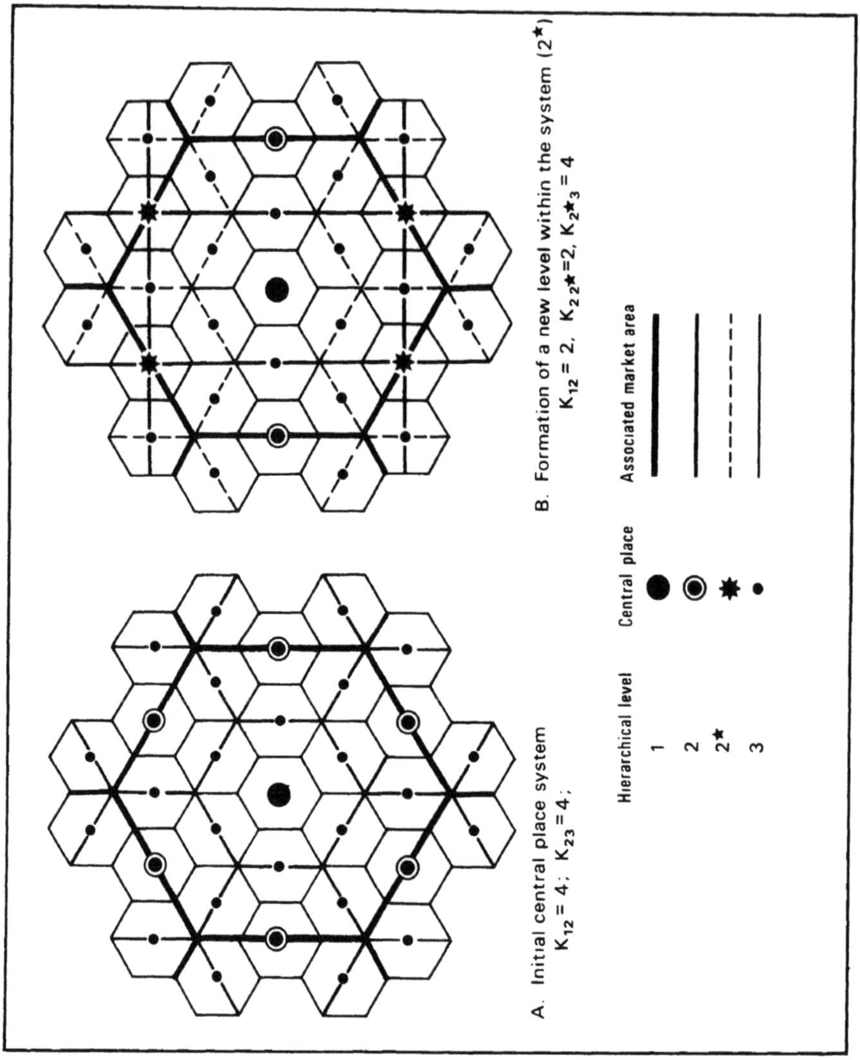

Figure 4.3: General hierarchical frameworks for planning (Source: Parr, 1980, 1981)

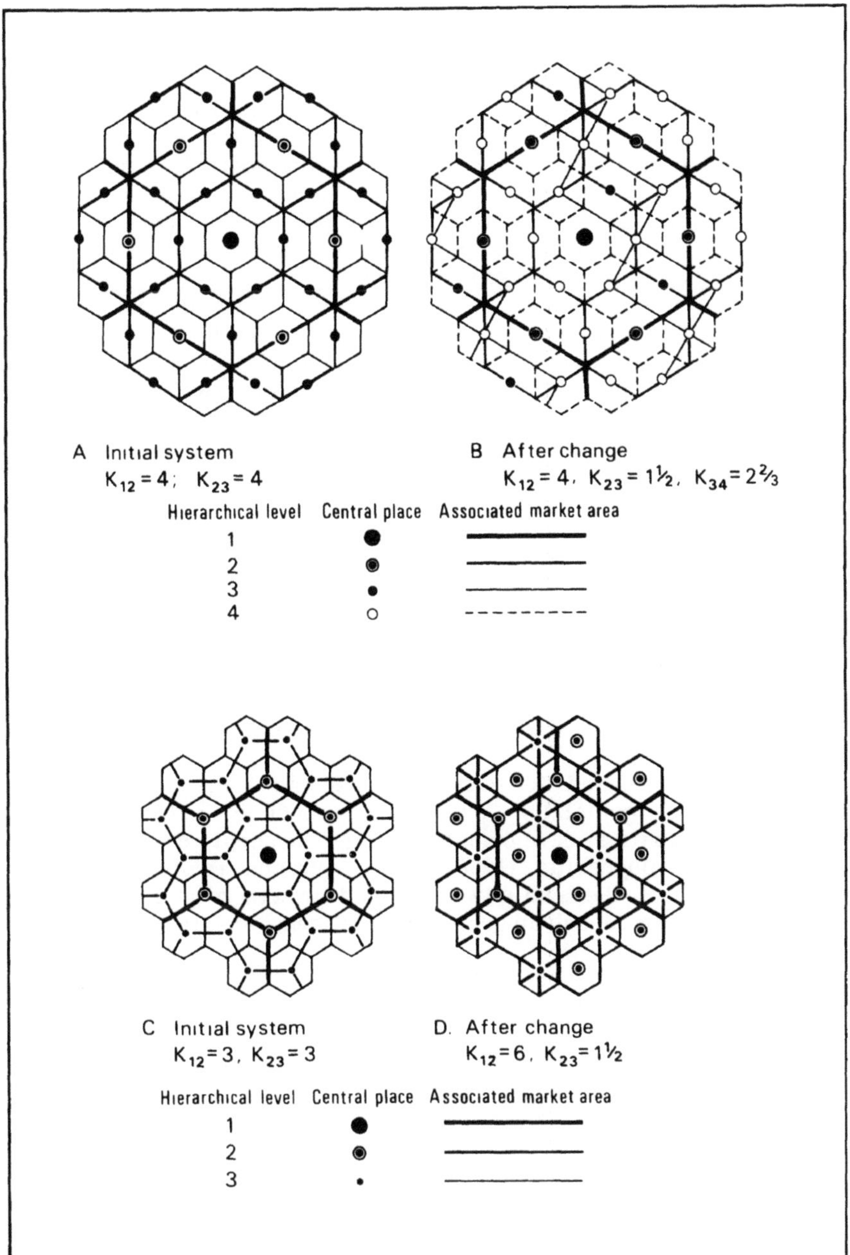

Figure 4.4: General hierarchical frameworks for planning (Source: Parr, 1980, 1981)

foreign aid and development programmes. With this it is argued that a policy emphasis on developing middle-sized cities has also emerged in many countries, in order to reverse or at least reduce polarisation, and to commercialise agriculture in rural regions. These ideas are clearly linked to those of Johnson, although Rondinelli envisages such 'intermediate' or 'secondary' cities as places with over 100,000 population, but which are considerably smaller than the largest city in the country. In particular, Rondinelli (1982) maintains that these can be used to promote a more equitable distribution of population and resources than can be achieved by establishing a few selected growth areas.

Mention of growth areas brings us to the strategy of "concentrated decentralisation", which became the main thrust of national and regional urban planning policy in many countries as from the 1960s onwards. This approach was mediated through the idea of growth poles and growth centres. As explained in Chapter 3, the original idea of the growth pole lay in the works of Schumpeter (1911) and Perroux (1950, 1955). It initially referred to a set of expanding industries located in an urban area, which thereby induce further innovations, linkages and economic development. It will be recalled that historically, Myrdal, Hirschman, Friedmann and Vance had all agreed that the early stages of economic development have normally been associated with spatial concentration in a single core region. This, of course, can be associated with the idea of the natural growth pole. On the other hand, Myrdal argued the need for intervention if growth is to be spread out. The notion of planned polarisation reversal gives rise to the planned growth pole. As Brookfield (1975) has noted, however, the original sectoral idea of growth poles was progressively widened to eventually become synonymous with an urban centre containing expanding economic activities. This geographical-spatial development of the concept owed much to the work of Boudeville (1966). The approach really represented the evolution of the spatially conceived growth centre, as opposed to the sectorally conceived growth pole (Richardson, 1978; Brookfield, 1975).

Growth centre and growth pole policies have been the subject of considerable academic debate and by now there are a number of detailed expositions (Hansen, 1967; Darwent, 1969; Lasuen, 1969; Kuklinski, 1972; Hansen, 1972; Parr, 1973; Moseley, 1974; Buttler, 1975). The employment of such techniques for practical planning purposes has been just as frequent, first in European countries, but later, increasingly in the developing world where it seems to dovetail well with the idea of reducing urban primacy. As noted by Richardson (1978), in such contexts the policy can serve to link national economic policy to intra-regional strategies by filling the gaps existing in the urban hierarchy. Examples abound, for instance, regional physical development plans in Kenya have incorporated the growth pole concept, as they have in Tanzania, India and the socialist states of China and Poland as well. Growth pole policies have also been employed frequently in Latin American countries, as shown in the review by Richardson and Richardson (1975). For example, in Chile in the late 1960s, a growth pole was established in each of the 11 main regions of the country. In the case of Venezuela, a planned growth pole based on heavy industry was developed at Ciudad Guayana and a secondary pole at Ciudad Bolivar also in the east of the country (see Turner and Smulian, 1971). Similarly, in Colombia, after 1969, the primacy of Bogota was planned for by means of the expansion of the next three urban places of Medellin, Cali and Barranquilla.

The growth pole concept has been applied to correct urban concentration in a large number of different ways and by promoting poles of widely different sizes. As a result, many argue that it has become a very diffuse and weak idea. Moseley (1973) has gone so far as to suggest that growth centres and poles exist only in the eye of the beholder. Similarly, Gilbert and Gugler (1982) have recently argued that the concept is flaccid, having been applied both by capitalist and socialist governments alike. The by now rather cloudy and ill-defined idea of the growth pole is exacerbated by the suggestion that in practice it has failed to live up to its promise (Richardson, 1978). Principally, it is argued that the strategy has not worked in the manner predicted. This partly reflects the varied interpretation which has been placed on the concept by different planners and academics, but it also results from the problem of identifying an appropriate population base size for growth poles. Another issue that is frequently raised is the long time horizon that is needed for the establishment of a growth pole or centre. More fundamentally, many commentators argue that the whole concept is basically Western, both in origin and philosophy. From a more radical stance, it is argued that growth pole policy is based on the dispersal of polarisation at the national and international levels, and not on its eradication. Hence, from a Marxist perspective, it is frequently posited that growth poles are dependent on principles of unequal development and the continuing dominance of the advanced countries. However, most commentators conclude that in spite of such formidable drawbacks, the policy is not without its potential merits, principally due to its flexibility, and that it should not be overlooked in urban and regional planning. A basic theme here is the need for a link to be forged between growth pole policies and rural-based strategies in order to spread growth more equitably in developing societies.

This brings the account to a recent and very important school of thought. A growing number of regional planners would suggest that the growth centre policy is deficient in that it is premised entirely on the principle of development from above. The concept of rural development associated with more fundamental change is connected with the doctrine of development from below, as noted earlier in this chapter. This is particularly associated with the work of Friedmann and Douglas (1975), and more recently, Friedmann and Weaver (1979), and is frequently referred to as "agropolitan development" or an "urban-based rural development" strategy. The starting point for such a philosophy is that polarised development and growth pole theory have acted as the "handmaiden(s) of transnational capital" (Friedmann and Weaver, 1979, p. 186) and have been "completely attuned to transnational corporations" (p. 188). Under such conditions, it is argued that development has been dominated by purely functional principles. In contrast to this, Friedmann urges for a return to a territorial basis for global development, founded on a "basic needs" approach. This partly reflects Johnson's move away from the "fetishism of growth efficiency" (Friedmann and Weaver, 1979, p. 195). A principal idea is that basic needs must first be satisfied within areas and this is associated with selective regional/territorial closure in economic terms, and also with the communalisation of productive wealth. It is envisaged that later, the economy can be diversified and non-agricultural activities introduced. The suggestion is made that when activities are bought in, an urban location is not mandatory and that hitherto agglomeration economies have been greatly over-stated. Thereby, it is envisaged that agropolitan development will lead to cities that are based on agriculture. The approach as a whole is designed to build on strength from within and is premised on the skills and resources of the local population, and not outside aid and capital. However, judicious use of the latter is not precluded subsequently. Such an approach, involving the recovery of territorial life seems very attractive given the overwhelming problems of urban-based development currently being experienced in Third World countries. It is argued that in the end, "large cities will lose their present overwhelming advantage" (Friedmann and Weaver, 1979, p. 200). It is clear that the new agropolitan approach advocated by Friedmann has political connotations and is linked in the main, but not exclusively, to socialist policies. Prime examples cited by Friedmann and Weaver (1979) include China, Vietnam, the Republic of Korea, Sri Lanka, Bangladesh, Pakistan and Tanzania. The involvement of the mass of the people by means of active public participation at the local level becomes a vital prerequisite of such an agropolitan approach, but with it comes the opportunity of influencing spatial perceptions and the urban-rural balance to development. These important themes will be developed in subsequent chapters.

EXAMPLES OF URBAN SYSTEMS PLANNING IN THIRD WORLD COUNTRIES

Given the controversial nature of the planning issues and development strategies outlined previously in this chapter, it seems clear that the political dimension is basic to the form assumed by urban systems planning in Third World countries. The principal contrast may be drawn between socialist and capitalist countries. As noted in Chapter 3, large cities have often been bolstered in non-socialist countries during the post-war period. When the economic virtues of large cities have come into question, some form of deconcentration has been attempted, but normally by promoting agglomeration elsewhere, employing growth poles and the like. Hence, Gilbert and Gugler (1982) maintain that outside the socialist countries, few if any real attempts have been made to control the growth of metropolitan areas and to reduce spatial inequalities. It is the socialist countries such as Cuba, China, Cambodia and Tanzania that have made a more concerted effort to control or reduce the growth of large cities. As Stretton (1978) notes, communist governments have generally been suspicious of city life and uncertain as to how they should respond. It is these countries that in the end have frequently moved toward anti-urban policies, or at least strongly rural-based strategies of development. China, for example, is a model of the agropolitan approach advocated by Friedmann.

The example of China is, in fact, often cited. Since 1949, predominantly anti-urban, pro-rural policies have been adopted (see Schenk, 1974; Wu and Ip, 1981). However, as Chang (1982) notes, policies have changed periodically since 1949. On the one hand, strong arguments have been presented for deurbanisation. On the other, some have stressed the importance of centralised planning and the development of major urban-based industries. As Wu and Ip (1981) observe, planning has, therefore, been successively bottom-up and top-down. The period 1949-58 witnessed quite rapid urbanisation. But from 1958 for the next 20 years, the accent was strongly anti-urban and owed much to the ideology of Mao Tse-tung (Ma, 1976). Urban areas were seen as centres of imperialism, surplus extraction and consumption; in short, they were increasingly regarded as the hangover of capitalism. After 1958, an extensive rustification programme was embarked upon, whereby youths and urban bureaucrats were sent out to work in the countryside, sometimes permanently. At the same time, every effort was made to promote the self-sufficiency of rural communes. The rural areas were encouraged to develop their own resources and to establish their own rural industrial plants. This is a prime example of the basic needs approach associated with increasing territorial economic closure. Thus, since 1958, the Chinese have basically renounced urbanisation and migration, and indeed, China's 13 per cent level of urbanisation is relatively low when compared with many countries of a similar overall level of economic development (Chang, 1982). However, urban growth has continued in China, for example, Shanghai now exceeds 10 millions, having increased from a population of around 5 millions in 1949. But it is estimated that as much as one-third of Shanghai's population is presently engaged in agricultural activities. Certainly, attitudes to urban living have been changed in China and the overall emphasis has been placed on the provision of basic needs and the promotion of social equity. Whether such fundamental policy shifts can be explained by ideology alone is, however, open to question (Ma, 1976). With in excess of one million new mouths to feed each and every month, at one level the transfer of people to the countryside can be seen as a necessary expedient in order to cope with the ever-increasing demand for food.

The socialist state of Cuba provides another interesting example. Prior to the revolution in 1959, Havana stood as the classic primate capital city. The country was characterised by high rates of rural-to-urban migration and the rapid growth of spontaneous settlements in Havana. Stretton (1978) notes that at this time, in the region of one-third of all homes in the capital were self-built. The 1953 Census showed that 55 per cent of urban housing was either insanitary or inadequate. Although not initially conceived as a communist revolution, Castro gradually developed his reforms along socialist lines. The leaders of the revolution came to regard Havana as imperialist, privileged and corrupt, and after 1963, it was positively discriminated against. No further work or investment was centred on the capital and the provision of housing and jobs there virtually ceased. This freezing of metropolitan Havana to let the rest of the country catch-up is often referred to as the "Havana strategy", whereby a city's physical fabric is left to deteriorate in order to reduce its attractiveness to potential in-migrants. Other policies sought to emphasise the growth of provincial towns in the 20,000-200,000 size range in an effort to counterbalance Havana, without establishing further large urban centres. At the next level down, the regrouping of village settlements into "rural towns" was effected in order to improve rural living conditions (Hall, 1981a). Notwithstanding these policies, urbanisation has reached 60.3 per cent, and the population is still heavily concentrated in Havana, although its share of the national total has fallen since 1943. On Lehmann's (1982) figures, the growth of Havana has declined from an average annual rate of 2.7 per cent in the period 1943-53 to 1.3 per cent in the period 1970-75. Currently, its population stands at the 1.75 million mark. The Government has implemented stringent migration controls and people are only allowed to move from jobs elsewhere to new jobs in Havana with ministerial permission, and overall there has been a marked decline in internal migration (Lehmann, 1982). Although as noted by Stretton (1978), planning has often tended to be technocratic and dictatorial, much has been achieved in correcting urban-rural imbalances in Cuba. Clearly, the central aim of policy has been to 'urbanise the countryside and ruralise the city'.

Other Caribbean examples can be cited which illustrate some aspects of the mixed-economy approach to urban planning in Third World countries. These examples also relate to the case study material included in the latter parts of this book. The history of settlement and socio-economic patterns in Barbados was considered in Chapter 3. Barbados is an archetypal open, small-island economy, long dependent on the export of primary products. Although only 430 k^2 in area, a population of 248,983 was supported in 1980, at an extremely high average density of 578 persons per km^2. However, nearly one-half of this total population lives in the Bridgetown urban area. The existence of marked socio-economic disparities between the southwestern coastal zone and the rest of the country was demonstrated in Figure 3.7. It may be posited that Barbados illustrates the suggestion that small territorial extent demands urban concentration for economic efficiency. But alternatively, it can be suggested that the country is small enough to be considered as a single urban region, so that such marked disparities should not occur.

There was little or no planning in the colonial period and the Barbadian Town and Country Planning Office was not established until 1959. In the same year, an Interim Act was passed establishing development control over the highly developed south-western coastal strip, and over a short stretch of the east coast from Belleplaine to Bathsheba (for locations see either Figure 4.5 or 3.8). Comprehensive planning control was not established until the Town and Country Planning Act was passed in 1965. This only became fully effective in July 1968, after the gaining of independence. The act introduced a planning system closely modelled on the British system in all respects. Accordingly, the Chief Town Planner was required to prepare the first ever physical development plan for the island within a period of four years. In the event, it was published in 1970 (Town and Country Planning Office, Barbados, 1970).

The plan has been reviewed elsewhere by the present author (Potter and Hunte, 1979; Potter, 1983b). Overall, it was regarded as a blueprint for the future settlement pattern of the island and its theme was an unequivocal call for decentralisation from Bridgetown. The settlement system proposed for the end of the 20 year period of the plan is shown in Figure 4.5. The elevation of Speightstown in the north, and Oistins in the south to the status of regional urban centres was envisaged, their projected populations standing at 6,900 and 7,800 respectively. As a third tier to the settlement hierarchy, seven district centres with populations generally falling between 3,000 and 5,000 were to be established at Holetown, Belleplaine, St Lucy, St John, St George, Welchman Hall and Six Cross Roads. Finally, some 58 village centres were to be bolstered to form a fourth hierarchical tier (Figure 4.5).

The plan thereby provides a clear example of the application of traditional central place notions to the planning of a national settlement system. However, the settlement pattern proposed seemed to take little heed of the existing distribution of population and economic activities, or indeed, recent government locational decisions. Thus, as indicated in Chapter 3, of the eight new industrial estates developed up to 1970, seven were located in the south-western metropolitan parishes of St Michael and Christ Church (refer back to Figure 3.8). Similarly, the plan took little account of the disposition of high grade agricultural land and water catchment areas. In short, the document appeared to be more publishable than practicable (Potter, 1983b). In many respects, it amounted to an idealistic application of central place notions, without any real effort being made to dovetail them to local circumstances. A somewhat more limited decentralisation to a few key areas might have been envisaged, especially to bolster the social and community facilities of the rural north and east. The plan demonstrates that there is no simplistic formula and warns against the uncritical adoption of western ideas. It was a pity that no alternative plans were produced for evaluation and that there appears to have been little public participation in the planning sequence. Matters relating to the physical development plan also serve to illustrate the crucial but sometimes conflicting or ambivalent statements of governments. In 1978, the United Nations conducted a survey of governmental attitudes to population distribution in Latin American and Caribbean countries. The results,

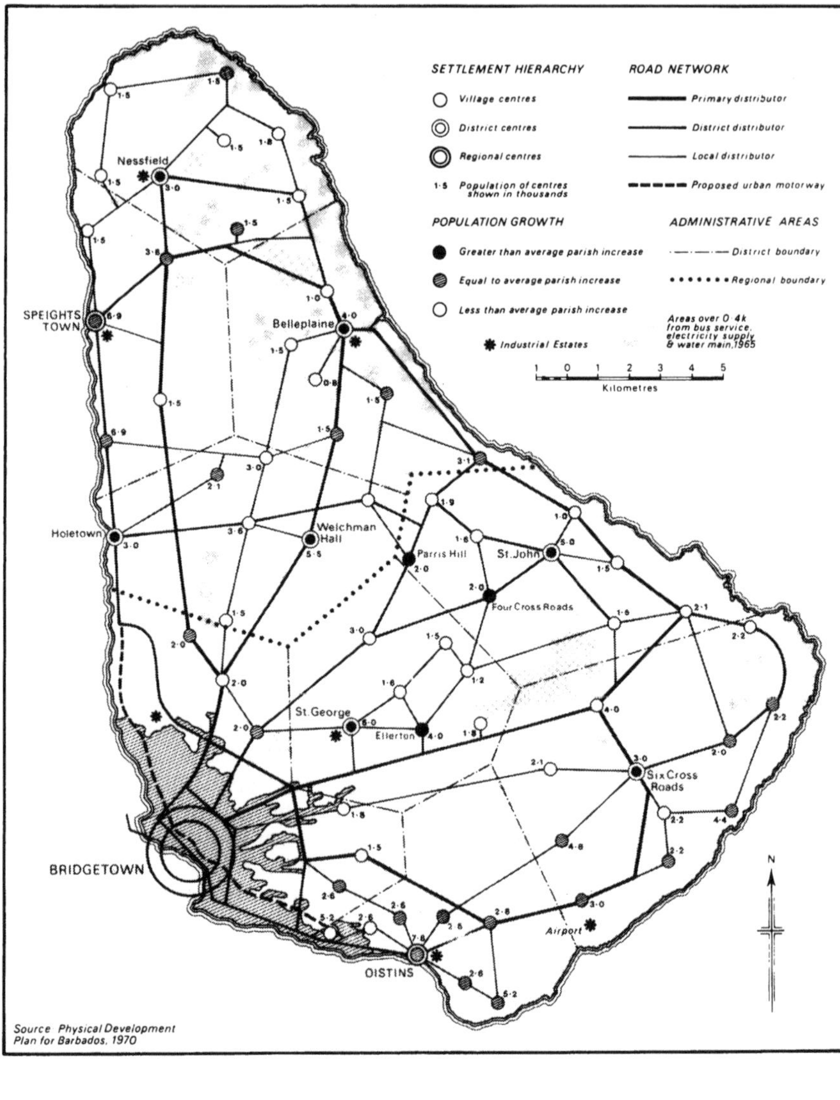

Figure 4.5: Physical development planning proposals for Barbados

reproduced in Peek and Standing (1982, Table 1, p. 2), indicate that of the 27 respondent countries, 21 reported that they regarded their population distributions as extremely unacceptable, four that they were substantially unacceptable, and one, Cuba, that it was slightly unacceptable. Only one country stated that it considered its existing population as being entirely acceptable. That country was Barbados, despite its espousal elsewhere of the singular aim of decentralisation since 1970.

The sequence of maps reproduced in Figures 4.6 to 4.9 summarise the basic philosophy of settlement planning in another West Indian nation, Trinidad and Tobago. The total national population was 1,114,800 in 1977, and over the past 70 years, oil has been the fulcrum of the economy, accounting for around 30 per cent of Gross National Product. A major characteristic of Trinidad is its highly uneven distribution of population and development. The recent phenomenon of rapid urbanisation has reinforced the inherited pattern of skewed development, which focuses on the north-west and west coasts. This is made clear by the mapping of basic infrastructure (Figure 4.6) which pinpoints the major urbanised zones, and in particular, Greater Port of Spain and San Fernando (Figure 4.7). The major physical plan for Trinidad and Tobago (Town and Country Planning Division, Trinidad, 1978) notes that this pattern of unequal development poses a dilemma. On the one hand urban areas offer opportunities for socio-economic advancement. On the other, their growth creates pressures which threaten the very bases of these advantages. Some 60 per cent of the national population of Trinidad and Tobago is now urban, and around 40 per cent is to be found concentrated in the sprawling east to west capital region. The growth and residential structure of Port of Spain were examined in Chapter 3.

The 1978 plan advanced four alternative national urban development strategies (Figures 4.8 and 4.9), and these can be related directly to Richardson's (1981) typology reviewed earlier in this chapter. The first two strategies, shown in Figure 4.8, suggest a pattern of concentrated urbanisation. The first strategy, that of "trends development" is basically the <u>laissez-faire</u> one which would witness the continued growth of eastern Port of Spain and San Fernando. This unplanned concentration may be set against the second strategy of "planned concentration". This would have the advantage of maximising existing opportunities for agglomeration economies, and would see the expansion of the Port of Spain area as far east as Sangre Grande (Figure 4.8). But the social costs of such a strategy would be high and would do little to help reduce the uneven pattern of development. Two explicit strategies of dispersal were therefore considered in the plan (Figure 4.9). The first envisaged a pattern of maximum "dispersal" to relieve congestion and stimulate the lagging regions. It was argued, however, that such a strongly welfare-based approach would serve to reduce economies of concentration to an unacceptably low level. In the fourth and final alternative, therefore, a pattern of "dispersed concentration" was advocated, wherein the stated aim was to combine the respective merits of the strategies of planned concentration and dispersal. It was argued

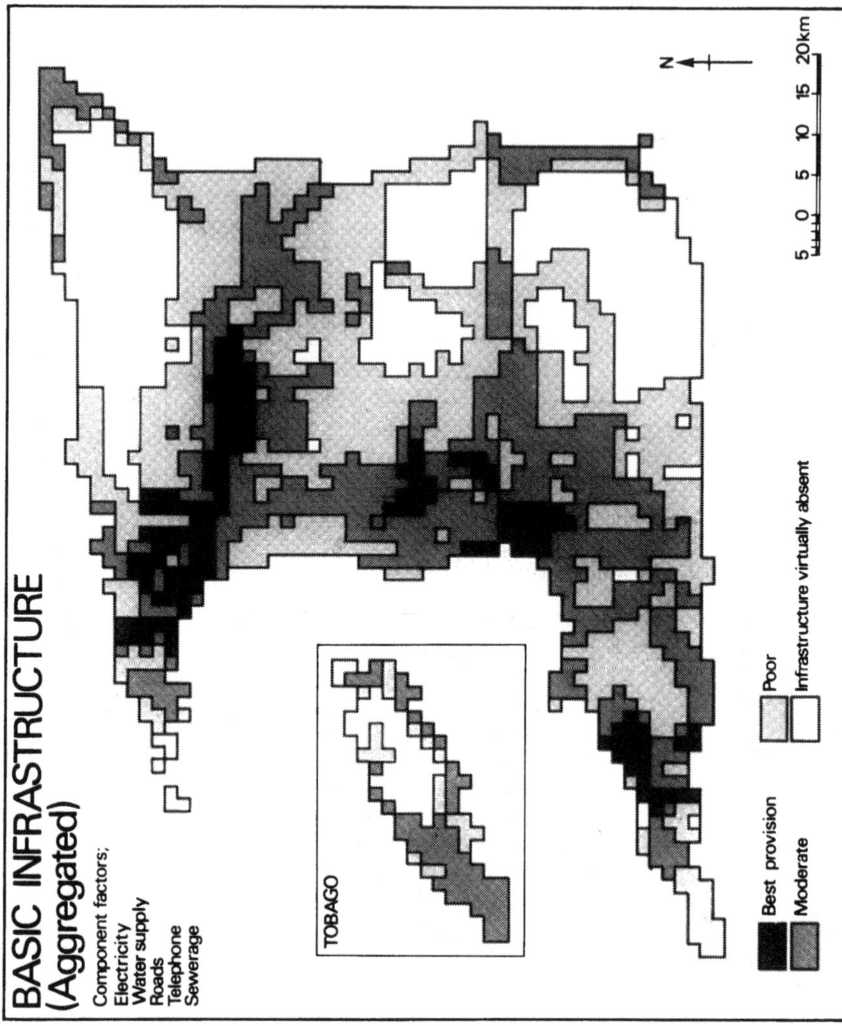

Figure 4.6: Basic infrastructure in Trinidad and Tobago

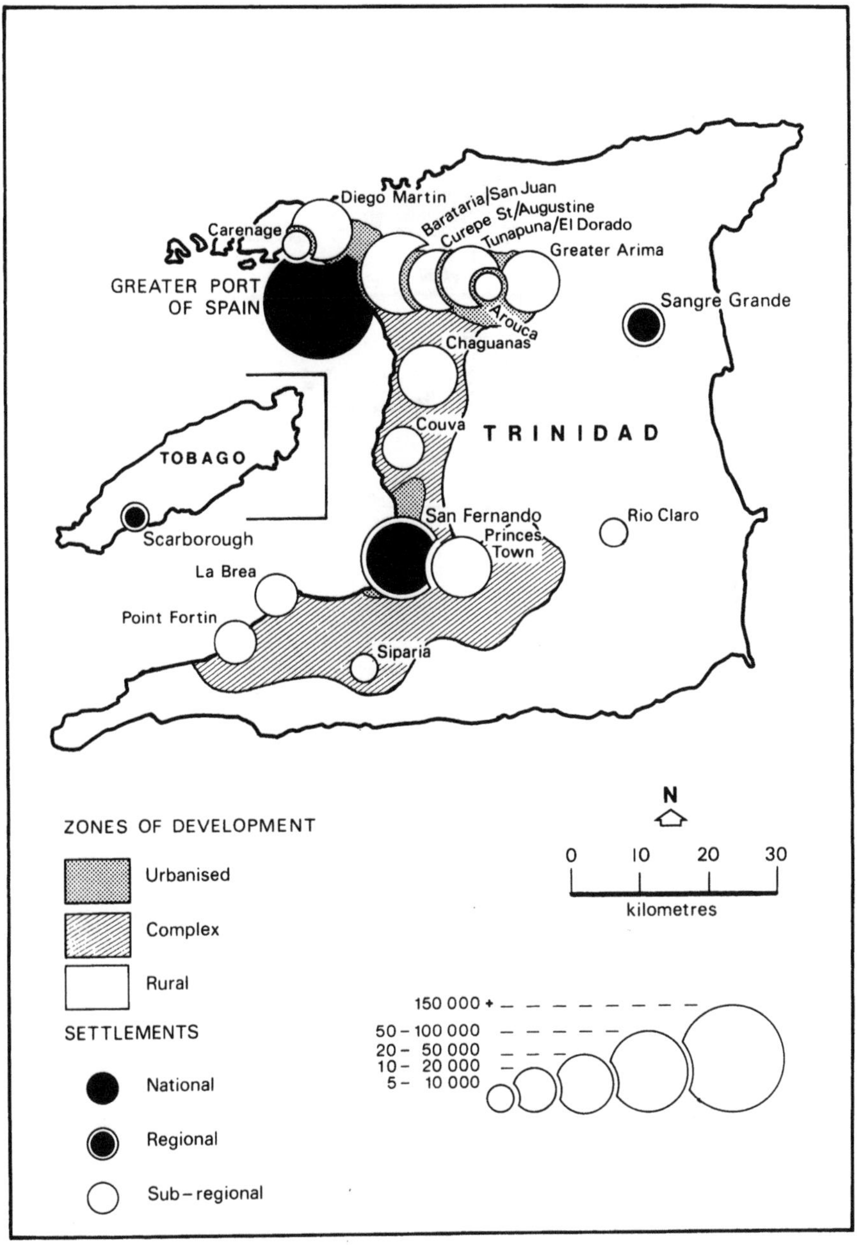

Figure 4.7: Zones of development in Trinidad and Tobago

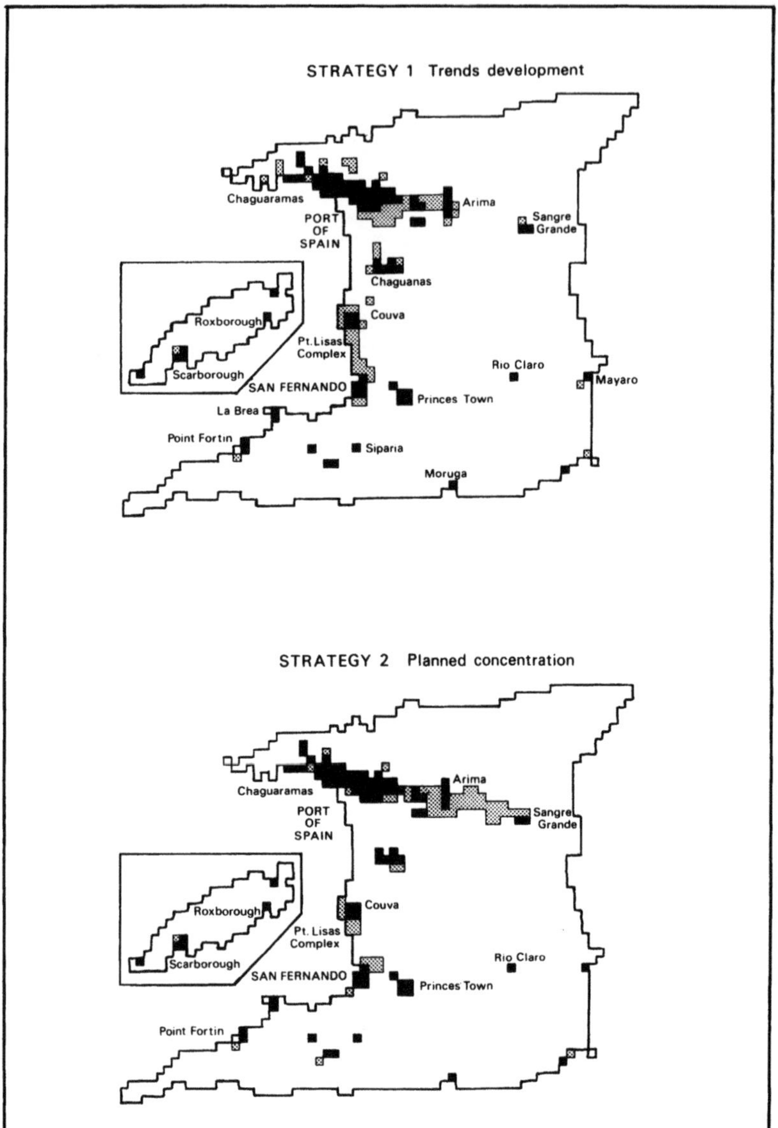

Figure 4.8: Alternative plans for Trinidad and Tobago: trends development and planned concentration

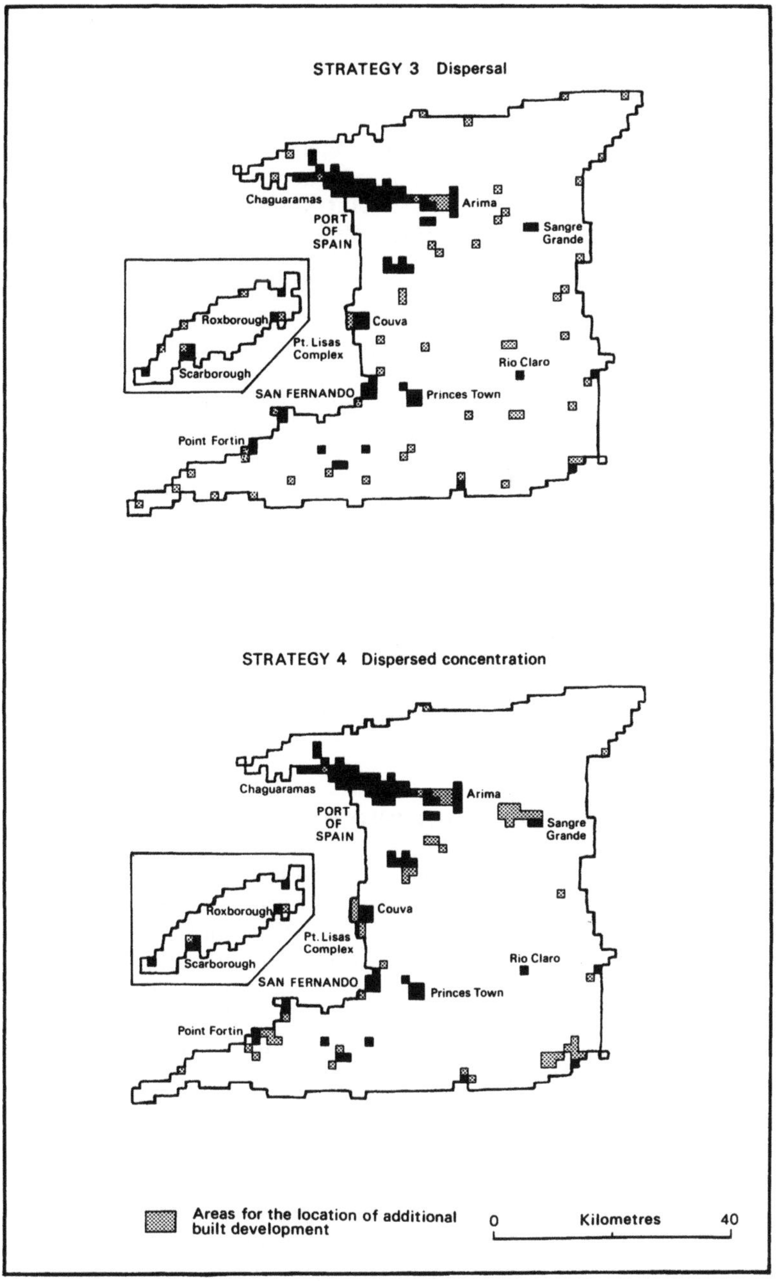

Figure 4.9: Alternative plans for Trinidad and Tobago: dispersal and dispersed concentration

that such a settlement system would harness the economies inherent in the existing settlement cores, whilst at the same time promoting development in the lagging rural regions.

It was this approach that was adopted in the final plan shown in Figure 4.10. The main proposal was the development of four "growth poles" between 1980 and 2000, respectively at Point Lisas/Couva, Point Fortin/La Brea, Sangre Grande and Guayaguayare/Galeota Point. The development of the first of these is already well under way, the Point Lisas Industrial Development Company having been established in 1966 and brought under government control in 1970. This multi-billion dollar complex is based on chemical and steel works and a major port development project (Mulchansingh, 1983; Pollard, 1984). The Point Fortin/La Brea growth centre is scheduled to come on stream between 1985 and 1995, and is based on the existing oil refinery at Point Fortin. Sangre Grande on the other hand is seen as the focus of more gradual growth, mainly in the form of residential overspill from eastern Port of Spain. The designation of Guayaguayare as a futuristic new town to reach a total population of 20,000 by the turn of the century already appears to have lost favour among politicians and planners.

Urban planning in Trinidad and Tobago offers a good example of the strategy of concentrated urbanisation and also illustrates the wide diversity of developments for which the term 'growth pole' has come to be used. More particularly, the trade off that exists between considerations of social equity and economic efficiency is highlighted, as is the inevitable socio-political nature of decisions made about the relative merits of different national settlement strategies.

PLANNING FOR THIRD WORLD HOUSING

The serious nature of the problems posed by the provision of housing in Third World cities was discussed in Chapter 3. However, as Turner (1980) stresses, it is not a question which simply involves 'housing'. At the same time, it is essential to make it possible for individuals to improve their overall life capacity. Thus, improvements in employment, transport, infrastructure and social facilities must accompany developments in housing. Without these wider changes, particularly the provision of employment, people will not be able to afford improvements to their housing. This argument has major policy implications, as is illustrated below.

In most developed countries, the state has taken a leading role in the building of homes. But as noted in the previous chapter, this has not been the case in most Third World countries. There are, however, exceptions and in Hong Kong, Singapore and China, government building programmes have to some extent succeeded in controlling spontaneous settlements. The case of Hong Kong is an interesting and well-documented one (see Dwyer, 1975; Drakakis-Smith, 1981). Since 1954, the Hong Kong government has been responsible for the construction of over 400,000 new domestic

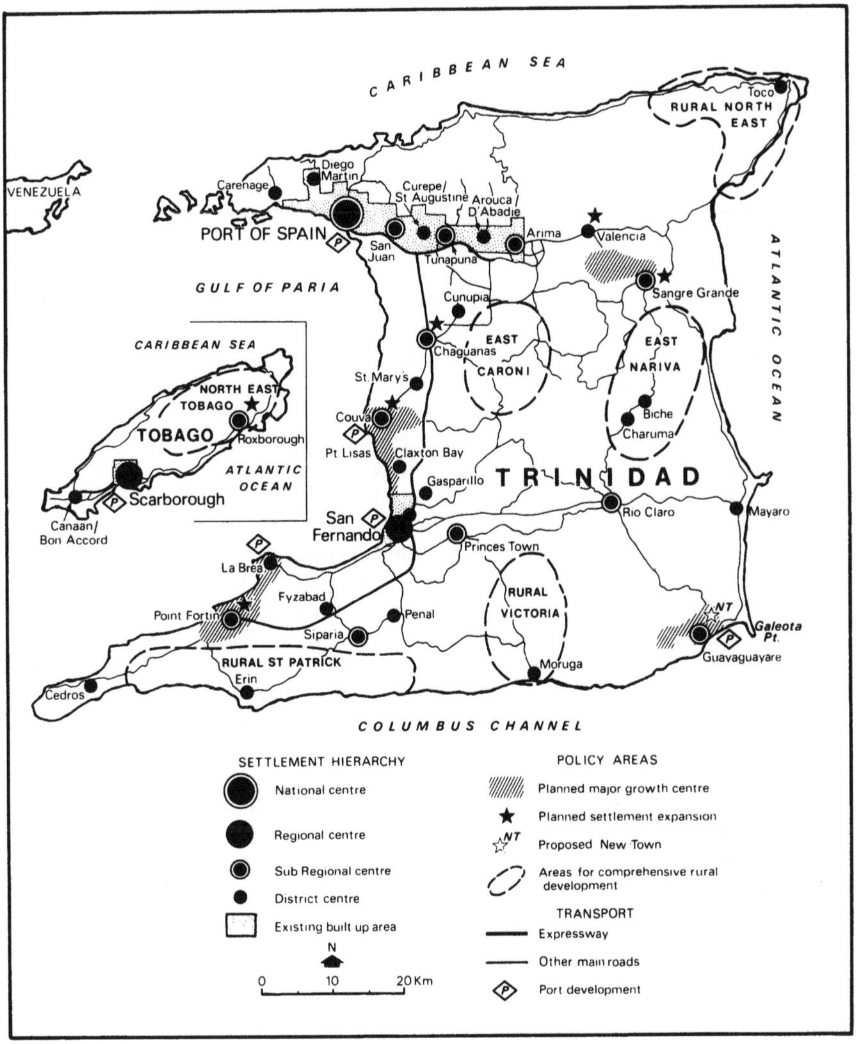

Figure 4.10: Urban and regional development plan for Trinidad and Tobago

residences. These now accommodate over 2 million residents, or approximately 44.5 per cent of the total city population. The scheme, which stands as one of the largest public housing programmes in the Third World, was basically designed to rehouse squatters and not as a means of assisting low income families per se. Both Dwyer (1975) and Drakakis-Smith (1981) argue that a strong motive on the part of government was to free sites occupied by squatters for more lucrative permanent development. The programme commenced in 1954, having been precipitated by a major fire in the Kowloon area, which had left some 53,000 squatters homeless in the previous year (Yeh and Fong, 1984). It was decided to embark on a large-scale programme of high-density accommodation of a very basic design. Thus, the standard unit was a seven storey H-shaped block. All facilities were provided on a communal basis, including water taps, bathrooms and toilets, and were located in the central crosspiece of the block. Some 240 blocks of this type were built up to 1964, and these housed half a million people. It is now generally accepted that the standards adopted in these early schemes were too low. Accordingly, a major effort to improve the quality of resettlement housing was made during the period 1964-73, during which time the building programme was extended and upgraded. After 1964, newly built blocks were larger, at least 16 storeys high, and the individual housing units were more spacious, having their own toilets, tap-water and kitchens. On the debit side, however, there was an increasing tendency for these new estates to be located on the urban periphery, away from the main employment opportunities. Notably, a special low cost housing programme was also introduced in 1961. Since 1973, squatter resettlement has been retained as the principal plank of housing policy, but emphasis has also been placed on the improvement of the early resettlement estates. Although Hong Kong's housing programme has not been without its critics, especially in relation to the very high residential densities involved, it stands as one of the most ambitious resettlement schemes ever attempted in the Third World.

Another example of a government housing scheme instigated with the specific aim of eliminating squatter settlements is provided by Caracas, Venezuela. This programme was mentioned in Chapter 3, when the housing structure of Caracas was discussed, including the prevalence of ranchos or squatter settlements, and the highly restricted site occupied by the city (Plate 4.1). The housing programme was carried out by Banco Obrero, the principal housing agency, at the instigation of the dictator, Jimenez. This crash programme started in 1954, and between then and 1958, some 97 high-rise developments, or "superblocks" were built, 85 in Caracas (for their location see Figure 3.16) and 12 in the nearby port of La Guaira. In all, in excess of 16,000 apartments were provided in these superblocks. Some of the centrally located superblocks are shown in Plate 4.2. Most are 15 storey blocks, and they are luxurious by comparison with the rancho areas (Plate 4.2). The headlong speed of the rehousing project is frequently attributed to Jimenez's desire to make an overt show of his power and his concern for the welfare

Plate 4.1: A view of Caracas, Venezuela

Plate 4.2: Caracas, Venezuela: Barrio area and high-rise response

of the people. It is now generally agreed that the outcome was little short of disastrous. Thus, many ranchos were bulldozed with little notice and the residents transferred to the superblocks, which had been extremely hastily designed. For example, a major fault was the interior location of the stairwells. Jimenez was overthrown in 1958, by which time the superblocks were in near chaos (Dwyer, 1975; Drakakis-Smith, 1981). For example, some 4,000 families had invaded empty flats, whilst rent arrears had reached staggering proportions. Squatter shacks had been built close to the blocks, notably on the green spaces provided between them. There had been little or no physical maintenance of the blocks, but far more damaging was the fact that adequate social facilities had never been provided. The largest group of buildings, located to the east of the CBD and known as the Urbanizacion 23 de Enero had come to be dominated by political factions. With the overthrow of Jimenez, the programme was suspended and subsequently, evaluation surveys were carried out. These served to emphasise that the low incomes of the residents made it impossible for them to pay the rents and maintenance charges without sub-letting. Such overcrowding further exacerbated the social problems of high-density living. A major recommendation was the need to build up community involvement by means of tenants' associations. In 1961, a large-scale social work programme commenced in the 23 de Enero area. As Drakakis-Smith (1981) observes, Caracas offers perhaps the most infamous example of an ill-planned and misdirected mass public housing programme in a Third World city.

Most authors are agreed on the lessons that are to be drawn from such examples. In a nutshell, apart from a few wealthy city states, most Third World governments cannot afford high-rise, monumental responses to their housing problems. But more significantly, nor can the mass of poor people in these countries. The headlong rush into high technology, Western inspired housing programmes seems singularly inappropriate, and frankly, incongruous when viewed both environmentally and socially. Such "oppressive housing" tends to filter up to middle class groups, if, that is, it benefits any social group at all. These facts tend to place a fundamental question mark against the efficacy of government involvement in the provision of housing in Third World countries. Turner (1983, p. 207), for example, has recently written:

> Speaking to those with responsibilities for housing policy and its implementation, Otto Koenigsberger says that "if government is to improve a low-income majority's housing conditions then it must not build houses".

In the past 15 years, there have certainly been major changes in housing theory and practice in many capitalist Third World countries. The lead has been taken from the central arguments put forward by Mangin, Turner and others, that the vast majority of low-income residents in less developed cities are resourceful and responsible, and that if the opportunity exists, they will build and subsequently improve their own homes via self-help. Thus, the new conventional wisdom in Third World urban housing has become the promotion of aided self-help. The pivotal argument is that the government should provide the missing facilities and infrastructure, but the poor should actually build homes for themselves. The state should also concentrate on providing jobs and social services, which by and large, the people cannot create for themselves. Three major types of aided self-help may be recognised: the upgrading of existing squatter settlements, site and services schemes and a variant of the latter, core housing. Since 1970, these have become the new orthodoxy, being actively supported by the World Bank, and as Laquian (1983) notes, almost 100 countries are at present pursuing community upgrading and site and services projects.

The upgrading of existing squatter settlements represents the most basic form of aided self-help. In order to improve existing slums, basic infrastructure such as water standpipes or sewerage connections are installed and/or improvements to the dwellings themselves are made. The principal aim of upgrading schemes is to minimise housing costs and to avoid the residential and job dislocation that inevitably results from large-scale government resettlement schemes. In many instances, merely removing the threat of eviction and adding basic missing infrastructure will serve to upgrade a longstanding slum area or squatter settlement. In such cases, housing improvements may still be left entirely in the hands of the residents themselves, by means of a gradual process of consolidation (Skinner and Rodell, 1983). This type of approach is distinct from more comprehensive schemes, where improvements are also made to the dwelling units. Given the irregular and essentially unplanned nature of spontaneous settlements, frequently the equalisation and squaring up of plots to give a regular layout and road plan is a prime requirement. A typical starting point might well be the hotch-potch of timber dwellings shown in Plate 4.3. However, there is a trade-off, for the more the effort to regularise the layout, the greater the destruction of houses that is likely to be involved. In parts of Asia, and in the Caribbean too, where lightweight timber or bamboo framed houses are common, this is often much less of a problem, as such houses can be moved relatively easily. A good illustrated example of what can be achieved by means of a slum upgrading project is given by Alan Turner (1980, pp. 256-67) in the context of Manila. Turner stresses in unequivocal terms the vital need for the full and effective participation of the public concerning new layouts and the facilities most needed. If the residents are not fully consulted on these issues, then there is little or no chance of successful upgrading.

Upgrading schemes go hand in hand with site and services programmes. Whilst the former are designed to make the best use of existing houses, the latter represent a strategy for providing new ones on the most economical basis. This is necessary both to rehouse the few who have been displaced by upgrading schemes, and to provide the homes necessitated by new in-migrants. Skinner and Rodell (1983) attribute the broad concept of "land-and-utilities",

Plate 4.3: Squatter housing in the Four à Chaud area of Castries, St. Lucia

Plate 4.4: A 'chattel-style' timber house on a rock- pile base, Holetown, Barbados

now generally referred to as site and services schemes, to the practical works of Abrams and Koenigsberger between 1955 and 1963. Site and services schemes involve providing entirely new homes by extending public services such as roads, water, sewerage, drainage and electricity to sites in preparation for construction. Such schemes vary from the mere provision of services and pegged-out building lots, to the erection of partly finished houses. This means that the approach can be tailored to suit a wide range of income groups, and its potential utility for very low income residents is an important merit. The lots themselves may be either sold or leased to individual residents. At best, site and services schemes offer a family a plot of land on which they can build at their own pace, with gradual servicing, access to credit, building materials and advice, all at a cost which is affordable given their income (Gilbert and Gugler, 1982). Core housing schemes essentially offer a minimum shelter unit, which can be occupied almost immediately and can be extended when the residents are able to afford it. Once again, of course, it is essential that the residents are owners and that finance is available to facilitate improvement and consolidation. Most core housing projects are linked to site and services schemes, and it is frequently envisaged that it will take up to 20 years for the houses to be fully completed.

There is by now an extensive literature dealing with both the theory and the practice of aided self-help housing. Recent general accounts, for example, are provided by Laquian (1977), Muller (1982), Siebolds and Steinberg (1982), Peattie (1982), Burns (1983), Kabagambe and Moughton (1983), Angel (1983) and Phillips and Yeh (1983); whilst Ward (1981) and Devas (1983) examine the many problems of finance which exist. The potential merits of aided self-help approaches to Third World housing problems are so great that there is the danger that the approach is regarded as a "cure-all". We should certainly at least remind ourselves of the arguments presented by Burgess (1982) against self-help, essentially suggesting that governments may adopt such policies in order to avoid the redistributional and structural changes that would be involved in housing people in a decent manner. This line of reasoning sees self-help as reinforcing social inequities and providing a cover for government non-action on urban reform, progressive taxation and land speculation. The critical role of permanent tenure is highlighted in all studies of Third World housing. The example of the chattel or movable housing system in Barbados as a response to the spot rental system was discussed in Chapter 3. It is only with the chance of owning the plots on which such houses are built that true progress is likely to be made (see Plate 4.4). However, it is notable that aided self-help has not been confined to capitalist countries. In Cuba, for example, since 1979, 60 per cent of all new housing units have been constructed by voluntary 'micro-brigades'. These take the form of squads of workers who second themselves from their jobs in order to construct housing for their fellow workers. In turn, these workmates are expected to work overtime to maintain production levels in the absence of the micro-brigade workers (Hall, 1981b).

There is much sense in the argument that aided self-help must not be regarded as a panacea, nor must we fall into the trap of feeling that all public housing is necessarily bad (Turner, 1980). The latter author has suggested that a balanced housing strategy for low-income housing must consist of a number of closely linked policies, including the upgrading of existing slums, ensuring an adequate supply of land for new site and services settlements, providing security for tenure and access to finance, along with establishing wider socio-economic development and the use of appropriate technology. Without progress on each of these fronts, little is likely to be achieved in tackling the pressing housing problems that exist in most Third World countries.

FINAL COMMENTS

The chapter has shown that there are no simple and universal answers to urban planning problems in Third World countries. Trends that emerge from an examination of evolving thought concerning Third World urban planning are the need to consider the perceptions and aspirations of those being planned for, and to increase their effective involvement in the planning process. This is implicit in the agropolitan approach to development suggested by Friedmann and others at the inter-regional scale. It is also a vital accompaniment to the process of in situ self-help and upgrading which in various forms is increasingly coming to be favoured for intra-urban planning in less developed countries.

Clearly, however, the political will for people-based planning must exist outside the ballot box. This point is crucial, for political factors play a decisive role in urban planning. By its very nature, planning frequently serves the interests of the existing political elites and economic power groups. This may be regarded as the reason for the prevalence of top down systems of planning and decision-making in so many countries. Taylor and Williams (1982), for example, observe that governments in Third World countries may frequently fear public participation, or indeed any form of power sharing. This may also reflect the colonial period, with its legacy of highly centralised decision-taking (Logan, 1972). This carries the grave risk that politicians become increasingly out of touch with the peasantry and the urban masses in general. This is perhaps well-illustrated by the frequent adoption of negative attitudes to squatters, self-help housing and informal sector employment initiatives. It is also conceivably mirrored in the frequent drawing up of over-sophisticated and technical plans, which are unsuited to local circumstances.

The mere act of calling for the establishment of participatory planning will however achieve little. Further, it must be recognised that its attempted implementation will involve severe difficulties. But against this, its adoption in principle will hopefully help to foster a more responsible and responsive attitude on the part of the planned and the planners. In the chapter which follows, this issue is investigated. It would, of course, be naive in the extreme to imply that these changes will occur overnight and without formidable problems being encountered. But such changes are urgently required, as exemplified by Taylor and Williams (1982, p. 20):

> There is an urgent need to reduce this conflict of interests and to involve all segments of society in the decision-making process as much as possible, not merely out of altruism or democratic principle, but because it will lay the foundations for greater stability within the society.

Accordingly, the challenges and prospects associated with the establishment of participatory urban planning in Third World countries are considered in the next chapter.

Chapter 5

Public Participation in Third World Urban Planning

A major concern over the past 20 years in both rich and poor countries has been with the establishment of greater public involvement in the process of environmental planning. This chapter explores the need for, and potential of public participation in the urban and regional development planning process in developing countries. Although it might be argued that public participation is essentially a preoccupation of planning in advanced industrial countries, it is increasingly coming to be accepted that it is just as important in less developed nations, albeit perhaps in a somewhat modified form (Conyers, 1982; Franklin, G.H., 1979; Potter, 1984b, 1984c).

Some analysts might be tempted to argue that because of the immensity and pressing nature of the environmental and socio-economic problems faced by many developing countries, public participation in planning is at best a luxury, and at worst, entirely unnecessary. In the latter regard, Koenigsberger (1983, p. 51), for instance, has recently claimed that in many poorer countries, the high levels of self-help activity mean that the question of public participation does not arise, as "it is the public that does the planning and the development; it is the planner who is not allowed to participate". Although this argument would seem to be somewhat extreme and to ignore the planners' responsibility of finding appropriate policy frameworks for planner-client interaction, it does reflect the extreme pressures exerted on planners in developing countries. Others may be inclined to comment that educational and literacy levels in many Third World countries are such that trained professionals are the only people that possess the requisite expertise to make judgements and reach decisions about environmental matters. The question as to whether planning should be technocratic or participatory thereby assumes special relevance in the context of the Third World.

The present chapter examines these issues, and it is concluded that despite the possible snares and pitfalls involved in encouraging greater participatory democracy in Third World urban planning practices, its promotion stands as a vital objective. As noted in Chapter 1, the theme of the present account is the pressing need for the development of soundly-based and appropriate urban and regional planning systems that seek explicitly to identify and act upon the perceptions and aspirations of individuals. The need for such an approach is well exemplified by the changes that have occurred in views of squatter settlements in the past 30 years, as detailed in Chapter 3. Viewed initially as blots on the urban scene testifying to the operation of the culture of poverty (Lewis, 1959), close scrutiny of the aims and aspirations of those living in many such settlements showed them to be areas of rational self-improvement, or "slums of hope" (Turner, 1967, 1968; Mangin, 1967). Thus, the basic argument presented in the present chapter concurs with Lloyd (1979), that it is imperative for planners and researchers to understand how the poor view their society, and that further, any policy, "whether it re-structures society or merely papers over the cracks will affect those who live in the shanty town and will imply their acquiescence and perhaps active participation" (Lloyd, 1979, p. 226). Hopefully, it will be active public involvement that is encouraged rather than mere acquiescence; and this is seen as a major prerequisite in steering emphasis away from Western-style planning to appropriate frameworks, and away from purely technical solutions to what are essentially complex social problems.

BACKGROUND TO PUBLIC PARTICIPATION IN PLANNING

First, we consider some of the wider philosophical and pragmatic issues that surround public involvement in the environmental planning sequence. By now, many countries in both the developed and developing world regions espouse the need for greater public participation in planning, although in practice, many may only pay it lipservice. The overall approach is referred to by an almost bewildering variety of terms, including 'bottom-up planning', 'grass-roots planning', 'public involvement', 'participatory planning', 'democratic planning', and 'collaborative planning' (Conyers, 1982; Fagence, 1977).

All of these terms and the approaches they describe, have in common the philosophy that in a democracy, members of the general public should have the opportunity of participating in the decision-making process between voting in elections. As Conyers (1982), has argued, there are at least three important reasons for encouraging public participation in planning. First, it acts as a means of gaining insights into local conditions and the needs of local people. This is essentially the pragmatic rationale for public involvement, for without such basic information to help identify public preferences correctly, plans are unlikely to be successful (Sewell and Coppock, 1977). Secondly, it is logical to assume that individuals are more likely to be committed to plans if they are involved in their preparation. Thirdly, there is the largely philosophical reason identified above, that it is considered to be a basic democratic right that people should be involved in their own development. In short, such views subscribe to the notion that "planning is for people" (Sewell and Coppock, 1977, p. x).

The operationalisation of public participation via formally established mechanisms and legislation is, however, quite recent (see Fagence, 1977; Hampton, 1977). From the birth of modern-day British town planning in 1947, for the next 20 years, public participation was regarded by local planning authorities as keeping the public informed of what was going on, and basically little more (Fagence, 1977). The Planning Advisory Group was set up in 1964, and its report appeared the next year (PAG, 1965). It suggested the establishment of a two-tier development planning system and stressed that this would necessitate greater public information and actual participation in the planning process. The Planning Advisory Group's recommendations were reflected in the Town and Country Planning Act, 1968 which set up the structure/local development planning system. At about the same time, the Skeffington Committee was established to consider the means of fostering greater public participation in the development planning process and its report. <u>People and Planning</u> was published in 1969 (Skeffington, 1969). This stressed the need for public participation and suggested various means to achieve it, although these have been described as "scarcely innovatory" (Fagence, 1977, p. 265). The Town and Country Planning Act, 1971, embodied the spirit of the Skeffington recommendations, so that local planning authorities were to secure adequate publicity for their plans and the public were to be afforded the opportunity of making known their views.

The Skeffington report defined public participation as the act of sharing in the formulation of policies and proposals. The public were not simply to be asked to judge a finished or near finished product (Hampton, 1977). Thus, the Skeffington report distinguished clearly between what it regarded as involvement as opposed to participation <u>per se</u> (Grimshaw and Briggs, 1970). Involvement may be regarded as merely the first step, for participation implies far more: that is, an active role within the decision-making process itself. This distinction is important and will be referred to when the means of enabling public participation are reviewed subsequently in this chapter.

At the very heart of questions concerning public participation in planning is the issue of the nature of the interaction that is established between planners and the public, who are effectively the planner's clients. Three main hypothetical models of planner-client relations can be identified, as shown in Figure 5.1 (Pocock and Hudson, 1978). In the first of these, the planner is seen as a leader (Figure 5.1a) and the dominant flow of ideas and information is from the professional planner to members of the public, with little or no flow in the opposite direction. Pocock and Hudson argue that it is this type of implicit model that has underlined planner-public relations in the UK since the Second World War. In an alternative formulation, the planner may be seen as a follower. In this reverse model, shown in Figure 5.1b, the public expresses its needs and the planner responds to these statements. Advocacy planning, where radical planners work to represent the views of deprived or minority groups would be an example of the operation of this mode of planning. In the

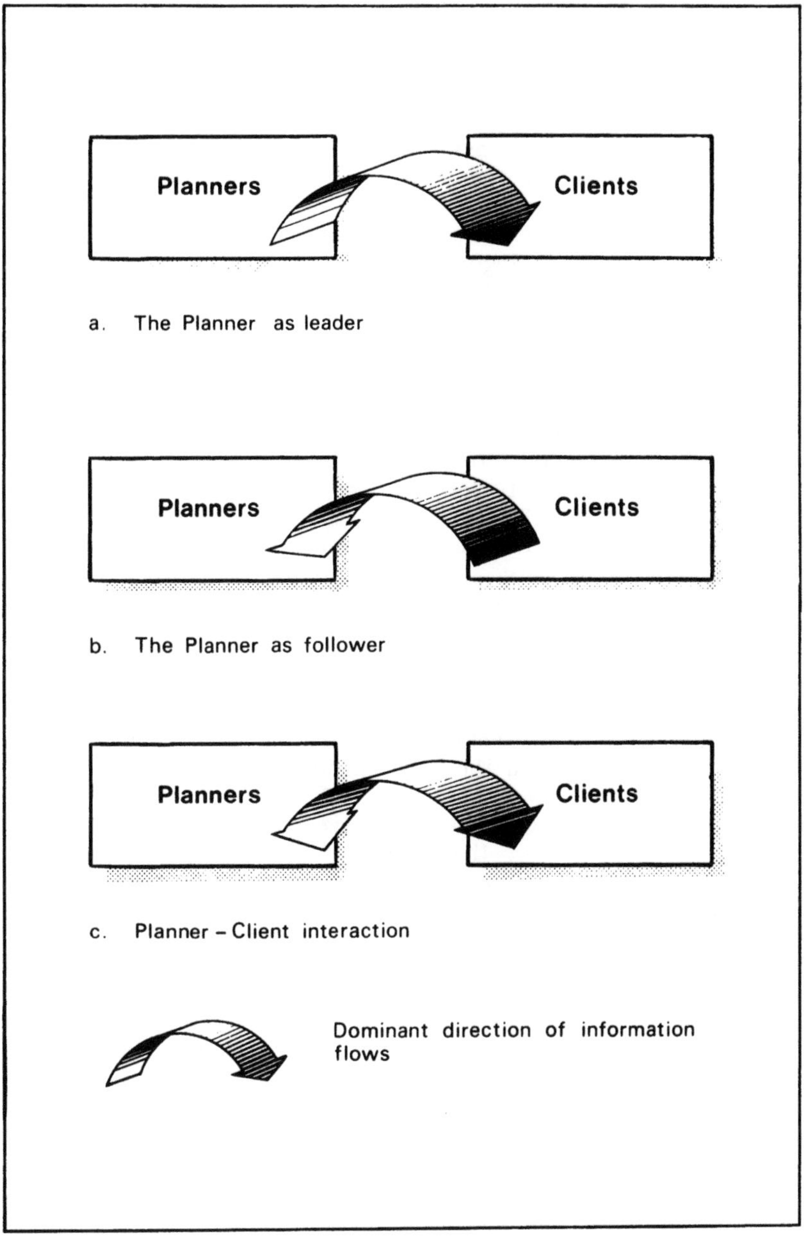

Figure 5.1: Planner-client interactions (Source: Pocock and Hudson, 1978)

third and final formulation, in many ways an idealistic compromise, there is continual planner-client interaction, with flows of ideas and information occurring in both directions throughout the planning sequence (Figure 5.1c). Here, planners explain the reasons for their proposed actions, whilst the public conveys its needs and aspirations to the planners. In this manner, a continued dialogue is established between planners and the public, so that a decision-making partnership is established rather than a process of sporadic consultation.

At a more fundamental level, these models and indeed the whole question of public participation are closely related to political philosophies of democracy (Fagence, 1977). The comment of Stretton (1978) that the motives for participatory planning are frequently radical, although its establishment often serves to conservatise the system is apposite in this regard. But as Fagence (1977) has stressed, there is a range of practices that can pass for public participation in the process of democratic government. These extend from a participationist view through to an elitist one, whilst Fagence regards the Marxist perspective as an intermediate one.

The traditional elitist view stresses the need for a ruling group within a democracy. This is often justified by the argument that the mass of society is either politically naive, apathetic or excessively volatile and intemperate. Normally, it is the former allegation of inertia or incompetence on the part of citizens that is cited. In the traditional elitist view, it is maintained that public accountability should be limited to periodic elections, so it is with such a system that the philosophy of the environmental planner as leader would come to full fruition. The restriction of public involvement to the ballot box would make environmental planning an expert and technocratic process. On the other hand, the strict participationist perspective, which owes much to classical nineteenth century European liberalism would place strong emphasis on the full and constant participation of the public. Finally, Fagence (1977) sees the Marxist approach as affording a bridge between the extreme participatory and elitist views. In particular, it can be argued that although the assumption that Marxist propositions are particularly pro-participation is generally correct, Marx in fact made many concessions to elitist preferences in terms of their influence in decision-making structures. Similarly, Marx envisaged stimulated revolutions from a working-class base, these having been fostered by the process of education.

Whatever the philosophy and practice of democracy that is subscribed to, it is certainly the case that the idea of public participation in planning is here to stay. In the words of Fagence (1977, p. 371), it has by now "endured sufficiently ... to be taken seriously: it need no longer be considered the current planning phobia". However, although it must be recognised that participation is highly desirable for a variety of reasons, from a pragmatic perspective it is equally important to acknowledge that there is no easy means of achieving it (Conyers, 1982). Accordingly, the problems involved in implementing public participation are reviewed in the next section, and subsequently, the specific issues concerning its potential application in Third World contexts are addressed.

PROBLEMS OF PARTICIPATORY PLANNING

Arnstein (1969, p. 216) has adroitly observed that "the idea of citizen participation is a little like eating spinach: no one is against it in principle because it is good for you". Hence the clear implication is that the concept is plagued by practical difficulties and impediments. Such problems are encountered in the context of both developed and less developed countries, although naturally, the present review concentrates on the latter. The fact that there are problems to be faced in implementing public participation should come as no real surprise if its establishment is indeed a cornerstone of true democratic government.

An issue that is frequently raised is whether or not members of the public really do wish to be involved in environmental policy-making. Clearly, advocates of elitist political systems would argue that often they do not. Indeed, as noted by Conyers (1982) there is much evidence that in many instances where individuals have the opportunity, they do not actively participate. Some regard this as a problem of particular relevance in Third World countries. Further, it is a general feature that peoples' perceived levels of satisfaction tend to be higher for those social domains which impinge most closely on their daily activities. This would appear to be especially the case for the lower income groups within society (Potter, 1985). These factors may well lead to a form of parochialism, so that some groups may be happy to defer to experts when confronted with non-local issues. The non-joiner problem is a near universal one and indeed, it must be recognised that it is a democratic right not to participate. The major point is that people need to be directly interested in a decision-making problem and to feel they can influence the outcome, in order for them to readily participate.

Ideally, all those concerned by a given issue should have the chance of participating, but it is not always clear who has legitimate interests. Further, in many instances not all those with a vested interest can participate, so that the issue of representation becomes a major one. If social surveys and interviews are employed to mobilise public involvement, then this problem is normally met by taking a representative sample of the views of the diverse groups involved. But in other instances, certain individuals or key groups may come to represent the views of wider groups (for example, via public hearings or consultations with community leaders). The problem of representation is, of course, closely allied to the fact that it is often those who are educationally, socially and economically better off that tend to participate most readily.

Another linked area of concern is frequently expressed. That is, whether the people know what they want and what is likely to be good for them. Perhaps even more fundamentally, it is asked whether members of the general public perceive the full range of possible alternatives that is open to them in a given situation. Conyers (1982) has stressed this as a matter of special concern in Third World contexts, particularly with respect to rural areas, where inherent conservatism may mean there is a fear of the unknown and untried. In this regard, it is also argued that perceived solutions to problems as seen at the local scale, may conflict with the dictates of national policies.

The potential communication gap existing between planners and the public is also seen by many as a vexing question. This is not just a difficulty relating to language and nomenclature, however, but rather points to differences in attitudes, expectations and, like as not, basic philosophies too. As will be stressed in the next section, the fact that many professional planners have been trained abroad undoubtedly exacerbates this problem in most Third World countries. This brings us back to the need to overcome the cult of the expert in planning: the express aim of public participation.

A lack of willingness on the part of the public may well result from past experiences of involvement where expectations have not in the event been fulfilled. Thus, if in the past, public participation has been no more than a cosmetic and largely spurious gesture, then this will obviously militate against continued participation in the future. This relates directly to the differences existing between consultation, involvement and participation. Even with regard to public participation in the preparation of structure plans in the UK since the 1971 Act, for example, Pocock and Hudson (1978) argue that this has amounted to little more than presenting the public with a very limited number of predetermined alternative scenarios, rather than any real degree of involvement in actual decision-making. Similarly, we shall see shortly that some techniques of public participation merely represent the flow of information and attitudes from the public to the planners, and not a genuine two-way interchange of decision-making views and preferences.

Finally, the argument that genuine public participation results in slower and less efficient planning is often cited as a further potential impediment to its establishment. It is indeed probably true that in most circumstances participatory planning is likely to involve more paper work, more soul searching, and will probably be both more costly and time consuming as a consequence. But it does not necessarily mean that planning will thereby be rendered less efficient. If a somewhat more ponderous and most costly decision results in a more workable long-term plan, then this is obviously more efficient than a quickly derived cheap solution that is in the end found to be wanting.

Undoubtedly, the problems involved in the establishment of effective public participation in urban and regional planning are great and may well be even more so in the context of the Third World. But such difficulties are by no means insurmountable. Obviously, efficacious planning is not likely to be easily achieved in the face of formidable environmental, social and political circumstances. The possible advantages of the establishment of participatory planning in Third World countries are likely to be great.

Its successful achievement may well depend on commitment and the careful selection of approaches that are suited to the local conditions prevailing in Third World countries. These issues are specifically addressed in the next two sections.

PUBLIC PARTICIPATION IN THIRD WORLD PLANNING: ASSESSMENT OF SCOPE

The discussion of problems involved with participatory planning gives no clear impression of the advantages which may follow its successful implementation. We must put to one side for just a moment the question of whether apposite techniques for effecting public participation in Third World contexts can be found. If they can be identified, what merits would they bestow on the planning sequence?

Conyers (1982) has argued forcefully that participatory planning is of particular relevance to less developed countries. The idea that development is a basic human right is a commonly accepted one. Conyers notes that this principle is strongly emphasised in many recently independent nations, as a result of the period of colonialism being regarded as a phase when democratic rights were repressed. As an example, the strongly pro-public participatory views of President Kaunda of Zambia are cited by Conyers (1982). Another rationale for the promotion of strong participatory democracy in Third World countries stems from a closely related fact - that of continuing dependency on Western countries - not only in the economic sphere, but often in the technical and academic arenas too. Thus, there may often be an implicit or explicit tendency to adopt tried and tested western solutions to economic and environmental problems. This type of response is well exemplified by the emphasis that was placed on industrialisation in early Third World national development plans (Dickenson et al, 1983). The same is likewise true of the dependency associated with the establishment of enclave industries and the development of tourism. It is often far from clear as to how appropriately such developments fit in with the goals and objectives of different groups in society.

Closely allied with these comments in the field of professional planning is the fact that many Third World planners have been trained abroad, often in the major planning schools of Western Europe and North America, located in major metropolitan centres. As cogently argued by Zetter (1981) this gives rise to the likelihood of what may be referred to as "cultural collisions" between the environmental needs of developing countries on the one hand and the conventional wisdoms and accepted solutions of industrialised nations on the other. These may occur principally via the importation of inappropriate and frequently over-sophisticated models and modes of planning. This is likely to be reflected in two fashions. Firstly, planners may come to regard Western-style solutions and standards as the norm and apply them in extremely divergent cultural, social and environmental circumstances with insufficient critical appraisal of their true worth. Secondly, a predisposition to adopt complex techniques for the apparent sophistication and scientific objectivity that they bestow on the user may also result. Hence, the syndrome of the Third World urban planner who wants to calibrate gravity models when even basic data concerning the provision of services have never been collated. Further, as Lloyd (1979) notes, having been trained abroad, planners become even more of an elite, and may unwittingly take an increasingly pejorative view of those they are employed to serve, or at the very least, become increasingly distanced from them. These fears, suggesting the particular need for appropriate forms of public participation in Third World urban and regional planning, are well-exemplified in the following extract:

> And so we rest the case of the People versus Mr. Urbano Planner y Administrador. We have charged the defendant with not taking people, especially the interests and outlooks of the ordinary poor, sufficiently into account in preparing his plans and launching his programs. Rather his master grid descends inexorably upon the city populace and the tao below must fit into its framework or be crushed. We further charge Mr. Urbano Planner y Administrator with making little effort to establish two-way communication with the urban masses, much less have them actually participate in the decision-making processes, outside of elections, governing their own lives. (Hollnsteiner, 1977, p. 319).

In an essentially similar vein, Fagence (1977, p. 319) has bemoaned what he describes as the:

> Use of the comprehensive, elitist master plan approach to decision-making, in which the process proceeds along a path dictated by the expert with perhaps a preconceived plan in mind, in ignorance (consciously or innocently contrived) of the aspirations or preferences of the client public.

If such accusations are true, it is perhaps little wonder that inappropriate and/or unfortunate planning decisions are often reached which seem poorly suited to the needs of the people and to the local environment. Numerous instances of inappropriate planning were given in Chapters 3 and 4; for example, the early unyielding presumption against spontaneous settlements, and the dependence on over-technical settlement plans based on the dictates of classical economic location theory. In such circumstances, the planner can become a remote figure of authority, telling the people what they must do, not working alongside them to produce solutions that are both feasible and mutually acceptable:

> Many modern planning techniques need to be augmented or replaced by an appropriate technology of planning. Just as high-technology hardware is generally imported, requires outside experts for its operation, and often is alienating, the same is true of much of the software of planning techniques. Too often,

> the 'sophisticated' methodologies become instruments of mystification, expanding the influence of the outside planner or expert while shrinking the influence of the purported beneficiaries. Technique can thus serve as an instrument of dominance (Kent, 1981, p. 318).

An appropriate technology of planning in Third World contexts must thereby involve well-articulated means of encouraging public participation in the planning process, which together with strong community-based planning foundations enable the aspirations and perceptions of individuals to be directly taken into account. In this light, self-help imperatives in the cities of developing countries must be construed as a form of public participation, albeit one that cries out for formal recognition and regularisation by planners and politicians. Hence, a basic and fundamental need in planning and development studies is to study the "cognitive maps" of low income groups. As defined by Lloyd (1979) the "ego-oriented cognitive map" of individuals embodies their personal and mainly self-centred view of society. Another perspective is described as the "externalized analytical structure", which represents an observer's wider overview of prevailing societal conditions. It is these types of public views of the world that need to be bought to the fore in Third World urban planning practices.

TECHNIQUES FOR PARTICIPATORY PLANNING IN THIRD WORLD COUNTRIES

The potential means of establishing public participation in planning are many and varied, reflecting the fact that no one technique is likely to be adequate and that a combination of several is normally required. This is also reflected in the highly varied listings of techniques that are reviewed by different writers. Those of Fagence (1977), Hampton (1977) and Sewell and Coppock (1977), for example, are reproduced in Figure 5.2. Although these show considerable consensus with regard to a number of commonly used methods such as behavioural and attitude surveys, public meetings, exhibitions, information documentation and task forces, there is considerable diversity at their margins. Fagence (1977), for instance, subdivides methods into those he regards as being conventional, those which are innovative and those based primarily on principles of self-help (Figure 5.2).

As Hampton (1977) has clearly recognised, techniques of participatory planning may conveniently be divided into three main groups, according to the principal direction of the information flows involved. First, some techniques are mainly concerned with dispersing information to the public. For example, use of detailed reports, specialist reports, leaflets, general publications, press and other mass media releases basically fulfils this function. Another clearly identifiable set of techniques, including behaviour and attitude surveys, questionnaires, study groups and kits and general comment

Techniques of public participation in Planning

a. Fagence (1977)

1. Conventional means

(a) Exhibitions
(b) Public meetings & Hearings
(c) Information documentation
(d) Questionnaire surveys
(e) Documentary reporting (Media)
(f) Other means-(ideas, competitions, referenda, public enquiries)

2. Innovative means

(a) Delphi method
(b) Nominal group method
(c) Charrette
(d) Other means- (gaming-simulation, scenario writing)

3. Means of self-help

(a) Self-Help manuals
(b) Planning Aid
(c) Task Forces

b. Hampton (1977)

1. Behaviour and attitude surveys
2. Existing political structure
3. Press and other mass media
4. Leaflets and other publicity
5. Detailed reports
6. Specialist reports
7. Consultative groups
8. Community forums
9. Community workers
10. Exhibitions
11. Study groups & kits
12. Public meetings
13. Co-options to committees
14. Comment forms

c. Sewell & Coppock (1977)

1. Public opinion polls & surveys
2. Referenda
3. The ballot box
4. Public hearings
5. Advocacy planning
6. Letters to editors & public officials
7. Representations of pressure groups
8. Protests and demonstrations
9. Court actions
10. Public meetings
11. Workshops or seminars
12. Task Forces

Figure 5.2: Techniques of public participation in planning

forms represent the gathering in of public views and opinions by planners. Finally, only a handful of methods involve general interaction between planners and the public. These include the existing political structure, community workers and co-options to committees. These approaches represent the direct participation of the public in the actual decision-making process, rather than their involvement in the collection and dissemination of information prior to decision-making. This three-fold grouping of means can be related directly to the three models of planner-client interactions shown in Figure 5.1.

In Fagence's (1977) categorisation of methods (Figure 5.2a), conventional means are recognised as being the least costly and the least demanding. All of the methods included under this heading are potentially useful in the Third World setting as elsewhere, although each has specific limitations and drawbacks. Exhibitions, for example, much praised in the Skeffington Report, suffer from the basically one-way flow of ideas that they involve. Further, their effectiveness depends on accessibility to all - a potential problem in areas of dispersed population and/or poor transport provision. Public meetings and hearings can also be employed to good effect, but as Fagence notes, such gatherings are often dominated by the articulate and are susceptible to professional control. Arnstein (1969) argues that meetings and hearings also frequently involve a basically one-way flow of information from planning agency to the public. Information documentation, in the guise of various publications such as reports of survey, policy statements and bulletins, represents another means of public involvement that is heavily skewed toward the literate and intellectually conscientious. However, the per capita costs of the method are high and it tends to exacerbate the non-joiner problem. Questionnaire surveys and interviews represent yet another method of mobilising public involvement in planning, and England (1974), in particular, has argued strongly for their use. Correctly constructed and implemented social surveys and interviews can act as sound vehicles for the analysis of the views and the preference structures of communities. As Fagence (1977) notes, their potential worth is all too frequently underestimated by planners and policy-makers and they can be used with individuals, groups or leaders of the community. Thus, surveys and interviews can form an important information gathering technique for community-based development programmes in the developing world, for example, in relation to squatter upgrading schemes. Finally, Fagence (1977) recognises documentary reporting such as media coverage and a miscellaneous set of other means such as ideas competitions, referenda and public inquiries. All of these conventional means of participation involve the predominant flow of information from the planner to the public. However, despite the possible shortcomings associated with such a feature, given the environmental and socio-economic conditions existing in most poor countries, these methods are likely to be of considerable potential. In particular, questionnaire surveys in association with public meetings and information documentation should do much to foster useful public participation at the community level in many Third World countries. The other methods reviewed are also likely to be relevant in specific circumstances, but these will have limited applicability given their dependence on generally high levels of educational attainment.

This same problem faces the application in Third World contexts of the methods of participatory planning described by Fagence as innovatory (Figure 5.2). Each of these endeavours to establish a better dialogue between planners and the planned, but requires an "above-average intelligence level of the participants" (Fagence, 1977, p. 292), as well as their careful selection. The Delphi technique aims to combine the knowledge and abilities of a diverse set of experts in order to arrive at a consensus. Its main aim is to eliminate round table panel discussions, which frequently involve problems of dominance and also bandwagon effects. The Delphi method normally entails four rounds of questionnaires which allow views to be expressed anonymously. First, general ideas on the topic under consideration are solicited and summarised. Secondly, the participants respond to these tabulations. These are again summarised and used as a response base for the third round. Fourthly, the list is refined yet again, and this time its contents are either voted upon or ranked. The nominal group method is not too dissimilar, in that it represents a situation where individuals work in the presence of others, but in which they do not interact. Thus, verbal skills are not essential to the method. A further advantage is that a large number of participants can be accommodated. The procedure employed often involves seven stages, as follows: (i) the convenor identifies the problems or issues under discussion and the results are tabulated and displayed; (ii) each participant lists his responses privately; (iii) the participants' responses are tabulated; (iv) the ideas and issues involved are discussed; (v) these are then ranked; (vi) further discussion occurs and (vii) the ideas or issues are finally ranked. The third innovatory method identified by Fagence is the Charrette. Here, local residents from a community form a steering committee, which itself sets up working groups to look at particular problems. Most if not all of these groups draw on expert advice from available professionals. At the end of their deliberations, each working group produces a report and these are then synthesised into an overall plan.

The Delphi and nominal group methods suggest themselves as potentially useful approaches in Third World urban planning, as they minimise dependence on verbal skills and allow a measure of anonymity. Both methods effectively represent a series of linked semi-structured questionnaire surveys. Against this, they require much organisation and are clearly not techniques for mass involvement. Further, the vital issue of representation is raised in relation to these methods. However, it seems that such approaches could well be useful and effective if used with respect to specific local scale problems, for example, in connection with an upgrading scheme for a squatter settlement (see Pfister, 1982 and Skinner, 1983).

Finally, as a third major category, Fagence recognises self-help means of public participation (Figure 5.2). Fagence argues that these are by and large a direct response to the inherent conservatism of planning. Thus, they provide the means for effective participation of the public in the complex planning sequence. For example, self-help manuals offer written guidance for local groups. Similarly, planning aid is basically a response technique designed to meet the technical requirements of disadvantaged and threatened groups. Task forces, on the other hand, represent the setting up of special teams to examine and report on matters involving complex problems. Given the prevalence of self-help imperatives in the Third World, all of these three approaches can be regarded as of potential in cities of the developing world. An example would be the production of manuals concerning housing standards, layouts or even building precautions against natural hazards such as earthquakes and hurricanes. However, in reality such methods all represent ad hoc ameliorative responses to existing planning problems, rather than the means of achieving comprehensive public participation per se.

Clearly, it would be wrong to think of these methods of participation as separate. To do so would be to reduce participation to the level of a purely technical exercise. Thus, as stressed repeatedly, effective public involvement will entail the use of many of these techniques in combination. This reflects the fact that public participation should represent an holistic approach, with its own ideology and philosophy. This is itself witnessed in some of the wider, more all-embracing forms of participation that are recognised by Conyers (1982). Thus, the use of extension services is seen as an important approach in the Third World. This involves technical officers from specialist governmental departments acting as a link with the public. Conyers stresses the potential of multipurpose extension staff in promoting development, and cites India's village level extension workers as an example. The problem can still remain that the people are not directly involved in the planning process itself, the method basically facilitating the effective communication of local ideas. Secondly, systems of decentralised planning and strong local government are regarded by Conyers as direct facilitators of public involvement in the planning process. It can be argued that in many colonies, local government was only encouraged to the extent that it fostered acceptance of the colonial regime (Conyers, 1982). The express aim of decentralised planning is the actual formulation of plans at the local level. Such an approach, subscribing to the development from below strategy, is often employed in socialist countries (Stohr and Taylor, 1981). However, Conyers maintains that the capacity to establish local planning systems is probably beyond many developing countries in the short-run period. The difficulties involved in the integration of local plans with the national development strategy is another area of concern. Further, such planning is still carried out by officials rather than the public. It is perhaps most realistic to regard these approaches as affording frameworks which facilitate the possibility of more effective public participation.

Finally, Conyers (1982) in common with other writers recognises community development as an increasingly important approach to development. Again this approach must be regarded as much more than just a method of obtaining popular participation in planning. The principal aim of community development is, of course, to raise standards of living, and self-help measures represent an important component in achieving this. However, the method also frequently involves the establishment of community development workers. The theory is that the main initiatives for development and change should come from members of the community itself, rather than from the development workers. In practice, this involves a delicate balance between local imperatives and expert professional advice. Conyers (1982) argues forcefully that community development programmes offer a number of advantages in the Third World. This view is also stressed by Kent (1981), especially if the approach represents community groups working on their own initiative. Kent argues that there are many problems inherent to mass public participation in planning, so that the aim should not be to maximise it in purely numerical terms, but rather to optimise it qualitatively. Kent's plea is thus for a radical form of participation in planning, namely that planning should actually be based in the community, not predicated by outside experts. Similarly, Kent pleads for the establishment of an appropriate technology of community-based planning, an argument previously covered in this chapter.

SOME EXAMPLES OF PARTICIPATORY PLANNING

In this section, some selected examples of participatory planning practices in developing countries are briefly cited. The aim is not to provide a detailed review of the array of techniques examined previously, nor an in-depth description of the experience of one particular country. Rather, the account seeks to give some idea of the diverse scales and circumstances in which public participation can be fostered. At one end of the scale it can represent an overall paradigm for planning practice, whilst at the other, it can merely mean the use of isolated techniques to articulate the voice of the people in relation to a specific planning task or issue.

At the wider scale, as noted by Kent (1981), there has been much community-based development planning, including the village level development programmes of India and China and the ujamaa programme of Tanzania. Such approaches have, however, been neglected by "modern technique-orientated, centrist planners" (Kent, 1981, p. 320). The Indian community development programme is perhaps one of the best known examples (Conyers, 1982). The scheme, launched in the 1950s, witnessed the establishment of local councils supported by teams of technical officers and multipurpose community development officers at the village level. As recently noted by Misra and Natraj (1981), the approach was clearly inspired by the ideas of Gandhi, who regarded the self-sufficient village as the basic unit of social and economic development. Gandhi did not ignore the fact that there would have to be linkage and interdependence between such villages. Similarly, Gandhi was not against industrialisation *per se*, but rather stressed the need for its progress at an appropriate human scale. Critics of the Indian experience point to the fact that ostensibly the approach has done little to reduce the gross inequalities that exist. Perhaps more fundamentally with regard to the present account, it is also argued that the approach has afforded only limited involvement of the majority of the population in the development process. Another example is afforded by China, which during the past three decades has practised both planning from above and from below simultaneously (Wu and Ip, 1981). But, as noted in the previous chapter, it is the Chinese rural development policies that have perhaps been most successful. Ma (1976) notes how Mao Tse-tung brought an explicitly pro-rural and anti-urbanism policy to China in the post-1949 period. A major approach has been based on decentralised planning and administration, linked to what are effectively principles of community development. The latter is effected by communes, brigades and teams in the rural areas and their equivalents in towns, with outside assistance and expertise being introduced where necessary (Conyers, 1982). Obviously, the experience of China has been very different from that of India, for in the former, such policies have been used to consolidate a social and economic revolution. As a part of this, a major effort has been made to reduce the rural-urban dichotomy and to encourage urban to rural population drift (You, 1981; Hoa, 1984).

The contrasting characteristics of India and China might be taken to imply that community-based public participation in planning is viable and effective in a wide variety of political and cultural contexts. Indeed, although it may be thought by some to flourish only in liberal democratic or socialist societies, evidence from the Philippines shows that public participation is sometimes encouraged even by dictatorial regimes. However, in such circumstances its role is likely to be amputated to that of a local palliative. Thus, Blunt (1982) looks at public involvement in the Philippines where all powers have been centralised in the President since the establishment of a martial law Government in 1972. Within the country, there has been an extremely fast rate of urbanisation and approximately 30 per cent of the local population of metropolitan Manila is now made up of squatters. The early attitude of the Government to housing was largely negative, being one of clearance and resettlement. In 1974, the National Housing Authority, financed by Central Government, took the first step in changing public housing policy to one of upgrading *in situ*. This started with the Tondo Foreshore Project. The upgrading of squatter settlements and the provision of site and services plots gained official acceptance with the 1977 Slum Improvement and Resettlement Programme. The principle of public participation was incorporated at an early stage in this change of planning philosophy. At first, it was principally achieved by holding public meetings and including preference questions on socio-economic surveys. In relation to the Tondo project, a mass meeting was held at which the overall plan was considered. At this stage, a general plan involving the overall framework of roads and infrastructure networks had been decided upon. However, within this configuration, alternative block plans were prepared by resident groups. Subsequently, each household was invited to cast its vote in favour of one particular design. Blunt, however, notes that there have been two major constraints to effective public participation. First, the lack of previous experience of participation among local groups and secondly, the political system which not only discourages criticism but imposes its own leaders right down to the local level. The recent creation of community planning councils, sometimes with elected officials, may be seen as a partial effort to overcome some of these problems. However, the overall success of these moves will largely depend on local implementing agencies responding positively to public views and initiatives.

The need for effective public consultation and participation in squatter settlement upgrading schemes has been stressed several times in this volume and by other writers elsewhere (Turner, 1980). The final example given here relates to the derivation of planning policies for the low-income housing areas of eastern Port of Spain, Trinidad. This zone of irregular housing on hilly terrain has already been discussed in Chapter 3. In 1971, as a part of the preparatory planning sequence, a social survey was carried out in 2,445 or 20 per cent, of the households in the area. However, a disappointing feature of this work was that it stressed almost exclusively factual topics in order to fill gaps in the planners' information base. Only a couple of questions genuinely sought to establish the needs and aspirations of the populace, asking for example, about their reasons for moving to Port of Spain and their preparedness to join self-help housing and general improvement groups (see Urban Redevelopment Council, 1972). The results of this survey have been strongly criticised by Conway (1981), and certainly it seems that the exercise represented a missed opportunity in terms of identifying and acting upon the overall views, aspirations and perceptions of the community. Perhaps as a corollary, despite all this effort and the subsequent publication of a redevelopment plan for the East Port of Spain area (Town and Country Planning Division, 1973), few if any actual improvements have been made on the ground.

CONCLUSIONS

The foregoing account has sought to demonstrate that public participation is an important potential ingredient of the urban planning process in Third World countries, despite the manifold problems that are likely to be involved in its implementation and subsequent operation. The enormity of the social and economic problems faced by the majority of less developed countries and their relative paucity of resources seem to make this more rather than less relevant in developing as opposed to developed countries. The review of techniques for effective public participation has shown that many are applicable in the less developed realm, albeit in particular circumstances and at different scales. The majority of commentators seem agreed that community-based development programmes are especially relevant, whilst social surveys and interviews can be particularly effective if used at the local scale, or with respect to the examination of specific issues at the national scale. It can be suggested that such approaches are of vital significance if the country-specific and people-specific nature of sound Third World planning and development are to be fully catered for. In short, such approaches do not provide a definitive solution to environmental and socio-economic problems, but rather a constructive and supportive framework for their appropriate discussion. Thus, it must be emphasised again that the role of public participation in the planning process cannot be considered in isolation from the wider social and political contexts, a point ably summarised by Fagence (1977, p. 331):

> Thus, it may be suggested that citizen participation in urban and regional planning should be seen as a part of a much wider spectrum of government activity in the social welfare field.

This is obviously an important point and from a structuralist perspective we should hardly be surprised if only a cosmetic form of public participation emerges under a repressive government or dictatorships of either the far right or the far left. The promotion of true development presupposes the existence of the will to improve the lot of the poorest groups in society and those with the weakest voices. The planner must plan with the people, seeking by the best means available to identify and act upon their perceptions and aspirations (Lloyd, 1979). Some individuals might regard these views as highly idealistic, but as Fagence (1977, p. 370) has commented:

> The commitment towards citizen participation in planning by planners should be derived from the social ethic of planning, one of the foundations of the planning profession. The social ethic or ideology in planning comes from the strong tradition of social commitment and idealism in planning.

Chapter 6

Perception Studies and Third World Urban Planning

If it is accepted that identifying the aims and aspirations of ordinary urban residents stands as a central task in the promotion of effective environmental planning, how can this be achieved? The next two chapters examine this question from a spatial-geographical point of view. Thus, the present chapter is primarily technically oriented, in that it aims to identify the methods that can be used to investigate the environmental perceptions of individuals and groups as part of the urban planning processes in Third World countries. A related theme is the evaluation of the strengths and limitations of the various specific methods identified. The somewhat wider philosophical, ideological and political issues involved in the employment of such techniques and approaches as a part of the planning sequence, and as a process contributing to public involvement in planning have been well-aired previously in this volume, notably in Chapters 1 and 5. Hence, in the present account, these important considerations are initially held in abeyance, but are reconsidered in the conclusion of this chapter, when a critique of the overall field of planning-oriented spatial perception research is provided.

Whilst every effort has been made to ensure that the present review is as detailed as possible, inevitably the chapter cannot provide a comprehensive outline of each and every potentially relevant technique, nor is it possible to dwell on any one in great detail. In short, the account should not be regarded as a manual. However, in examining sets of methods, an effort is made to pinpoint the main bibliographic references for those who might wish to take them further, perhaps ultimately into the realms of academic and applied research. Secondly, examples are provided throughout the chapter to illustrate the scope and potential of the methods reviewed. Some of these illustrations come from the author's own research, whilst others are drawn from the wider literature in the field. Following this, two major case studies of planning-related perception research based on the author's work in Trinidad and Tobago and Barbados are presented in Chapter 7.

In referring to illustrative case studies it is worthwhile reiterating a point made several times previously in the present text, this time via the comment of Brookfield (1973, p. 12) that "Thus far there have been few formal attempts to apply the methodology of the behavioural movement in the social sciences to problems of the Third World". Although over ten years on, this comment still applies today. Thus, the principal theme underlying this chapter is that developing an understanding of how people view the world, both in terms of their information levels about places and their spatial patterns of emotional attachment and detachment is essential in breaking into the links existing between settlement systems, space preferences, migratory fields and planning imperatives.

The essence of these complex relationships has been well-summarised, again by Brookfield (1975, p. 51), who firstly notes in discussing spatial inequalities in development, growth and modernisation how "Dichotomies, or polarising constructs, are basic to the simplest structure of human perception into comprehensible order. Inevitably they grow into stereotypes: things that are". The inference here that inertia may frequently characterise spatial images and preferences by way of the existence and perpetuation of stereotypes clearly offers a major challenge to both the environmental planner and the development expert. This notion has been further exemplified by Brookfield (1975, pp. 98-9) when he comments that:

> Entrepreneurs - ... - will thus concentrate at the growing points; they will underestimate or simply ignore opportunities elsewhere. Space preferences will therefore become distorted, and growth in the developing region - 'North' in Hirschman's terms - will be paralleled by retardation elsewhere - in the 'South'.

Such an argument suggests that uneven development in space may well be sustained and perpetuated, if not initially generated, by a form of 'perceptual inertia' on the part of governments, firms, institutions, groups and households. If a situation of self-sustaining cumulative causation pertains, perhaps due to the impact of uncertainty on location together with associated perceived economies of scale and agglomeration (Webber, 1972), it may well be that a lack of realisation of full growth potential will be characteristic in the middle- to long-terms. From a structuralist viewpoint, it must also be recognised that certain groups in society might wish for the perpetuation of a sharply dichotomised map of space preferences, so as to maintain a steady flow of resources, investment capital and labour into the existing core regions. Whatever its causation, however, such continued 'polarisation' or 'backwash' may lead to an array of damaging psychological, economic and social effects in the retarded or peripheral regions. In particular, feelings of economic exploitation and cultural-social betrayal may result in calls for regional or national separatism. Any group of politicians, administrators or planners that is prepared to openly consult the public to identify and analyse their environmental perceptions and preferences has at least taken the first step toward honestly acknowledging that the existing distribution of settlements and resources may be unacceptably uneven, and thereby socially and economically divisive. Surveys and analyses of spatial perceptions and cognitions should thereby be regarded as vital pre-requisites to sound and effective resource management and physical planning in Third World territories, whether at the inter-urban or intra-urban scales.

THE STUDY OF SPATIAL PERCEPTIONS AND COGNITIONS: INTRODUCTORY PERSPECTIVES

As noted in Chapter 1, behavioural geography developed as a strong and important branch of the subject during the late 1960s and early 1970s. It has also been observed that the field must be regarded as providing a perspective which may be applied to all areas of traditional geographical enquiry, rather than as a strictly defined new sub-branch of the discipline. Early geographical work along behavioural lines was seen essentially as a reaction to the quantitative revolution, emphasising as it had scientific objectivity via the paradigm of logical positivism, and the search for links between variables measured at the aggregate level. In contrast, behavioural approaches offered the potential of focusing on the essential subjectivity of human decision-making and subsequent actions, and thereby, the importance of the individual as a basic unit of analysis. Even at this stage, the word potential should perhaps be stressed, for commentators such as Ley (1981) and Cox (1981) would argue strongly that this promise has not been fulfilled, a theme to which we shall return later in the text.

The rise of behavioural geography essentially matches the rise of 'behaviouralism' as a major theme within the social sciences as a whole, a development that without question owes much to the seminal works of Boulding (1956) the economist, Simon (1957) the polymath and Lynch (1960) the architect-designer. Within all of the social sciences in the 1960s there was a strong call for the study of 'images' in order to explain individuals' personal worlds (Pocock and Hudson, 1978; Gold, 1980), a call that aligned well with a humanistic approach (Ley, 1981). Hence the development of behavioural geography and environmental psychology (Lee, 1976; Proshansky, Ittelson and Rivlin, 1976), and parallel movements in the fields of architecture, design, planning, anthropology, ethnology and sociology.

As with any specialist field of academic enquiry, even one with a strong policy orientation, perception studies have acquired their own mystique which may appear almost impenetrable to the layman. But the basic ideas and concepts and the overall rationale behind them are really quite straightforward. Gold (1980) argues that there are four main features of the behavioural approach as applied to geography. First, it is posited that the environmental perceptions and cognitions on the basis of which people act may well differ significantly from the 'true' nature of the 'real' world. This is central to the field and essentially recognises that decision-makers act in space not on the basis of some standard and entirely objectively perceived environment, but rather with regard to their unquestionably subjective and more often than not highly simplified and/or distorted interpretation of that structure. The implication of this distinction has been neatly summarised by Brookfield (1969, p. 53):

> Decision-makers operating in an environment base their decisions on the environment as they perceive it, not as it is. The action resulting from the decision, on the other hand, is played out in a real environment.

In much the same vein, by the early 1960s, Kirk had drawn a sharp distinction between the "phenomenal" environment of natural and man-influenced physical features on the one hand, and the "behavioural" environment, regarded as a "psycho-physical" field of human structuring, on the other (Kirk, 1963). Second, Gold suggests that work in behavioural geography explicitly recognises that the individual influences the physical and social environments just as much as responding to them. This takes us away from a mechanical-deterministic view of the behaviour of people within environments, an approach which unfortunately is still commonly adopted by some. Thirdly, Gold stresses the essentially multidisciplinary outlook of work in the field, a point that has already been touched on here. Finally, Gold stresses that behavioural geography tends to focus on the individual rather than the group. This has already been described as a contentious issue in the present account, but it is a point that has some considerable relevance with respect to the employment of perception studies as a means of effecting public participation in planning. Thus, it is recognised here that examining individuals' views is the corner-stone of behavioural work, with all the uniqueness and idiosyncracy that this implies. However, in so far as generalised views and consensus images exist, it is argued that they should be studied. There are, in fact, a plethora of perception studies, often highly quantitative in nature, that deal with aggregate groups and their impressions. In fact, many of the methods examined in this chapter fall into this category. Also, an insistence that the individual should always be taken as the unit of investigation might well blind us to the fact that the perceptions of governments, administrators, professions, institutions, committees and other aggregate groups may have a fundamental bearing on environmental decision-making and policy foundation.

A MODEL OF ENVIRONMENTAL PERCEPTION-COGNITION AND DEFINITION OF TERMS

How can we conceive of perceptions and cognitions mediating between people and the 'true' environment, and what precisely is meant when we refer to spatial cognitions and perceptions? These two important and closely related questions must be considered before turning to review techniques of behavioural research.

Perception is a word that is used to describe the process by means of which individuals become aware of and interpret their surroundings via the medium of the senses, and also the product of that process (Skurnik and George, 1964; Goodey, 1973). Strictly defined, perception relates to the act of interpreting and coding the sensory stimulation that is experienced at the present time. But all individuals have received previous stimuli which influence and colour the here and now of everyday experience. This totality of knowing is referred to as 'cognition'. Such a summation of an individual's past, present and perhaps even future anticipated stimuli is obviously not dependent on direct first-hand personal experience alone and will be greatly influenced by the mass media and interpersonal communication (see, for example, Walmsley, 1982; Gould and Lyew-Ayee, 1983). It is precisely because past perceptions influence peoples' immediate cognitions that spatial stereotypes and perceptual inertia frequently occur. Hence, in common parlance, many individuals "know what they like and like what they know". Thus, when considering spatial cognitions, it is tempting to invert the old adage that "seeing is believing" and to suggest rather that "believing is seeing" (Pocock, 1981). Once established, regional images are likely to be remarkably resistant to change and often become a part of general parlance, even becoming enshrined in fictional literature (Pocock, 1979). Some researchers have gone so far as to refer to "myth maps" of countries due to the frequent occurrence of such distortions in the public's regional consciousness (Gould, 1969b; Jones, 1978), a theme which is investigated later in this chapter. Myth maps may be the result of chance, unfortuitous misinformation or even deliberate dissemination by those who stand to gain by their promotion and maintenance. Thus, it would be naive to assume that spatial images are not coloured by political expediency and/or commercial interest, as witnessed by regional policy and tourist resort advertising respectively.

A highly simplified model of environmental perception and cognition is presented as a framework for discussion in Figure 6.1. At a basic level, choice and subsequent overt spatial behaviour are mediated within the environment via the processes and products of environmental perception and cognition. As already stressed, this general notion applies whether we are talking of individuals, corporate groups, institutions, firms or indeed governments. It must be remembered, however, that choice is never absolute, for constraints serve to limit the possibilities that are open to different decision-makers at the same time, or the same ones at different times. Foremost amongst these are political and legal restrictions, financial and general economic conditions and the activities of urban managers such as planners and architects (see, for instance, Pahl, 1970; Gray, 1975; Hamnett, 1977; Williams, 1978). The intangible product of environmental perception-cognitions is referred to in a number of different ways. Among the most commonly employed descriptors are "cognitive" or "mental maps", "images" and "schemata", as indicated in Figure 6.1.

Let us ignore for just a moment the complex and important debates that have emerged of late regarding the overall efficacy of examining the products of spatial perception-cognition (Bunting and Guelke, 1979; Lloyd, 1982), their naming and whether indeed they can be measured in spatial terms at all (Tuan, 1975; Graham, 1976, 1982; Murray and Spencer, 1979; Downs, 1981). Whatever its precise form and nature, there are three major components of environmental perceptions/cognition that are generally recognised as making-up the image, and these are indicated in Figure 6.1. First, the

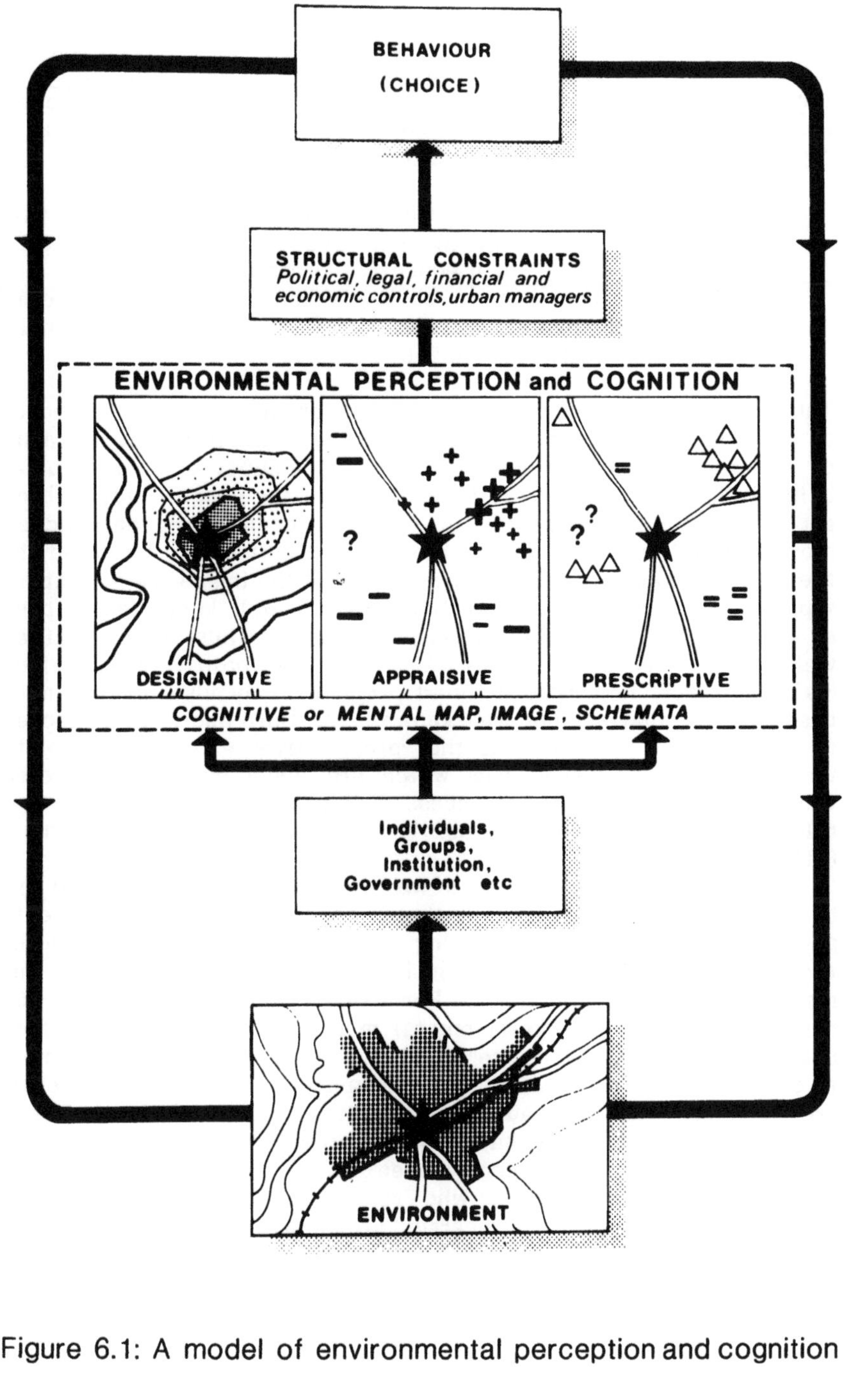

Figure 6.1: A model of environmental perception and cognition

"designative" component signifies the basic information possessed about places, that is, the "whatness and whereness of the image" as described by Pocock (1973, p. 256). Thus, we may envisage an isopleth map distinguishing between shades of terra incognita and terra cognita. It is obvious, however, that all knowledge concerning location is tempered by emotional reaction, thereby rendering an "appraisive" aspect to spatial cognitions. As Pocock observes:

> The appraisive aspect is emotional, concerned with feeling, value and meaning attached to the perceived - a response concerned with evaluation and, consequently, preference. It forms that part of the response, more obviously sensuous than intellectual, concerned with appraising the multifaceted urban personality or aesthetic (Pocock, 1973, p. 256).

Thus, to use an analogy, the designative or informational component of the image is the canvas upon which the brushstrokes of appraisive or emotional perception are worked. The analogy of a piece of art is highly appropriate, for the final product - the image - may be broad and impressionistic, symbolic or ethereal, or minutely detailed and more ordered and regular than reality itself. It should be noted that the appraisive and designative components of the image are not entirely separate. For example, areas that are not well known may appear as question marks on the appraisive schema, whilst areas that are well-known often stand out as being appraised positively (see Figure 6.1). However, less rational blind prejudice may also prevail in the case of areas that are not well-known. The third and final component of the image is described as the "prescriptive" element and relates to inferences and predictions which are made beyond the present time and/or known circumstances. For example, the response may relate to areas of expected stability and change in the future (Figure 6.1).

In the account that follows we shall review methods that can be employed to examine all three components of environmental perception-cognition Before doing so, however, it is worth noting that a good deal of discussion has surrounded the meaning and measurement of mental maps. For example, Graham (1982, p. 251) has observed that "The term mental map is perhaps one of the most unfortunate in the literature" having been employed, for example, to describe things as different as group space preferences and freehand sketch maps. The same author then states that "surely images being mental phenomena, cannot have physical properties" (P. 251). Certainly similar objections have recently been voiced by Tuan (1975) and by Bunting and Guelke (1979), who remind us that images are not representable objects that are carried around in people's heads. Kuipers (1978, 1981, 1982) in discussing what he refers to as the "map in the head metaphor" has observed that spatial knowledge often falls into entirely disconnected spatial components and further, can be asymmetrical in the sense of a route being known in one direction, but not in reverse. All of which is to say that mental maps often possess non-maplike properties. Similarly, Downs and Stea (1973) have posited that in certain instances, a mental map may be no more than a set of stimulus-response links. The argument that caution should be exercised when studying spatial images and mental maps is certainly valid, but it is ventured here should not dissuade us from seeking to examine them from a primarily pragmatic perspective. Certainly, it can be argued that most components of environmental images must have locational attributes, even if these have to be deduced by the investigator rather than being given freely by a subject, in much the same way that a person has a personality even if they are unable to describe and measure it themselves. Thus, elsewhere, the present author has argued that:

> ... whilst we should not be blinkered by a slavish belief in the acuity of mental maps and spatial images, equally we should endeavour to avoid falling into the trap of becoming so esoteric as to deny their very existence or possible role in guiding overt spatial behaviour. It is argued here that such a response would result in the premature dismissal of a set of techniques which may yet prove to be of particular use to environmental planners and policy-makers (Potter, 1983d, p. 264).

This quotation largely expresses the overall spirit of the review of techniques and methods that can be used to investigate appraisive, designative and prescriptive aspects of environmental cognitions-images that is provided in the remainder of this chapter.

THE STUDY OF APPRAISIVE PERCEPTIONS IN THIRD WORLD URBAN PLANNING

As explained, the appraisive response reflects emotional reactions to environmental circumstances. It is in this regard that we must fully acknowledge the aspatial nature of some parts of the image, in the sense that some reactions may not be mappable or may not be held in a map-like form by individuals. Perhaps we can most aptly think of this as the image component that is added to the spatial-informational mental map to form the rounded and complete schemata. However, as will be shown, much data collected under this heading are capable of cartographic representation. Obviously, finding out how people feel about places and spaces is of considerable potential relevance to urban and regional planners, administrators and politicians in explaining locational choices and decisions. This point will be exemplified throughout the following review.

Space Preferences and Residential Desirability

The study of space preferences is at once both fundamental and quite straightforward; fundamental in the sense of generating useful and germane data and straightforward by virtue of necessitating only relatively simple techniques of analysis. The basic idea is to examine which areas are regarded as most desirable by individuals, holding other things constant. Thus, we are basically concerned with appraisive perceptions of specified areas, that can be mapped as "residential desirability surfaces".

The examination of such surfaces was largely pioneered in an important paper by Gould (1966; reprinted in Downs and Stea, 1973). In particular, Gould stressed the potential influence of spatial perceptions on human decisions and also that the views of individuals are unlikely to be totally disparate, so that it should prove possible to partition unique personal elements from the overall general viewpoint. The methodology adopted by Gould to achieve this separation was as follows. The respondents, mainly school and university students, were asked to place the areas making up the country in rank order according to residential desirability. The rank orderings of areas by different individuals were then intercorrelated in a pairwise fashion, and subsequently, the matrix of rank correlations subjected to principal components analysis. Thereby, the first derived component was taken as a statistical expression of the "consensus image". Essentially this method was employed by Gould and others in studies of space preferences in the UK, USA and Europe (see, for example, Gould, 1966, 1967; Gould and White, 1968, 1974), and by others at a variety of scales ranging from that of the world to the individual town. Such work certainly became popular, especially under the heading "mental maps", as popularised by the title of Gould and White's (1974) book. It is clear, however, that the use of the title in this context has been somewhat unfortunate, given the basically appraisive connotations of the surveys conducted. Indeed, Tuan (1974) complained bitterly that the products of such exercises were little more than cartographic renditions of public opinion surveys.

From a Third World perspective, Gould and his associates were clearly fully aware of the potential applied relevance of such research. Gould (1969a) looked at the case of Tanzania, as discussed in Chapter 1. The work indicated the primacy of the coastal areas and the principal axes of "modernization" (Gould, 1969a, 1970) in nationals' perceptions (see Figures 1.2 and 1.3). The level of income with which respondents would be 'reasonably satisfied' was also examined for each administrative area to give some idea of the financial inducements that would be necessary to smooth out the highly irregular perception surface. A similar analysis of space preferences in Ghana indicated the enhanced standing of the coastal districts containing the major urban centres and ports, and the south of the country in general (Gould, 1966; Gould and White, 1974). An exploratory study set in Nigeria highlighted two distinct areas of high residential desirability, referred to as the northern and southern cores. As was the case with both the Ghanaian and Tanzanian maps, there was a common theme of "peripheralism", with low levels of desirability being related to areas of low road network density (Gould and Ola, 1970). Finally, Gould and White (1974) examined the perceptions held of Malaya by Chinese and Malayan students and found significant cultural differences between them. In the execution of this early research, Gould was particularly aware of the potential applied relevance of such studies to development and planning in Third World countries, commenting that:

> ... there may be an area of enquiry here that is not only geographically intriguing, but capable of crossing the line between pure and applied research ... In much of the underdeveloped world the allocation of social investment is still of critical concern as many countries try to forge the basic infrastructure of transport, education, sanitation and health facilities. Are the areas that are already "mentally bright" going to receive a large share because they are prominent in the minds of men? Would an awareness and self-knowledge of this tendency have any beneficial influence? The stricture "Unto them that hath shall be given" seems to describe the tendency of a system of allocation with strong feedback features to produce the agglomerations and clusters of goods and people that are the main features of the urban revolution (Gould, 1966; as reprinted in Downs and Stea, 1973, p. 216).

Despite this exhortation, few studies until very recently have actually taken up the theme of space preferences in Third World countries. This very point is stressed at the outset by Weinand and Ward (1979) in their recent examination of space preferences among tertiary level students in Papua New Guinea. In particular, they stress that the link between migration and mental maps has not been fully explored (Fuller and Chapman, 1974). The respondents were asked to rank order 18 districts from the viewpoint of their residential desirability and later to rank seven towns. A consistency was exhibited throughout the rankings, and it was suggested that when asked to assess a district, respondents probably had major urban areas in mind. In overall terms, Port Moresby the capital, was poorly assessed, however, probably due to problems of civil order and the cost of living. The authors conclude by asking whether existing spatial attitudes can be altered radically by government initiatives. In another study, Honour (1979) looked at the 'perception of opportunity' of Jamaican school leavers in five different locations, using a grid square basis and a rating scale. Overall, the maps showed a strong local bias, plus the marked pull of Kingston, the capital. In contrast, there was poor regard for the traditional agricultural areas. More recently, the present author has studied space preferences in connection with planning proposals in Barbados (Potter, 1983d, 1983e, 1983f, 1984a, 1984b, 1984c), Trinidad (Potter, 1983c) and St Lucia, and this work is referred to subsequently in this chapter and the next.

These studies are in accord in that they do not employ the multivariate statistical routines originally recommended by Gould, but rather, much simpler techniques of analysis. Also, the works of Honour and Potter employed interval rather than ordinal data to scale residential space preferences. In fact, Gould (1969b) has specifically stated his strong predilection for ordinal measures of space preferences. However, this approach seems inappropriate given that research has shown that people find it extremely difficult to rank order any more than a very few objects. In the same paper, Gould also advocated the use of multidimensional scaling to explore the structure of space preferences. It is suggested by the present author that the adoption of such complex high technology multivariate methods is largely unnecessary and inappropriate, particularly in a Third World setting.

These arguments are illustrated by a final example of space preference research drawn from the author's own research work in St Lucia, West Indies. Some aspects of urban planning in St Lucia have been discussed previously in Chapter 3. A major feature is that over half the national population of 119,000 in 1980 lived in two areas: the northwest capital region extending between Castries and Gros Islet and the south, based on Vieux Fort (see Figure 6.2). Development planning envisages continued industrial growth in these two areas, and this dovetails with continued tourist developments in these zones, and also in Soufriere.

A full account of the survey work is not provided here, only a brief illustration of the approach used. A total of 186 respondents were interviewed in the five principal settlements of Castries, Gros Islet, Dennery, Vieux Fort and Soufriere (Figure 6.2). The interview schedule is reproduced in full in Appendix C. Each interviewee was asked to assess the residential desirability of the 11 administrative quarters comprising St Lucia. Responses were converted to a seven point rating scale, ranging from 1 if respondents said they would not like to live in an area at all, to 7 if they said they would like to live there very much. It is suggested that it is much easier for an individual to evaluate areas on an absolute basis such as this rather than on a relative, ordinal one. Further, sophisticated statistical methods of analysis are not needed initially, rather the scores recorded for areas can be averaged to give an overall mean residential desirability score. These are shown mapped for the five location sub-groups of St Lucian nationals in Figure 6.3.

Each of the perception surfaces shows a strong local dome of desirability, although such parochialism is less marked in the case of residents of Castries (Figure 6.3a). Noticeably, Castries residents also like the Vieux Fort area, but all the other quarters appear in shades of relative disenchantment. Residents of Gros Islet are a little more parochial and they assess all ten other quarters negatively. However, this general antipathy is a little less strong with regard to far away Vieux Fort and Soufriere. For the other three survey locations in the south and east of the island (Figures 6.3c to e), the capital region does tolerably well, but the outstanding feature is undoubtedly the enhanced perception of Vieux Fort and Soufriere.

Rather than using multivariate analyses, the separation between local and generalised national perception surfaces can be effected by simple bivariate correlation and regression analyses of mean residential desirability scores against distance from the point of survey. The notion that the further places are from a particular location the less they will be desired may be taken as a starting point.

Such a hypothesis of perceptual distance-decay is tested in Figures 6.4a to 6.4e. There is indeed an overall negative association between residential desirability and distance, ranging from -0.22 in the case of Castries to -0.36 for Gros Islet, again reflecting differential parochialism in space preferences. Obviously, the overall relationships are highly influenced by the strong liking expressed in each case for the home quarter, but the regression line may be taken to express

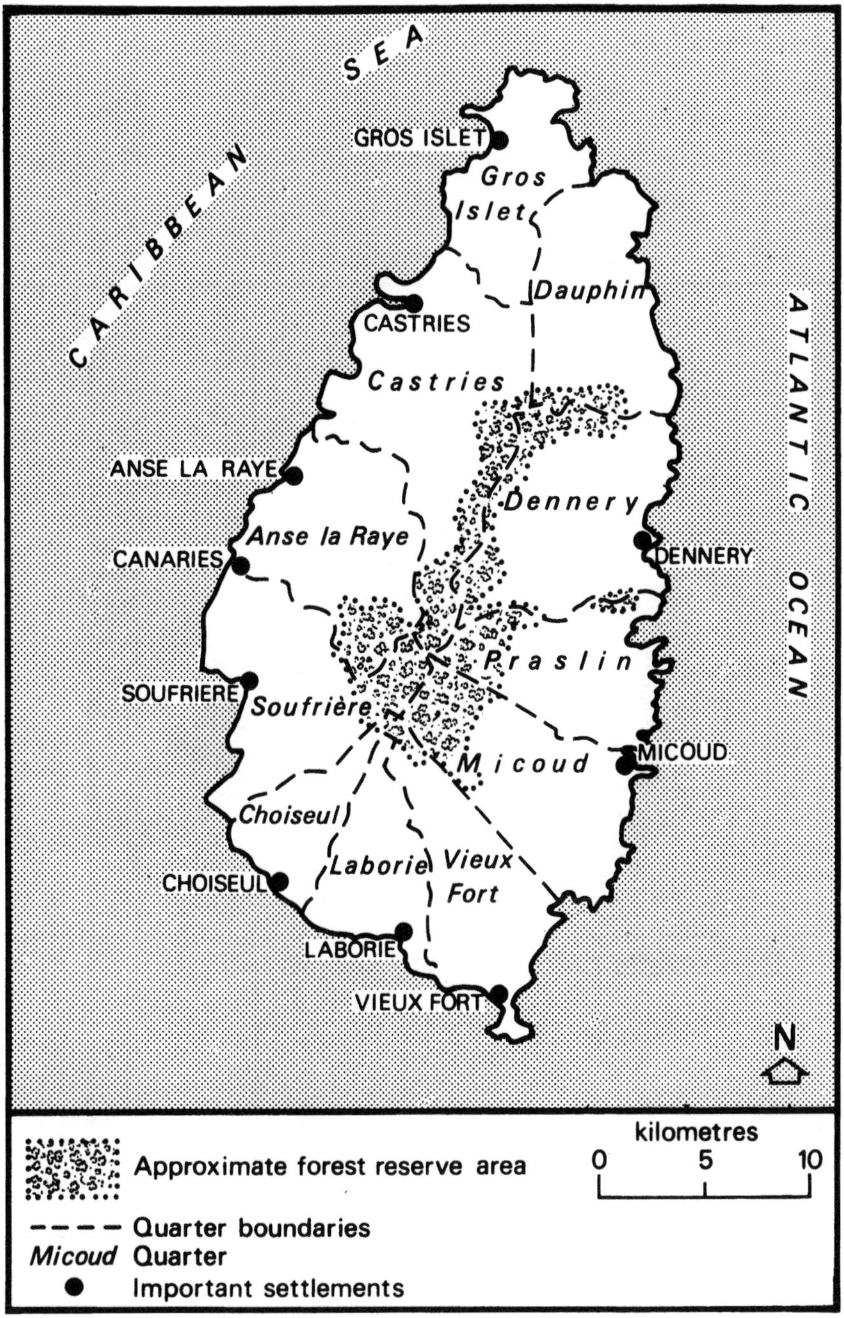

Figure 6.2: St.Lucia-principal settlements and administrative areas

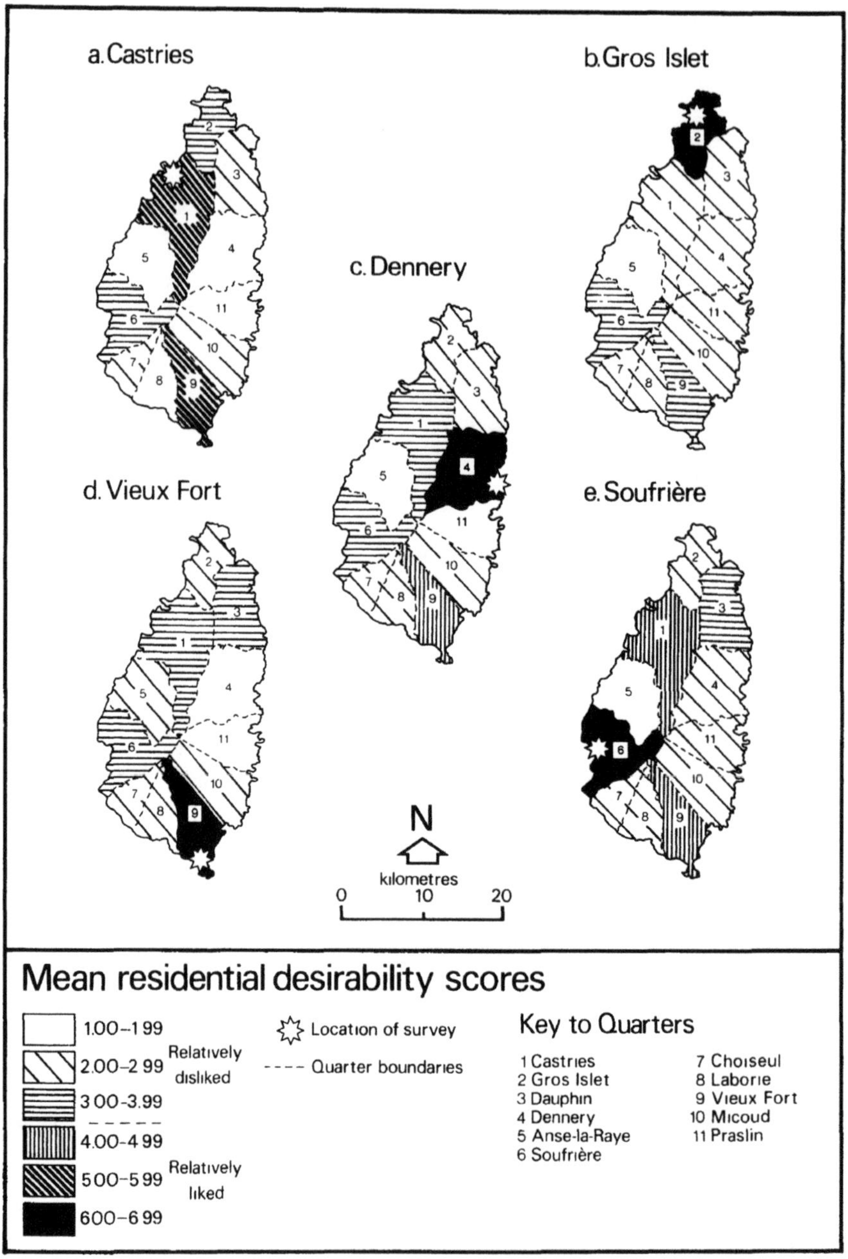

Figure 6.3: Mean residential desirability maps of St.Lucia

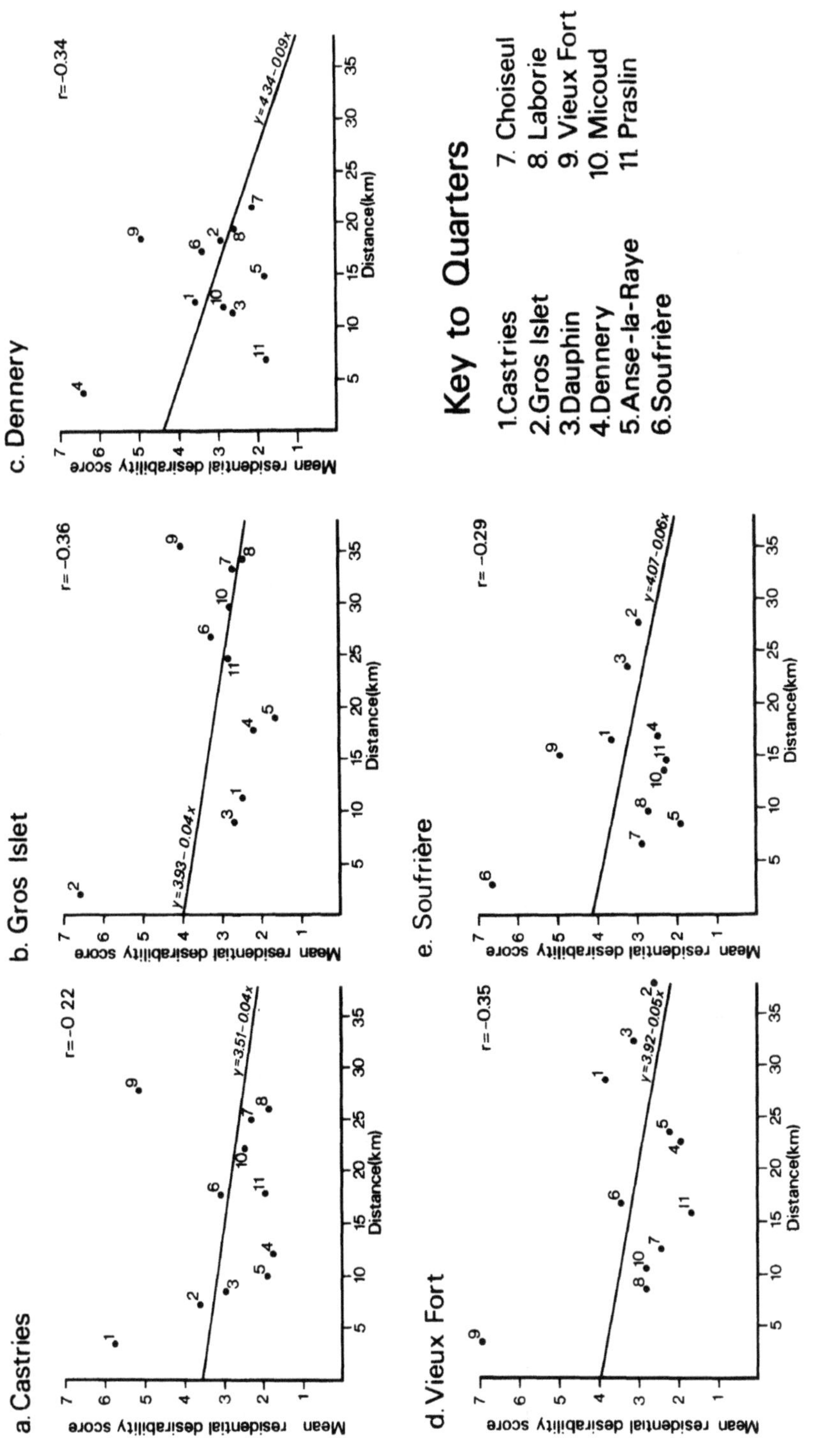

Figure 6.4: Correlation and regression analyses of mean residential desirability scores

the general trend if distance alone were effective throughout in a linear fashion. Thereby, the regression line effectively represents the local view in each case, whilst residuals from it should give some indication as to any national trend that exists. For example, for the Castries residents, Vieux Fort, Gros Islet and Soufriere are desired much more than would be predicted on the basis of their distance from the capital. These positive residuals may be mapped for all five survey locations (Figure 6.5). In each case, the home quarter appears as a positive residual from regression and apart from this, the next most apparent feature of the five maps is their similarity. Thus, the quarters of Vieux Fort, Soufriere and Gros Islet appear in each and every map, whilst Castries occurs in all but that pertaining to the residents of Gros Islet. Thereby, the perception surface clearly picks out the quarters which contain the major urban settlements. On the other hand, the much less developed areas of Dennery, Praslin, Laborie and Anse-la-Raye are held in poor regard in the consensus space preference map of St Lucians.

It would seem highly appropriate that basic inventories such as these are regarded as an essential part of the spatial planning process, revealing as they do the principal dimensions of nationals' environmental perceptions. This is especially so given the ease with which surveys such as the St Lucian example can be carried out. Such analyses can be seen as a potential form of public input into the planning process, albeit at the pre-proposal stage and of a rather non-specific nature. Similarly, comparable analyses can be carried out at the intra-urban scale in relation to residential desirability and housing quality, as has recently been done for Calabar, Nigeria by Ebong (1983).

Myth Maps

Stated in the simplest terms, myth map analysis addresses the fascinating question of whether space preferences reflect the actual socio-economic differences existing between areas. The basic idea was again pioneered by Gould (1969b), when he looked at the degree of association exhibited between the space preferences of students for American states and their level of composite social welfare to see "whether the over-all mental map for the group reflects reality" (Gould, 1969b, p. 38). In this enquiry the correlation was high ($r^2 = 0.61$). The residuals from the linear relationship pinpointed the rural states of New England and Florida, for example, as being greatly overestimated. In contrast, the Southern states, Great Plains and Minnesota were all consistently underevaluated. The possible role of myth map analysis over a period of time in analysing patterns of internal migration and development in Third World countries should be apparent.

For example, in the St Lucian analysis presented in the last section, it was ventured that positive residuals to the perception surface appear to be related to the main urban settlements. This association between urban based opportunities and images can be tested quite simply. In another section of the interview, respondents were asked to state

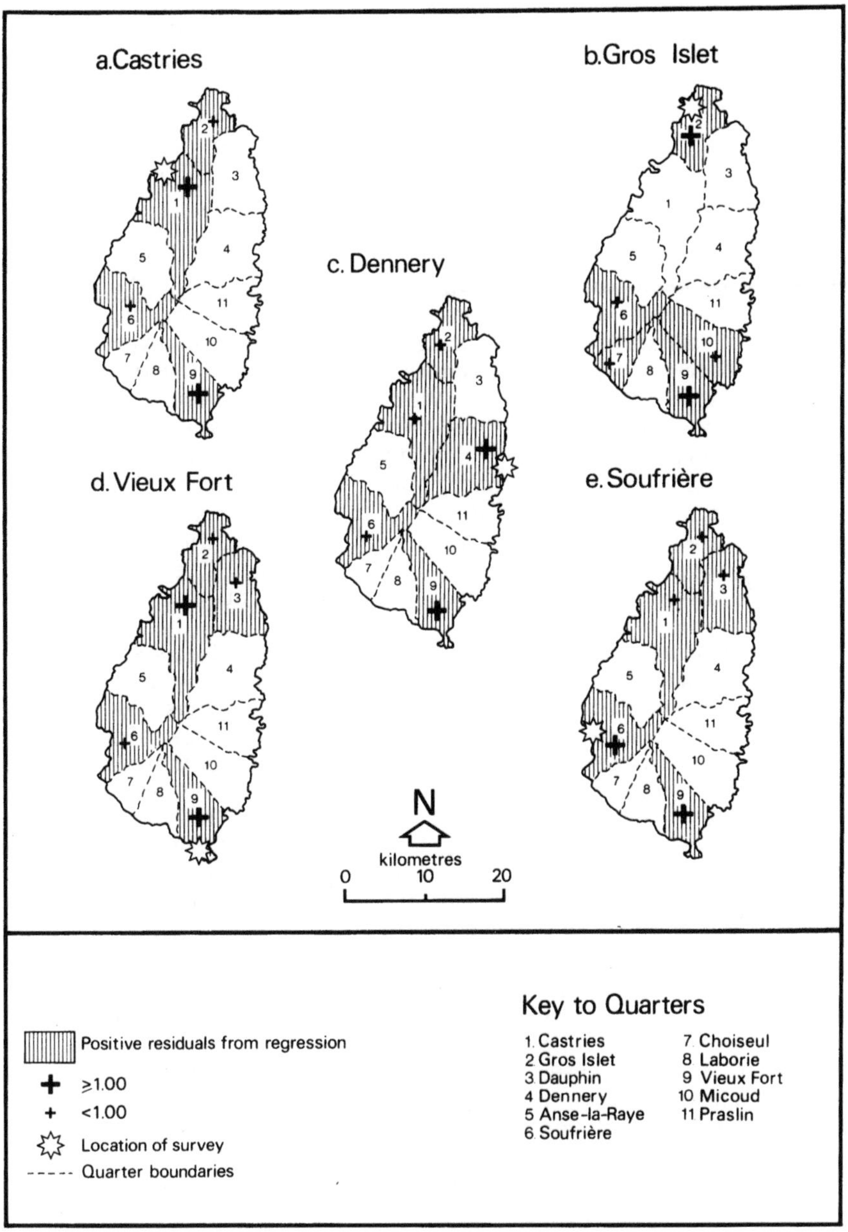

Figure 6.5: Positive residuals from regression of desirability on distance

their single most preferred area for residence (see Appendix C, question 3). Locatable responses are shown mapped in Figure 6.6. The figure confirms the findings of the previous analyses, in that very strong preference is expressed for the home locality. In each case, outside of the home quarter, only the main urban centres are cited as preferred areas. This pattern can be summarised by counting the total number of times settlements in each quarter are mentioned as being most preferred by respondents from outside that area. This gave the following rank ordering of quarters by popularity: Castries, Vieux Fort, Gros Islet, Soufriere, Anse-la-Raye, Micoud, Dennery, Laborie, Praslin, Dauphin and Choiseul. The quarters were then ranked according to their industrial development, as measured by the total number of separate manufacturing installations. The overall match existing between space preferences and levels of industrial activity was apparent. Thus, Castries, Vieux Fort and Gros Islet occupied first, second and third positions on both counts, whilst Dauphin and Choiseul were at the bottom of both lists. But contrasts in ranking were of equal interest. Dennery and Praslin, for example, recorded low perceptual rankings relative to their industrial infrastructure. On the other hand, Soufriere, Laborie, Anse-la-Raye and Micoud were all 'overrated'. Such mis-matches between "perceptions" and "reality" offer interesting clues to development and demographic change.

An excellent and much more detailed example of this sort of work in a Third World country is provided by Jones (1978) in his article on myth maps and migration in Venezuela (see also, Jones and Zannaras, 1978). Jones conducted 191 interviews in five areas of persistent out-migration in Venezuela, assessing how young potential migrants from these zones perceived the quality of life and economic opportunities provided by the 30 major cities. The background to this work was the generally held view that in Latin America the benefits of large cities are overperceived, and rural areas underperceived, so that "through the analysis of misperceptions of urban social and economic opportunities one may more fully understand the continued high in-migration rates" (Jones, 1978, p. 75).

The young respondents were asked to rank-order the 30 Venezuela cities according to first, job availability and second, overall social and physical desirability. Principal components analysis was then applied to both of these data sets, as in Gould's methodology. A high correlation was shown to exist between the two assessments (r^2 = 0.86). Both showed perceptual peaks at major urban centres such as Maracaibo, Merida, Caracas and Valencia. The perception surfaces were then compared with "objective opportunity", with the latter again being measured by a composite index, this time based on eleven socio-economic variables. The myth maps of residuals showed that the southeast (eastern Llanos and Eastern Interior) was strongly undervalued, as was the capital region of Caracas-Maiquetia. In contrast, urban areas along the central Andean cordillera and those serving as growth centres on resource frontiers in the West were consistently overperceived relative to the composite welfare index. Jones argues that such mis-matches are largely attributable to the regional images

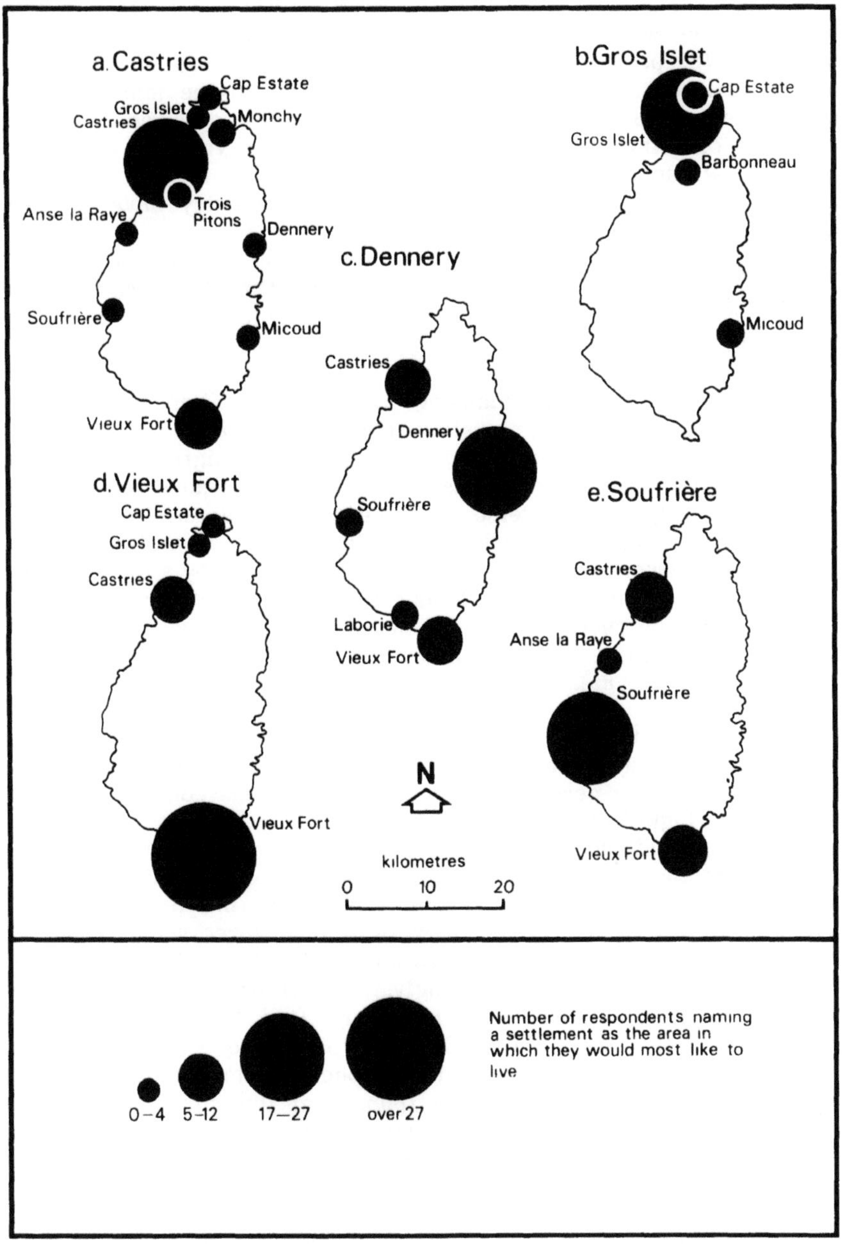

Figure 6.6: Residential preference maps of St.Lucia

propagated by Venezuelan national newspapers (Jones, 1975, 1978; Jones and Zannaras, 1978), with the rural Llanos and Eastern Interior being "viewed with a mixture of paternalism, condescension and pity" (Jones, 1978, p. 85). The poor performance of Caracas relative to actual socio-economic conditions is attributed to the joint influence of the Venezuelan government's decentralisation policy and the daily coverage of crime, traffic congestion and barrio violence in Caracas by the newspapers. In contrast, Jones argues that industrial progress in the east and western peripheries of the country have received wide acclaim in the press.

The potential of myth map-type analyses of this sort as a means of opening up debate concerning the settlement, preference, migration and policy space cycle (Figure 1.1) is considerable and Jones concludes by arguing that:

> National and regional planning authorities in Venezuela ... are somewhat stymied by existing migration models which do not get at the underlying reasons for urban in-migration. Ignoring migrant perceptions, their plans are forced into an objective mold - ... - designed to reverse the objective characteristics of urban places. Perhaps the question should be raised whether popular attitudes toward specific areas might be changed by a direct attack on misperceptions. Such a question begs another: would the correction of such perceptual biases be worth the time, effort, and cost, relative to direct investments in physical infrastructure? (Jones, 1978, p. 89).

This is an interesting and very complex issue. Jones' work on Venezuela has shown that urban in-migration is strongly influenced by migrant contacts and space perceptions (Jones, 1980). This conclusion is also reached by Lightfoot and Fuller (1983) in their study of six villages in northeast Thailand. Here it was shown that the strong movement of people to Bangkok from this area was heavily dependent on personal sources of information, thereby resulting in a stable and self-perpetuating pattern of migration. A stereotyped image of Bangkok has thus built up (Sternstein, 1971; Lightfoot, Fuller and Kamnuansilpa, 1981). Interestingly, an information programme concerning job opportunities in northeast Thai towns was initiated in three of the six study villages. Jobs were simply listed on noticeboards in the villages, whilst special committees publicised the project. Strong evidence was found that this simple programme encouraged a much higher rate of movement to the Northeast towns and a reduction to Bangkok. Migration management of this type stands as an interesting component of the development issues involved in the myth map equation.

The present author has also used a form of myth map analysis to compare nationals' space preferences for the administrative divisions of Barbados with the demographic histories and present-day socio-economic characteristics of these areas (Potter, 1983d), and this case study is reviewed in the next chapter. It can only be concluded here that the analysis of myth maps affords a relatively straightforward, flexible and potentially extremely rewarding method for all those concerned with urban and regional planning issues in Third World countries.

Attitudinal Surveys

Many basic aspects of individuals' appraisive environmental perceptions and cognitions can be investigated by means of simple attitudinal surveys. The techniques that have been developed for such purposes are legion and interested readers are referred to the standard texts in the field, particularly Oppenheim (1966) and Moser and Kalton (1971). In the present brief account, no effort is made to review all of the techniques available or indeed any one in detail: the aim is to give some brief impression of their overall scope and application and perhaps more importantly, to assess their possible employment in Third World urban settings.

Attitudes toward the environment, space and place may all be investigated by unstructured interviews. Such an approach has the merit of flexibility but involves problems of data recording and memory decay for the interviewer. More usually, standardised sets of questions are put to respondents, either in the form of structured interviews or self-complete questionnaires. In either case, questions can be of two types. First, they can be open-ended, where free answers are elicited and which have to be recorded in full. This approach provides the respondents with much freedom, but presents some difficulties of analysis. The responses are normally coded into categories and a form of content analysis carried out by counting the frequency of mention of particular answers. Such methods may be of considerable utility in developing countries, especially for researchers from outside the country and those not wishing to prespecify possible answers to the respondents. For example, in surveys carried out by the present author in Trinidad and Tobago, Barbados and St Lucia, a number of open-ended questions were included (see Appendices A-C). Prompts such as "what do you think are the three best (or worst) things about Port of Spain?" can be most instructive (see Potter, 1983c). Such methods have frequently been used in eliciting appraisive perceptions of Third World urban areas. Thus, Barker and Ferguson (1980, 1983) asked rural Kenyan students to state what they regarded as the best and worst features of Nairobi, and found that employment opportunities were most frequently cited on the positive side and theft and robbery in a negative context. Similarly, Lightfoot, Fuller and Kamnuansilpa (1981) elicited perceived advantages and disadvantages of Bangkok and found that young migrants were drawn by non-economic factors as much as by purely economic ones.

Alternatively, questions may be put in a closed manner, in the sense that the respondent is provided with a set of pre-specified alternative replies. Thus, the interviewee may be asked to tick an appropriate box, to agree or disagree with a statement, or at the simplest level, merely to state yes or no. Closed questions embody the advantages of quick and easy analysis but tend to limit and direct respondents. Such methods thereby depend greatly on the appropriateness of the questions asked and the response categories provided. Checklists, rating scales, ranking procedures, inventories, indices and Likert scales are among the many forms of data collection that may be employed (Oppenheim, 1966). A typical example of their use in urban planning-related cognitive study in a Third World context is provided by Lee Fong (1980) and Wong and Lee Fong (1980). This project examined the perceptions of residents of Tai Po new town in Hong Kong, a settlement that is projected to reach a population of 220,000 by 1985. Check lists and ten-point rating scales were used in an effort to evaluate residents' satisfactions with various aspects of the planned environment, for example, housing, shopping, services and environmental quality in general.

Finally, one other frequently employed technique of behavioural research must be mentioned, namely the semantic differential, developed by Osgood, Suci and Tannenbaum (1957). The method normally involves the use of bipolar seven-point rating scales, the extremes of each scale being defined by contrasting adjectives. For example, in asking people to assess settlements or administrative regions, scales specified by the adjectives urban/rural, developed/underdeveloped, modern/traditional might be used. The examination of the degree of consensus in the evaluation of the study objects by the respondents on the scales can be very informative, as can assessing the extent of intercorrelation existing between the scales. The full range of multivariate statistical techniques is frequently employed in connection with this task.

The efficacious employment of attitudinal surveys always demands careful preparation and attention to detail in execution. Put bluntly, silly questions will always generate silly answers and without forethought and precision, attitudinal surveys can amount to little more than spurious and over-sophisticated academic indulgence. Such misuse is to be bemoaned, especially as it may have led in the past to a feeling among planners that such techniques are of secondary or even minimal importance in stimulating public involvement in planning (England, 1974; Fagence, 1977). There are, however, other particular issues which must be addressed if attitudinal surveys are to be implemented in developing countries. One problem which must be faced in all environments, but which may be particularly acute in the Third World is that since the methods involve the pre-specification of possible answers, we have no means of knowing that the most apposite response categories or factors have, in fact, been included. There is an associated semantic difficulty here, too, in that we cannot be sure even when the respondent employs the terminology provided, that it is construed in precisely the way that was intended. This may be exacerbated in a Third World context, for there may be a cultural filter operating if expatriate researchers conduct such surveys, or even a social class one if indigenous planners are involved. Likewise, whilst it is obviously imperative to avoid leading and loaded questions in all surveys, Conyers (1982) has argued that another type of cultural problem occurs in developing countries where interviewees are particularly prone to give the answers they feel are expected of them. Conyers has pinpointed other difficulties involved in Third World surveys, such as identifying suitable sampling frameworks and the special need to keep surveys short and simple in order to ensure co-operation among all groups.

Personal Construct Theory and the Repertory Grid Test

There is an operational technique that can be employed to study individuals' environmental cognitions that does not necessitate the imposition of variables by the investigator, which is highly flexible and has the added advantage of possessing an accompanying theoretical base. The method is known as the repertory grid, and is grounded in personal construct theory.

Personal construct theory was developed by G. A. Kelly (1955) as a method of psychotherapy, and essentially seeks to identify a person's "constructs", that is the basic reference axes which they use as the dimensions of appraisal. Each person possesses a system of constructs which in total make up his or her personality. It follows that there is no such thing as objective reality, only a range of alternative ways of construing events and phenomena (Lee, 1976). The basic model that Kelly adopted was that every person is a scientist. In order to make sense out of the world, each of us develops an implicit theoretical framework - our personal construct system. Thus, Kelly argued that we are all scientists in the sense that we derive "hypotheses" (expectations) from our "theories" (personal constructs). Such hypotheses are then subjected to empirical testing via actual behaviour and the theories modified as a consequence (Fransella and Bannister, 1977).

The repertory grid test is the operationalisation of personal construct theory in the form of a structured interview. Kelly (1955) posited that individuals interpret objects by categorising them as either similar to, or different from one another in various respects, and these are the constructs (Barker, 1977). Lee (1976) stressed that constructs are not just perceptions but "perceived anticipants", in that they carry implicit predictions about future events, especially in terms of their associations with other constructs. During the repertory grid interview, the "elements" or study objects are normally presented to the interviewee in sets of three, or "triads". In the original use, the elements would be people or roles, but in an environmental context we may think of settlements, regions, housing types or even alternative planning strategies. The interviewee is asked to specify the way in which two of the elements are alike and thereby different from the third. The aspect of similarity defines a basic "construct", or dimension of appraisal, and the difference prescribes the "contrast". When a construct has been elicited, all of the remaining elements can then be 'assessed' or 'scored' on it. Triads can be presented until no further constructs are forthcoming, or all combinations are exhausted. In describing the resultant matrix of elements and constructs, Fransella and Bannister (1977, pp. 3-4) provide an interesting geographical analogy:

> The results of the grid have often been looked at as a map of the construct system of an individual, a sort of idiographic cartography as contrasted with, say, the nomothetic cartography of the semantic differential. To the extent that a grid gives us a map of an individual's construct system, it is probably about as accurate and informative as the maps which Columbus provided of the American coastline. At that, it may be a good deal more sensitive ... than the kinds of psychological instruments we have tended to use to date.

Let us now turn briefly to an example of the application of the approach in a Third World settlement study carried out by the present author in Barbados. The elements were selected by the investigator and comprised the principal settlements in each of the eleven constituent administrative parishes of Barbados. These are listed in columns 1 to 11 of the grid, shown in Figure 6.7. The names of the places were put on cards and the first triad randomly presented to this particular respondent was Speightstown, Belleplaine, St Lucy, as denoted by the three circles shown in row 1. The subject stated that he considered Belleplaine and St Lucy to be similar, as shown by the two crosses, as they are both agricultural centres. Speightstown was given as the "odd town out" (circle remains blank), because it is seen by the respondent as a commercial/fishing centre. These descriptions become the construct and contrast poles of the first elicited construct respectively (Figure 6.7). Each of the other settlements is now assessed by the subject on the basis of the construct. Wherever a settlement is regarded as conforming with the construct a tick is placed against the relevant cell. Thus, in the present example, settlements 1 to 4 and 7 are not seen as agricultural centres, unlike 5, 6 and 8 to 11 which are. The next triad elicited the construct "tourist resort", as shown in row 2, and the procedure was continued, with eleven constructs being derived in all.

Even this single example gives some hints as to the potential of the method. The first insights are gained from the number and type of constructs derived. If a set of grids are completed, frequently employed constructs are of interest because they may well point to the main areas where policy changes could profitably be made. This type of soft and subjective analysis also tells us about individuals' cognitions of single places. For instance, reading down any of the columns in Figure 6.7 provides a pen-portrait of the settlement. Thus, the subject in this case sees Bridgetown as a non-agricultural tourist centre/attraction, which is a commercial/fishing/boat building centre, has a dense population and many products available. Further, it is regarded as being developed, urban and a non-local focus - in short, therefore, the national capital.

But there is another fascinating dimension to the grid, and that is the degree to which the constructs are interrelated. Quite simply, the question can be posed as to whether there are summary dimensions or "superordinate" constructs which group together hierarchically subordinate ones. One way of addressing this issue is to apply principal components analysis, factor analysis or multidimensional scaling

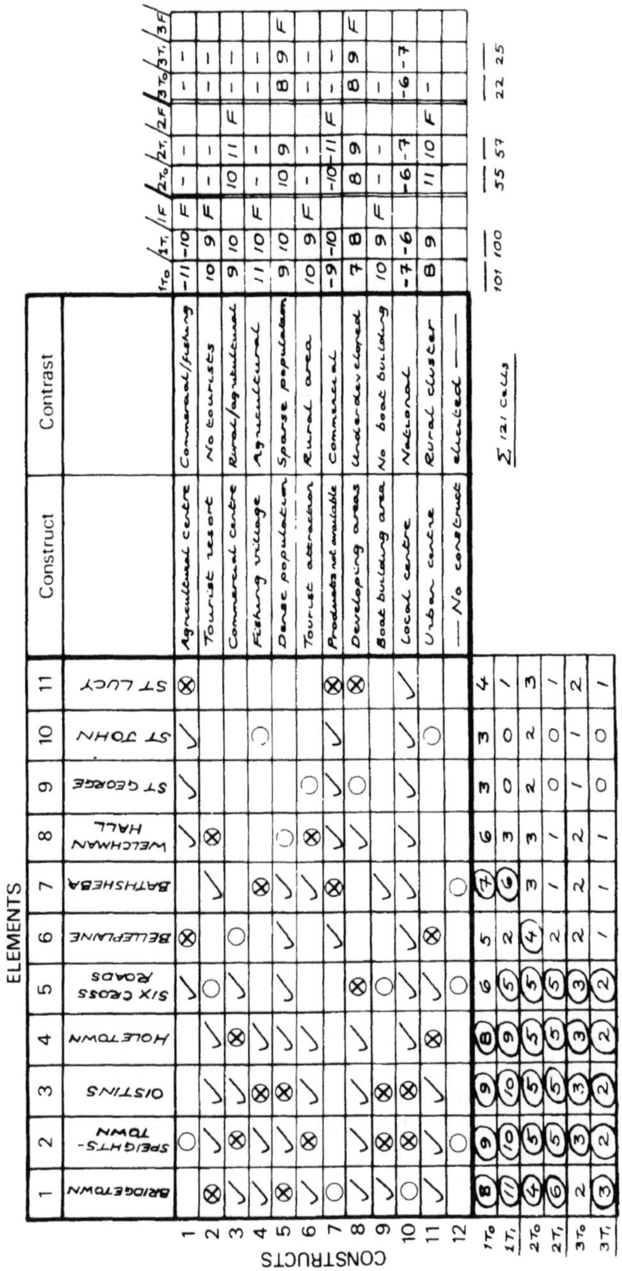

Figure 6.7: A repertory grid of Barbadian settlements

to individual grids. If both the elements and constructs are standardised, individual grids can be aggregated and multivariate analyses applied to this "supergrid".

It is worth pointing out however that essentially the same analysis can be performed by a non-parametric hand method of factoring grids, which was suggested by Kelly in 1955. However, few workers have used this simplified approach. It is argued by the present author that it represents a form of "appropriate technology" for Third World urban cognitive research (Potter, 1984b). For full details of the method, either Kelly (1955) or Potter and Coshall (1984) should be consulted, but the grid in Figure 6.7 has been hand factor analysed and gives some indication of the procedures involved. Even by means of cursory evaluation it is clear that certain constructs are highly related. For example, the construct 'tourist resort' in row 2 is highly positively correlated with the construct 'developing centre', for they are matched on 8 out of a possible 11 occasions. On the other hand, the 'tourist resort' construct is strongly negatively associated with the construct 'agricultural centre' (row 1), there being only one match between the two over the entire array of settlements. The extent of matching between rows may thus be used as a measure of association. To start the procedure proper, a summary row is derived by counting the number of ticks in each column. The totals are entered in a new row at the foot of the grid signified as 1To. Approximately one half of the column totals with the highest scores are then circled. The circles are seen as ticks and the non-circled numbers as blanks, and this pattern is now treated as if it were an entirely new general row or construct. The number of matches between it and each and every construct is counted, and these are entered in the new column 1To to the right of the grid proper. Thus, construct 2 is highly matched with the trial construct (+10), whilst clearly construct 1 is not (-11). In total, it is found that the trial construct agrees with 101 out of a possible 121 cells in the grid. It is possible, however, that there is a slightly better summary construct. Notably, three of the rows show negative relationships with the summary row, and these are "reflected", that is, treated as if the construct and contrast poles were swapped around. The column totals are adjusted accordingly for each settlement in 1Ti, and the same procedure applied to give new totals in column 1Ti. In this case, the second new row accounts for 100 matchings, one less than previously, so 1To is used as the first factor. The original constructs 'fishing village', 'tourist resort', 'tourist attraction' and 'boat building area' load strongly upon it, whilst 'agricultural centre' is entirely negatively related to it. We thereby have strong evidence to suggest that this is a tourist/coastal dimension. Other factors are then derived using the same methodology (see Potter and Coshall, 1984). In the Barbados study, the second factor appears to record an urban-commercial dimension and the third is a measure of population density and development (Figure 6.7).

A major characteristic of personal construct theory and the repertory grid is its great flexibility. In a sense, each application of the method is an experimental structured conversation which can be changed to suit the aims and circumstances of the work. This adaptability makes it a potentially very useful method in Third World settings. For example, the elements can be scored on the constructs in a variety of ways, ranging from presence/absence to ranking and rating scales (see Bannister and Mair, 1968; Fransella and Bannister, 1977). Although the method can take up to two hours to administer, it has been used with success, for example, in an investigation of Colombian peasants' perceptions of farming systems (Townsend, 1976, 1977), and of indigenous environmental knowledge among Nigerian farmers (Barker and Richards, 1978). Another merit of the method is that the elements can either be elicited from, or given to subjects. Similarly, the element triads can be presented to them as named cards or as photographs, or even on maps etc. Flexibility also exists in that some, or indeed all of the constructs can be supplied by the researcher. However, the chief hallmark of the method is that it allows people to express their own constructs with the minimum of investigator intrusion, so that constructs are normally elicited rather than given. There are other features that make the method particularly apposite in many Third World settings. The ability of the method to allow planners and administrators to "stand in the shoes of others" is particularly salient in this respect. For example, the method could be used to monitor changes in attitudes before and after major environmental planning projects, or to help evaluate the public's views of various suggested alternative plans. Hence, the repertory grid methodology can be used in connection with public participation in planning or ethnoscientific work looking at group cultural values and knowledge.

Projective Techniques

Finally in this section, mention is briefly made of a whole battery of projective techniques that are commonly employed by clinical and social psychologists to investigate appraisive perceptions and attitudes, although as yet, relatively little use has been made of them by geographers and planners. Projective techniques are ones whose purpose is hopefully not immediately obvious to the interviewees, a feature that helps to ensure that the investigator does not impose his own ideas and preselected categories on them. Further, such techniques are useful if the investigator wishes to delve somewhat deeper than the social facade and needs, therefore, to overcome barriers of awareness, irrationality, inadmissiblity, self-incrimination and politeness (Oppenheim, 1966).

Many of the methods work by association, whilst others involve the use of ambiguous stimuli. A popular technique and one that could find useful application in many geographical and planning situations is that of sentence completion. Sentence starters such as "In the south of the country ..." or "Port of Spain is ..." can be used to give a nondirective prompt to respondents. The normal form of analysis would be to code and group responses into categories in order to summarise perceptions. The technique has recently been used by Turner, S. D. (1982), in a geographical study of agriculture in Lesotho, using 18 introducers among 335 subjects. Turner sees the technique as a valid methodology for eliciting basic attitudes and perceptions using the phenomenological paradigm: that is, identifying essences of experience by means of allowing subjects to speak for themselves. Turner also stresses the cross-cultural empathy of the method which reduces interference from the researcher and his cultural background.

There are other projective techniques, several of which might find very useful geographic and planning policy applications, especially in Third World contexts. Included among these are cartoons, where simple drawings or pictures depicting a situation germane to the investigation and with empty 'speech balloons' are given to the respondent for completion. Another such method is that of picture interpretation, where respondents are invited to compose a story about the scene represented. The pictures are customarily ambiguous; for example, in an urban planning situation, an undisclosed person with a map or plan might be shown standing and looking over a low-income or high-rise housing district. Game techniques can also be used, where objects relating to the question at hand are shown to the subject in order to encourage conversation. The subject can then be set a game or simulation-type task to complete; for example, they might be asked to locate housing, jobs and services in an urban area, or to allocate new investments to regions. This type of model-building approach has recently been used in urban planning-related research by Murphy (1978) and Dijkink and Elbers (1981), and offers much potential in the investigation of appraisive environmental cognitions.

THE STUDY OF DESIGNATIVE PERCEPTIONS IN THIRD WORLD URBAN PLANNING

As already explained in this chapter, the designative component of urban and regional imagery relates to the geographical location, size, shape and disposition of the places and features making up the environment. At the most basic level, it connotes the quantity and quality of information that is possessed by individuals, groups and organisations about their surroundings. The designative component of spatial cognitions although different in tenor, is often closely linked to the appraisive component. The old adage that people know what they like and like what they know has already been cited earlier in this chapter, and is accurate in pinpointing the close reciprocal link that often exists between designative and appraisive imagery, that is between knowing and liking. One fascination in studying designative imagery is that it is highly likely that distortions, inaccuracies and biases will frequently occur, and these may well be related to environmental processes of change and to possible planning initiatives in some systematic manner.

Mental Maps

Although the term 'mental map' has often been used to describe the product of space preference and residential desirability surveys (Gould and White, 1974), its use is more fittingly linked with the outcome of free-recall sketch map drawing exercises. Such work was pioneered by Lynch (1960), an architect and city designer in his book The Image of the City. In this work, Lynch took samples of middle class people in three American cities and asked them to draw sketch maps of the central urban area, provide descriptions of various routes and to list distinctive elements. The structural elements depicted by these means were shown cartographically according to frequency of citation and were grouped by means of a five-fold structural typology of paths, edges, districts, nodes and landmarks. Lynch was basically interested in the legibility or visual clarity of the townscape, so that the consensus images derived were compared with the 'objective structure'. The clear planning orientation of the initial work on mental maps is of note. The Lynchian methodology, perhaps as a result of its simplicity has been used by many researchers in varied circumstances (see Pocock and Hudson, 1978, Chapter 5). In evaluating the methodology, Saarinen (1976, p. 110) has remarked that:

> The great advantage of the draw-a-map portion of Lynch's technique is that it is essentially a projective test that allows for and demands a maximum of structuring by the subject. There are no clues given, as would be the case if the subject were asked to outline areas on a prepared map. In such a case the individual's ignorance of connection would be overlooked because the map would already provide the structure. The use of a blank sheet of paper means that the only available source for the map produced are the subjects' minds.

This, however, presents the active researcher with a dilemma, for whilst it is true that the blank paper routine maximises the projective potential of the method, it must be recognised that map drawing on a clean sheet involves many inherent problems. Quite simply, some people are likely to be more able to draw than others, although they may in fact be characterised by broadly equivalent "mental maps" (Rieser, 1972). This particular problem is of course likely to be heightened in a Third World setting where semi-literate and illiterate groups may be common. Murray and Spencer (1979) in investigative research completed in a developed world setting found that basic drawing ability was related to the ability to produce an environmental sketch map. In some studies, it has even been posited that the size and shape of the sheet of paper used can influence the rendition of mental maps (Pocock, 1976).

These methods have now been used in Third World studies from the scale of continents to that of the single urban area, although such work is as yet in its infancy. Lewis, Degani and Hudson (1980) asked 104 Nigerian students to draw sketch maps of the continent of Africa. Similarly, at the intra-metropolitan level, Gulick (1963) had 35 students produce sketch maps of Tripoli in Lebanon. Notably, however, both of these exercises were carried out with samples drawn from educated elite groups. There is an example par excellence of the employment of free-recall sketch mapping for urban planning purposes in the Third World and this is the work of Appleyard (1969) in connection with the new city of Ciudad Guayana, the planned growth pole for eastern Venezuela. Some 300 or so subjects from four main areas were interviewed regarding their reactions to the design of the new city. As might be anticipated, the citizens were found to have quite parochial images, but perhaps more significantly, failed to appreciate many of the grander designs dear to the hearts of the planners. An interesting and useful outcome of Appleyard's research in Ciudad Guayana was his paper on styles and methods of structuring a city (Appleyard, 1970). In this, he distinguished between sequential sketch maps emphasising paths and nodes and spatial ones, which concentrate on buildings, districts and landmarks. In the Ciudad Guayana case study, sequential maps were found to predominate. Another Lynch-type study using a non-student sample has recently been conducted by Karan and Bladen (1982) in the Katmandu-Patan metropolitan area of Nepal. A total of 120 people were interviewed in three areas, although in the end, only 103 produced sketch maps that were sufficiently detailed for analysis. Notably, as "most respondents were unfamiliar with maps and had difficulty drawing one" (p.229), the respondents were provided with a map blank on which the north axis, a number of major squares and rivers were given. However, the findings of the work were rather limited, principally demonstrating that educational attainment was closely related to the accuracy and complexity of sketch maps.

In work carried out in the summer of 1982, the present author examined the mental maps of a sample of Barbadian school children, partly in order to compare them with planning zones, but also to evaluate the technique in a Third World setting. A preliminary report of some of the findings is given in Potter and Wilson (1983). The sample comprised four groups of children, with modal ages of 11, 13, 16 and 17 years. Members of each were provided with an outline map showing only the coast and the boundaries of the eleven administrative parishes of Barbados. They were then asked to add in 30 minutes all of the features/places they deemed important. The maps witnessed clear and interesting changes between the age groups, notably in terms of the mean number of separate places/features depicted, but perhaps more saliently, with regard to the style of spatial representation used. Single examples of completed maps for each age group are shown in Figures 6.8 and 6.9. Each of these is close to the average map pertaining to the entire age group. In the map of the 11 year old, areas are shown exclusively by names, arrows and simple symbols and there is a pronounced coastal orientation. The 13 year old's map contains roughly the same number of elements and shows a very similar style of depiction, but here an attempt has been made to draw in the Scotland district in St Andrew, although the result is highly confused. Turning to the map of the 16 year old respondent, a change is witnessed in the content and style of the cognitive map. First, it contains a larger number of features overall and second, an appreciation of regions or areal zones is indicated by the inclusion of tourist areas and the now well-defined Scotland District. This increasing sophistication is further evidenced by the

17 year old's map. In this, there is a close integration of coastal and inland features and a heavy bias toward the depiction of two-dimensional areas. The zones depicted include the densely populated west and southwest coasts, Bridgetown, the main agricultural parishes, tourist coasts and even relief features such as the Christ Church dome and St George Valley. As a summary, it was clear that the maps could be grouped into those in which features were shown only as dots, names and point symbols ("punctiform") and those which depicted two-dimensional zones ("areal" maps). Punctiform cognitive maps predominated overall, but exhibited a statistically significant decrease through the four ascending age groups (Potter and Wilson, 1983). Looking at the cognitive maps in total, approximately 32 per cent of the features shown were settlements and there was a close correlation between settlement size and frequency of citation. However, the low imageability of some settlements and the high imageability of others was particularly interesting in relation to published urban planning strategies (Potter, 1984b).

The study of mental maps is of interest, and further is of some potential relevance in relation to spatial planning studies, although the method is clearly not without its problems. In order to obviate some of these, however, non-graphical means of elicitation can be employed. For instance, the iconic method of photograph recognition can be used although factors such as the orientation of the scene can be crucial in such exercises. Further, verbal methods can be used, such as check lists of places known or even numerical rating scales of familiarity with entire zones or areas. Finally, listings of distinctive areas or of particular routes may also be employed, as in the original work of Lynch.

Cloze Procedures

A further means of investigating designative spatial images has been suggested by Robinson (1974). The method is that of the cloze procedure, which derives from the Gestalt concept of 'closure', whereby there is a tendency for individuals to complete missing or obscured parts of a visual image. This approach has been used in linguistic comprehension tests where words are systematically omitted from prose passages and must be fitted in by the subject. Robinson (1974) describes the headway made in studies of mental maps since Gould's work as minimal and argues that cloze tests might be developed to aid work in the field. One advantage is of course that the method gets away from the need for map drawing by respondents. The basic spatial methodology is simple: a map of the area of interest is used with some form of grid square or rectangle overlay superimposed. On this, some of the areas are systematically blanked out, say every fifth one on the grid. Interviewees are then asked to name settlements or some other stipulated feature located in the blank areas. The number of correct closures can be aggregated by distance and direction, or in relation to respondent characteristics (see Robinson, 1974; Porter, Hart and Machin, 1975). It can only be recorded that from the point of view of application in the Third World, the method still suffers

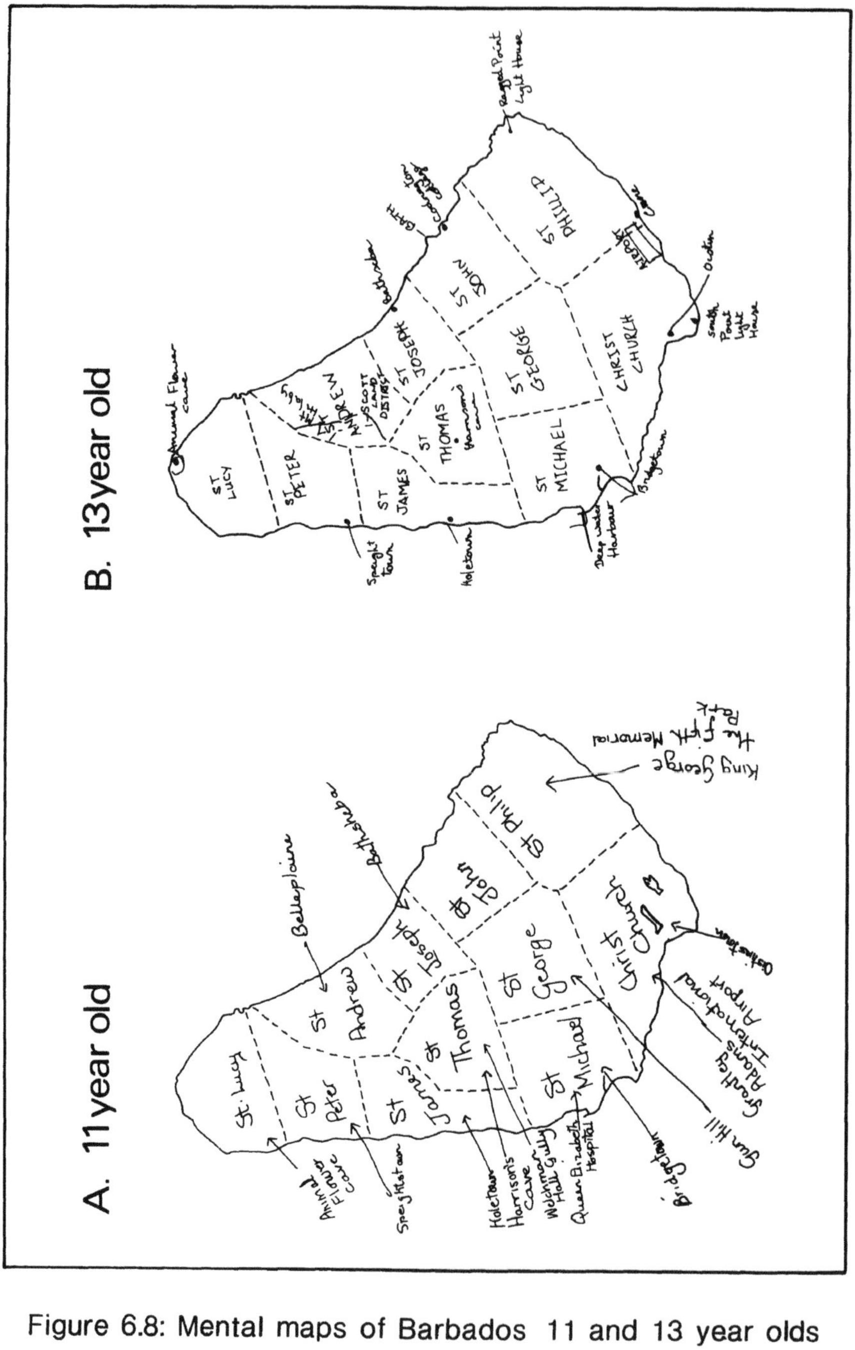

Figure 6.8: Mental maps of Barbados 11 and 13 year olds

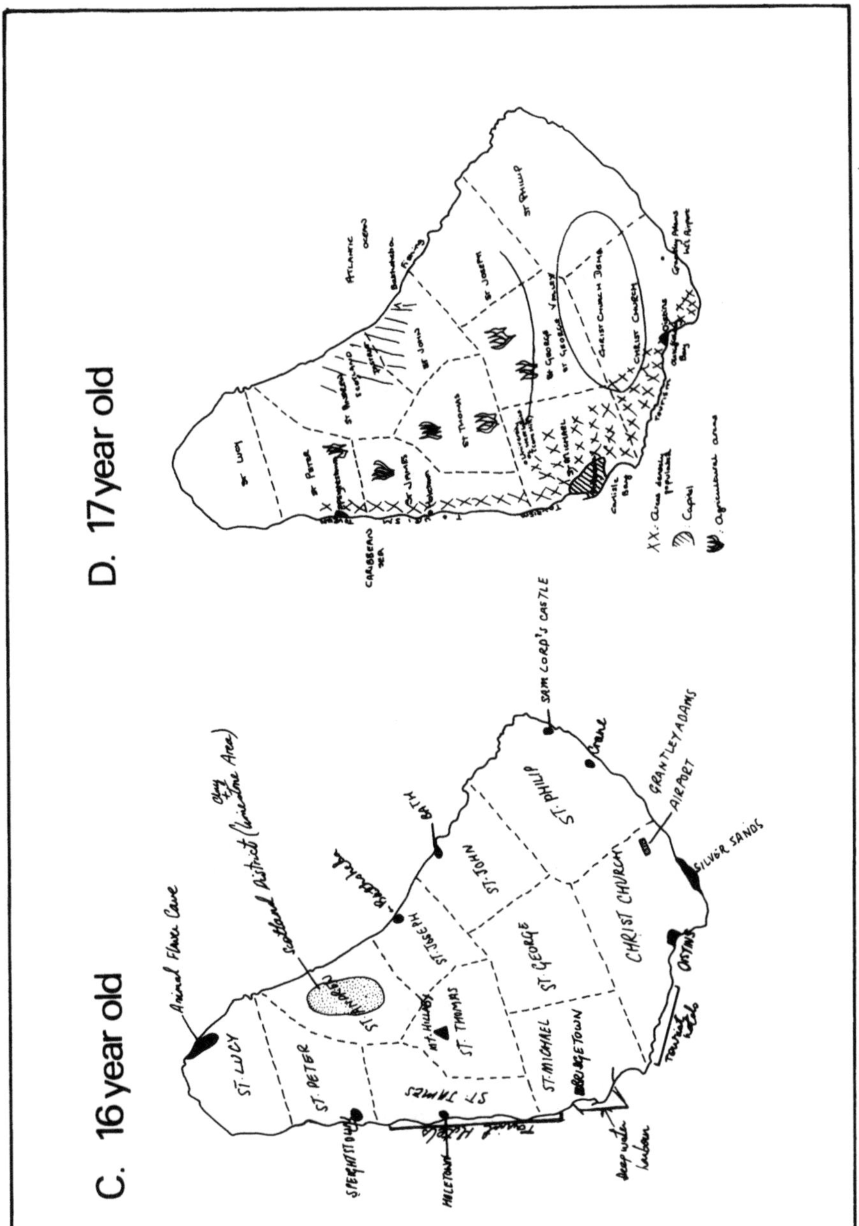

Figure 6.9: Mental maps of Barbados: 16 and 17 year olds

from the disadvantage of depending heavily on maps, but its evaluation in such contexts might prove to be an interesting and rewarding exercise.

Distance Perception

At the intra-urban level within behavioural geography, a great deal of research has examined the relationships between actual and subjectively perceived distances. The general finding is that distances within urban areas are consistently over-estimated (Lee, 1970; Canter and Tagg, 1975). Although any differential mis-perception of distances is likely to be reflected in behaviour, because over-estimation appears to be almost universal, it is difficult to see the direct practical application of such findings for planning purposes in Third World cities. In fact, virtually all of the work carried out on this topic has been set in Western countries. However, one notable recent departure is afforded by the work of Ferguson (1979) in Nairobi, Kenya. This research was based on samples drawn from the university student population and looked at the development of urban cognition as measured by distance estimation. The over-estimation of intra-urban distances was strongly confirmed for all groups of students, but estimates were shown to become much more accurate with increasing length of urban experience. Whilst studies of distance and directional biases might be of value to planners interested in the process of assimilation of in-migrants to urban areas, it is difficult to identify other areas of serious practical application. In fact, a closely related aspect of space perception may be of greater applied significance in Third World urban planning; that is the identification of perceived boundaries. Work carried out in Western urban areas has shown that mental map studies can be an interesting way of looking at how individuals define their community and neighbourhood areas (Herbert and Raine, 1976). Lee (1963), again in a Western study, asked respondents to draw their neighbourhood boundaries on maps and found the modal area accommodated approximately 2,500 people, much lower than the 10,000 level customarily employed by planners. This type of approach might well prove to be of some utility in studies of housing areas and communities in rapidly growing Third World cities.

THE STUDY OF PRESCRIPTIVE PERCEPTIONS IN THIRD WORLD URBAN PLANNING

In so far as environmental perceptions and cognitions are used by individuals and other decision-makers to draw inferences and make predictions about the future, they can be described as having a clear prescriptive dimension. Naturally, this component of the image dealing with anticipated change is one that can be tapped to advantage by environmental planners. Whilst the techniques reviewed in this chapter are particularly relevant at the early or basic survey stage of the planning sequence, they can also be employed when alternative planning proposals are being evaluated. Thus, there is not a separate range of methods for the examination of prescriptive environmental perceptions, but rather appropriate reformulations of the ones that have already been reviewed.

For example, space preference analyses and myth maps embody a strong prescriptive element in that they measure where people desire to live, so other things being equal, they can be used to infer where people and businesses might reasonably be expected to migrate. Specific examples of this type of work will be provided in the next chapter. Such analyses may be of direct relevance in decisions regarding resource allocation and regional incentives. Similarly, other appraisive attitudinal questions can be used to examine prescriptive perceptions. For example, in work carried out by the author in the Caribbean, respondents were asked whether they felt it would be a good idea to develop new towns, and if so, where they felt such new settlements might best be developed (see Appendices A-C). Personal construct theory and repertory grids, in particular, could also be used to good effect in prescriptive studies. For example, different planning strategies could be used as the elements and hypothetical strategies such as "the best plan imaginable", "the worst plan imaginable" also included. It should be remembered that constructs are perceived anticipants, so that even ostensibly straightforward repertory grid tests of the type presented as an example in the present chapter should bring forward regions, settlements, house types or whatever is being studied, in a prescriptive manner. Further, key constructs such as 'developed/not developed', 'changing/not changing', 'would like to live there/would not like to live there' can of course be specified by the researcher.

Projective techniques are obviously also well-suited to the examination of prescriptive imagery, particularly sentence completions, games and simulation exercises. The latter have been used quite frequently in developed world planning-oriented research. For example, Murphy (1978) used a three-dimensional model of "No-town" to investigate the planning preferences of five contrasting urban decision-making groups: property developers, local politicians, a citizens' pressure group, architects and planners. Likewise, Dijkink and Elbers (1981) asked children to build a city using 116 elements comprising houses, factories and shops. It is not hard to imagine the effective use of such techniques in a Third World urban planning context. Finally, as indicated previously in this account, free-recall mental mapping exercises and other designative techniques can be instrumental in identifying areas of low imageability on the perception surface and such features can be related to future patterns of urban development. However, it must always be remembered that cognitions and perceptions alone can never explain nor totally predict future spatial behaviour, which is the outcome of constraints and structural features, as much as preferences. However, the judicious recording and analysis of environmental cognitions should provide data which give a strong indication of likely behavioural trends, or at least a sound basis for their discussion.

EVALUATION

It is suggested that the analysis of individual and group designative, appraisive and prescriptive cognitions of environmental and other spatial circumstances constitutes a potentially extremely rewarding, although as yet, frequently neglected theme in Third World urban planning practice and research. This neglect appears somewhat surprising, given the likely cost effectiveness of such methods in most circumstances and the fact that they can be placed alongside other more conventional forms of public participation in the planning sequence. The techniques are also relatively simple to implement and analyse and further, it is posited that those detailed in this chapter represent a form of "appropriate technology" for the analysis of people, space and development in poor countries. It should be acknowledged, however, that it has become almost a cliche to suggest that studies of environmental perception and cognition are of immediate and direct relevance to planning and environmental policy formulation. In this connection, we should do well to heed the warning of Gold (1980, p. ii) that:

> Behavioural geography is (rightly) a strong policy-oriented field of study, but it has been all too common for geographers to make the mistake of claiming too much too early. It is not satisfactory to lecture the planner on his responsibility to adopt geographical concepts and techniques when the material that is being produced contains few usable recommendations.

The central argument behind the present chapter, however, has been that perception-based studies can provide a useful input into Third World urban and regional development planning, particularly by providing a forum for public involvement in planning. Although, as shown in Chapter 5, it is obvious that there are formidable obstacles to public participation in urban planning, devising mechanisms to overcome these stands as an important task. First, not every member of the public can participate for it is just not possible to take into account simultaneously everybody's views, nor indeed do all individuals wish to decide for themselves, as discussed in Chapter 5. The maxim has to be that "whilst everyone cannot have their way, all should have a chance to say". Public involvement must, therefore, be encouraged by all the means detailed in Chapter 5, but from a spatial-geographical viewpoint, interview and questionnaire surveys of spatial and more general environmental perceptions are of vital importance. However difficult it may prove to integrate these into the planning sequence, endeavouring to do so stresses the importance of the view that what people think matters in the planning process (Lloyd, 1979; Conyers, 1982). In particular, stereotypes, myths and prejudices in spatial perceptions must be established as such, and their causation and consequences fully investigated.

It is, of course, recognised that the application of such perspectives alone will do little or nothing to eradicate poverty, inequality and injustice. In this context, there is inevitably some truth in Cox's

(1981) recent claim that behavioural geography is "just one more instance of bourgeois thought" (p.256), as a result of it being grounded in the *status quo* ideology of capitalism. However, in terms of the wider aims of social development, this basically structuralist argument seems to be pessimistic, harsh and rather inflexible. The present author has argued that the inclusion of cognitive surveys in the planning process in Third World countries has the highly desirable effect of bringing the professional planner face-to-face with the views and aspirations of those being planned for (Potter, 1984b). This function may be especially important in many Third World countries where planners are likely to have been trained abroad, as a direct consequence of which they may well be inclined to bring inappropriate solutions to indigenous problems. Further, the cognitive methodology is important in so far as there are many situations where politicians and administrators whilst espousing the need for change in the public arena, remain stubbornly inert in the realm of actual implementation. The collation, analysis and publication of perception surveys should at least help to reduce the likelihood of this happening. As concluded elsewhere by the present author (Potter, 1983d, 1984b), although cognitive studies should not be viewed as a panacea, given the social, economic and political will, they can be harnessed to help promote more effective socio-economic planning and a more equable distribution of resources and development potentials in Third World countries.

Chapter 7

Case Studies of Planning Related Perception Research

Following the lengthy review of techniques and methods provided in the previous chapter, and notwithstanding the fact that a substantial range of examples was included in this account, the present chapter provides specific and detailed case studies of the application of planning-related perception studies in two Third World countries, namely Barbados and Trinidad and Tobago, West Indies. These studies are set in the context of the accounts of urban development and planning systems in these two countries that were provided in Chapters 3 and 4 respectively. The overall aim is to provide an impression of the way in which sets of cognitive-behavioural techniques may be applied in an holistic manner to examine planning-related themes and issues, in both the historical and contemporary contexts. This is important, for in the previous chapter, although a large number of separate techniques were reviewed, little indication was given of the manner in which these can be combined together to form an overall research methodology.

The case studies are based on the author's research work carried out in the two countries in 1980 and 1982. It must be appreciated, therefore, that the work was carried out by an academic researcher interested in planning policies and residents' perception of these. This represents a somewhat different approach from the much more specific and like as not highly pragmatic perspective of the practising planner. Hopefully, any such difference is one of degree and not of kind, so that the chapter should hopefully impart some impression of the possible contribution that these methods can make in planning-related research. Once more, the overarching theme is that the incorporation of such approaches into the planning sequence should help to prevent the syndrome of "the people versus Mr Urbano planner y administrador" (Hollnsteiner, 1977). To this end, some idea is given of the ways in which these studies can be used in the planning sequence and of the likely costs involved in the conduct of such surveys, in terms of both time and personnel. It can be argued that the more urban planners assume the mantle of researchers as well as administrators, the better. Such a view, however, must be based on acceptance of the fact that planning is never ideologically neutral, for it either strengthens or weakens existing spatial and social patterns. Thus, people-oriented planning requires the soliciting of information concerning the views of citizens, a fact all too easily overlooked behind a bureaucratic maze of legal and professional requirements, or an entrenched status quo view of the role of spatial planning. In short, effective democratic participatory planning can only be achieved if a dialogue is established between planners and the public. Perception based studies afford one means of effecting such communications. Again, although the examples presented in this chapter are mainly drawn at the inter-urban scale, it is stressed that the same type of approach is just as applicable at the local or intra-urban scale of analysis.

SPATIAL COGNITIONS, SOCIO-DEMOGRAPHIC CHANGE AND PLANNING IN BARBADOS

The background to the Barbadian field research has already been set out in previous chapters, but a brief resume of the history of settlement evolution and spatial planning is provided here to place the cognitive survey work in context.

Barbados, the most easterly of the Caribbean islands is a small (430 km^2) independent nation, although it supports a population of 248,983 (1980) at a very high density of 578 persons per km^2. Traditionally, the economy has been based on sugar monoculture, but since 1960, manufacturing industry and tourism have become increasingly important as noted in Chapter 3 (Potter, 1981, 1983a). Both of these activities have tended to perpetuate a dependency on overseas firms and foreign capital and have done little to reduce the open nature of the economy. The existing settlement pattern is highly skewed, primarily reflecting the course of British colonial development since the first landing of settlers in 1627 (see Chapters 3 and 4). The early settlements were all developed on the sheltered leeward or western coast of the island, first at present-day Holetown, then in sequence at Bridgetown, Oistins and Speightstown (see Figure 7.1). As shown in Chapter 3, this pattern, which is highly reminiscent of the mercantile model of settlement evolution suggested by Vance (1970), persists today. By the middle of the nineteenth century, however, Bridgetown had developed as the chief settlement and seaport, by that time accounting for some 28 per cent of the national population. Through the second half of the nineteenth and early part of the twentieth century, Bridgetown became still more primate, accounting for 30 per cent of total population in 1900 and 42 per cent by 1970. Increasingly over this period, high density ribbon-type development occurred along the west coast main highway (Lowenthal, 1957; Potter, 1983b). This highly concentrated settlement pattern was therefore the outcome of totally piecemeal and unco-ordinated processes of locational decision-making(Potter and Hunte, 1979; Potter, 1983b).

Town and country planning was not established until 1959, when an Interim Act introduced development control over the west coast and a short section of the east coast, thereby reflecting the existing pattern of population and infrastructure. As noted in Chapter 4,

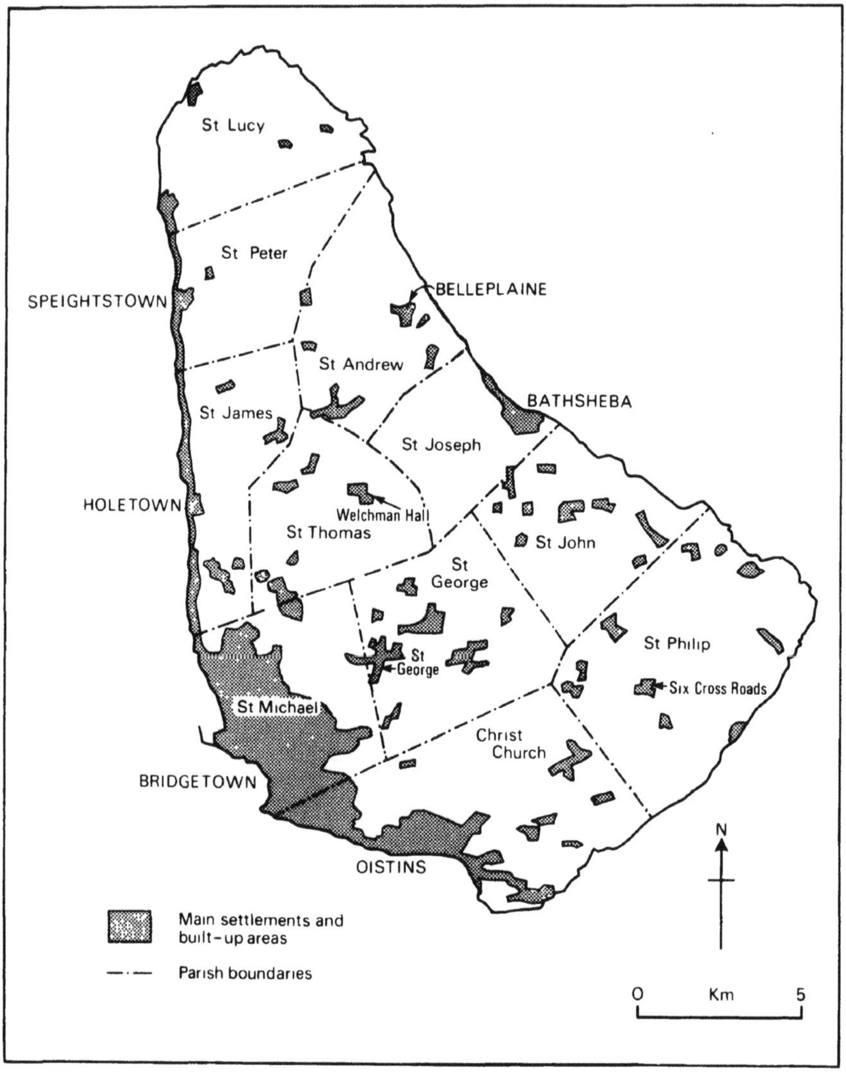

Figure 7.1: Barbados-principal settlements and survey areas

it was not until two years after independence, in 1968, that a national system of planning was introduced. This most notably led to the preparation of the first, and as yet only physical development plan for the island. The Physical Development Plan for Barbados, which was produced with technical assistance from the United Nations, was published in 1970 (Town and Country Planning Office, Barbados, 1970). Naturally enough, the chief remit of the plan was seen as the reduction of the primacy of Bridgetown by means of deconcentration. Thus, as noted in Chapter 4, the plan advocated a rationalisation of population, economic activities and services in the form of a planned hierarchy which was clearly inspired by classical Christallerian central place theory (Potter, 1983b). It may be recalled that two regional urban centres were designated for the northern and southern halves of the island at Speightstown and Oistins respectively. Additionally, seven district and 58 village centres were planned, these being located throughout the island (see Figure 4.5). No alternative planned settlement strategies were provided whereby the aim of decentralisation might be effected and thus the physical plan was largely presented as fait accompli. Given this, how sound was its overall advocacy of a planned settlement hierarchy and how well did it fit the existing settlement pattern and physical realities of the country? But just as significantly, did the identification of problems and the solutions proposed in the plan correspond with those recognised by different groups of Barbadian nationals? Do the people regard Bridgetown as being too large and do they feel that decentralisation and the development of new urban areas is a sound policy?

These were just some of the questions addressed in an interview survey carried out over a five-day period in 1980. A copy of the interview schedule used is reproduced in Appendix A. Three interviewers were employed and one full day was spent interviewing in each of the five main settlements of Bridgetown, Speightstown, Oistins, Six Cross Roads and Belleplaine (see Figure 7.1). Thus, the field survey entailed a total of 15 person-days work. Interviews were generally conducted in the street and the sample was to comprise a random selection of 45 residents from each of the areas surveyed. It was not feasible to quota sample, but respondents from the different settlements were found not to be unduly different with regard to age, sex and length of residence (see Potter, 1984a for details). In the final event, some 207 respondents were interviewed. Ideally, for planning-related purposes, the survey should have been carried out using larger quota samples in each of the eleven administrative areas making up Barbados, rather than just five of them. Even so, the manpower and financial resources involved would have been relatively low. The results of the survey are not presented here in their entirety, only an illustration of their possible relevance in the planning sequence.

As a part of the interview, respondents were asked to indicate their residential space preferences for each of the administrative parishes of Barbados (see question 6, Appendix A). Their responses were converted into a residential desirability score ranging from 1 (low preference) to 7 (high preference), and mean residential desirability scores calculated for parishes for each sub-sample. Gould's multivariate approach using ordinal assessments was regarded as unnecessarily sophisticated. The results of the residential desirability survey are shown mapped in Figures 7.2A to F. The view of the residents of Bridgetown, St Michael witnesses the existence of a pronounced dome of local desirability which extends to the neighbouring coastal parishes of St James and Christ Church (Figure 7.2A). However, in strong contrast, the northern and eastern parishes, especially the rural-agricultural ones of St Lucy, St Andrew, St Joseph and St John are all relatively disliked. In fact, the desirability surface exhibits a strong west coastal orientation, focusing on the principal built-up and tourist development areas. The space preferences of residents of Christ Church parish are shown in Figure 7.2B. Again, a clear local bias is revealed and once more the immediately adjacent coastal parishes of St Michael and St Philip are the only others which score relatively highly. From this coastal axis, desirability levels decline in a broadly northeasterly direction. A similar pattern emerges when we turn to the mental map of St Philip residents shown in Figure 7.2C. Only St Philip and Christ Church parishes are liked to any real degree and the entire northern two-thirds of the island falls squarely into the disliked category. Thus, a clear north versus south spatial divide exists, at least in the eyes of residents of the southern part of the country and a strong preference for the western coastal area is also revealed.

Are these patterns matched in the eyes of denizens of the northern half of Barbados? The residential desirability surface of residents of St Peter (Figure 7.2D) shows a strong preference for the home parish and for the neighbouring coastal area of St James. However, noticeably the other nearby northern parishes of St Lucy, St Andrew and St Joseph are picked out as being relatively disliked. In contrast to the maps of southern Barbadians, it is the areas a relatively long way away, in particular Christ Church, that are the next most liked. In fact, once again, there is a pronounced west and south coast bias in the space preference map. The picture becomes more complete if we turn to the map for residents of St Andrew parish shown in Figure 7.2E. In this, a strong reciprocal antipathy for other northern parishes is displayed in the form of a mirror image from that held by residents of St Peter. Thus, just as denizens of St Peter are uninspired by the habitat offered by St Andrew, so inhabitants of St Andrew appear to think equally poorly of St Peter. Saliently, however, Christ Church parish in the far south is far more favourably perceived than many of the areas located much closer to the survey parish.

Each of the maps clearly shows a local component with high desirability being recorded for the home parish and a general diminution with distance away. But each map also shows a non-distance related component and these, taken together, seem to represent a national consensus view. First, the western and southern coastal parishes are generally more favourably perceived than their distance from respondents would imply. In particular, the parishes of Christ Church and St Philip fare rather better than might be anticipated. The capital region of St Michael, however, does not do nearly so well. On the

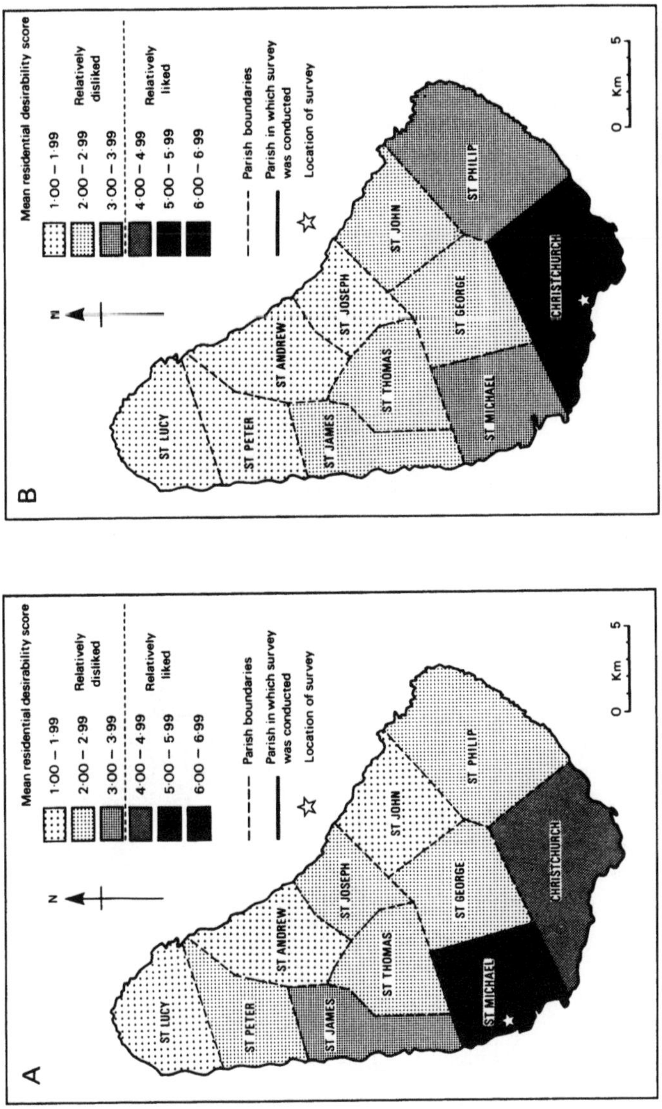

Figure 7.2: Residential desirability maps of Barbados

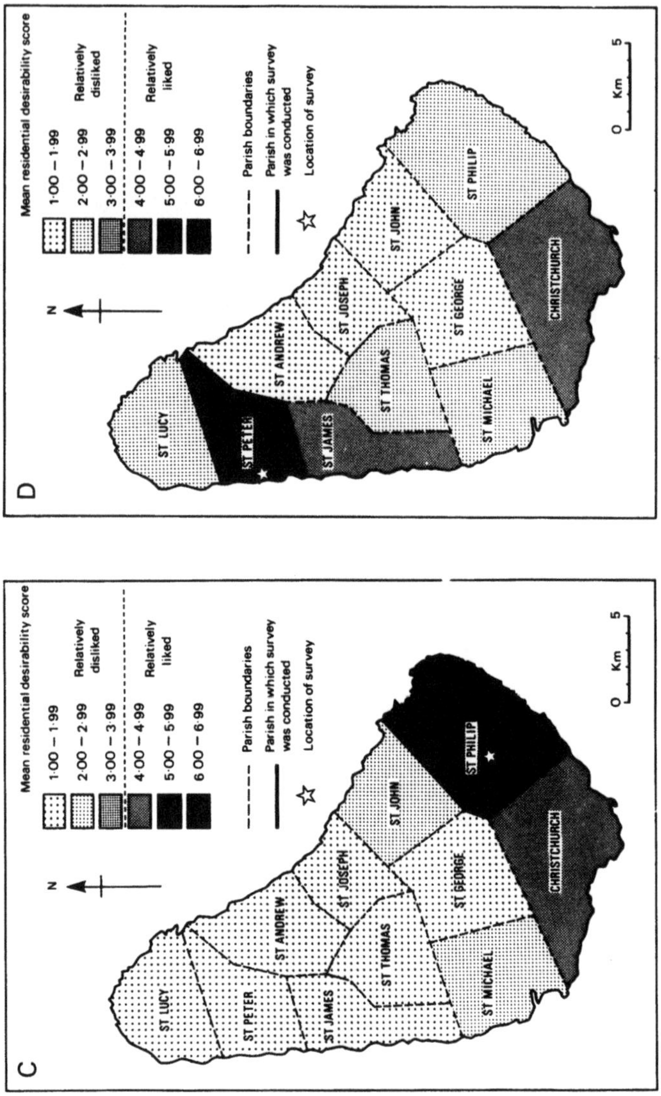

Figure 7.2: Cont.

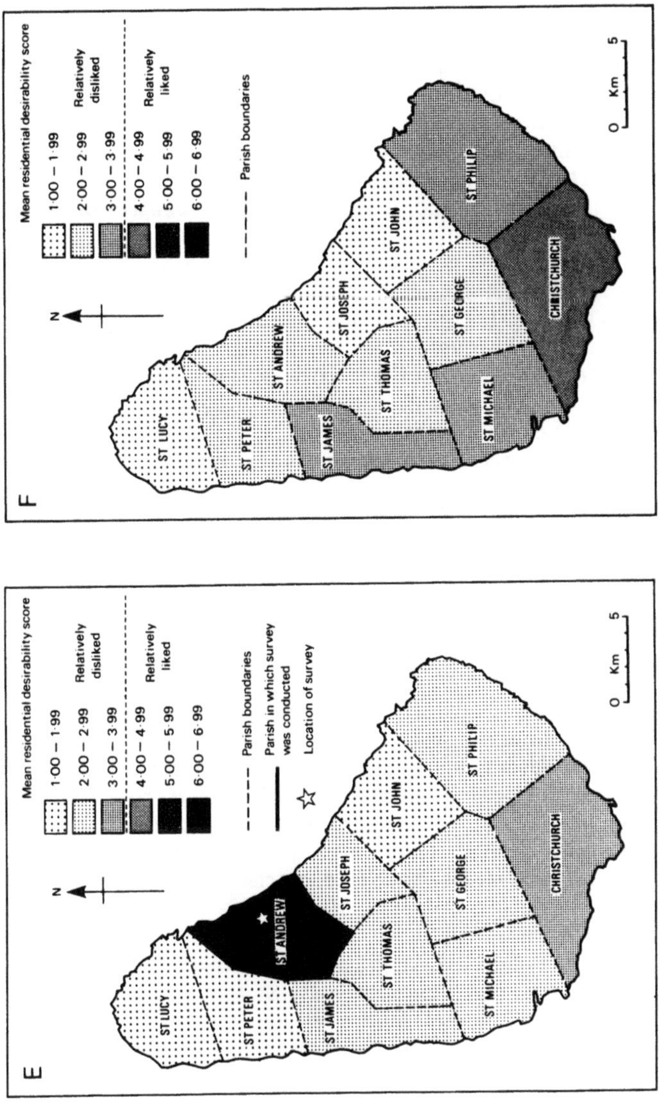

Figure 7.2: Cont.

other hand, the northern parishes are uniformly poorly perceived by Barbadians. Some of these points are summarised on the composite map of mean residential desirability scores recorded for parishes among the entire sample of respondents (Figure 7.2F), although this composite pattern is of course strongly influenced by the preference of respondents for the five survey areas selected.

These visually derived conclusions can be tested more rigorously by correlation and regression analyses. First, the hypothesis of metropolitan dominance of perceived residential desirability levels was examined. The mean residential desirability scores recorded by parishes were correlated with the distance of the centre of gravity of the parish from Bridgetown. If this is done for the entire sample, a clear negative association is recorded (Figure 7.3). Again, the pattern is influenced by the five locations selected for the survey, but even so, Christ Church (3) and St Philip (9) record noticeably high desirability scores. However, the hypothesis of a consistent distance-decay in perceived desirability with distance from the capital clearly breaks down if analysed for the five locational sub-samples (Figure 7.4). In the case of Six Cross Roads, Speightstown and Belleplaine, near to zero correlation coefficients are recorded. Once again, however, the analysis indicates the enhanced perceptions held of the parishes of Christ Church (3), St James (5) and St Philip (9). In fact, as might be expected, residential desirability is better explained by distance from the survey location itself (Figure 7.5). Overall, negative correlations are recorded in all 5 cases, ranging from $r = -0.36$ for Belleplaine to -0.81 for Bridgetown. The residuals from the regression lines indicate the enhanced desirability of Christ Church (3), St James (5) and St Philip (9) yet again.

In question 5 of the interview survey (Appendix A), the respondents were asked to imagine that they had entirely free choice and to state where in Barbados they would most like to live. Preferences stated for each parish other than by its residents were totalled to give an alternative measure of residential desirability. The significance of this method is that it negates the influence of the strong home locality bias shown for the five parishes in which the survey was conducted. The rank ordering of parishes according to this criterion is shown diagrammatically in Figure 7.6. Christ Church predictably appeared as the first ranking parish, followed by the western coastal parish of St James and then jointly by St Philip and St Thomas. Clearly, the main contrast is between the southeastern coastal parishes and the northeastern ones. The capital region of St Michael is perceived in middling-to-poor terms.

In another section of the work a combined index was constructed to reflect the socio-economic standing of the parishes. This was based on variables such as levels of car ownership, the number of high-income workers, agricultural employees, university educated, professional workers and the number of tourist beds (Potter, 1983d). The ranking of parishes according to this socio-economic index is shown alongside their ranking by residential preferences in Table 7.1. Comparison of these two listings allows us to perform a straightforward ordinal type of 'myth map' analysis. The overall relationship

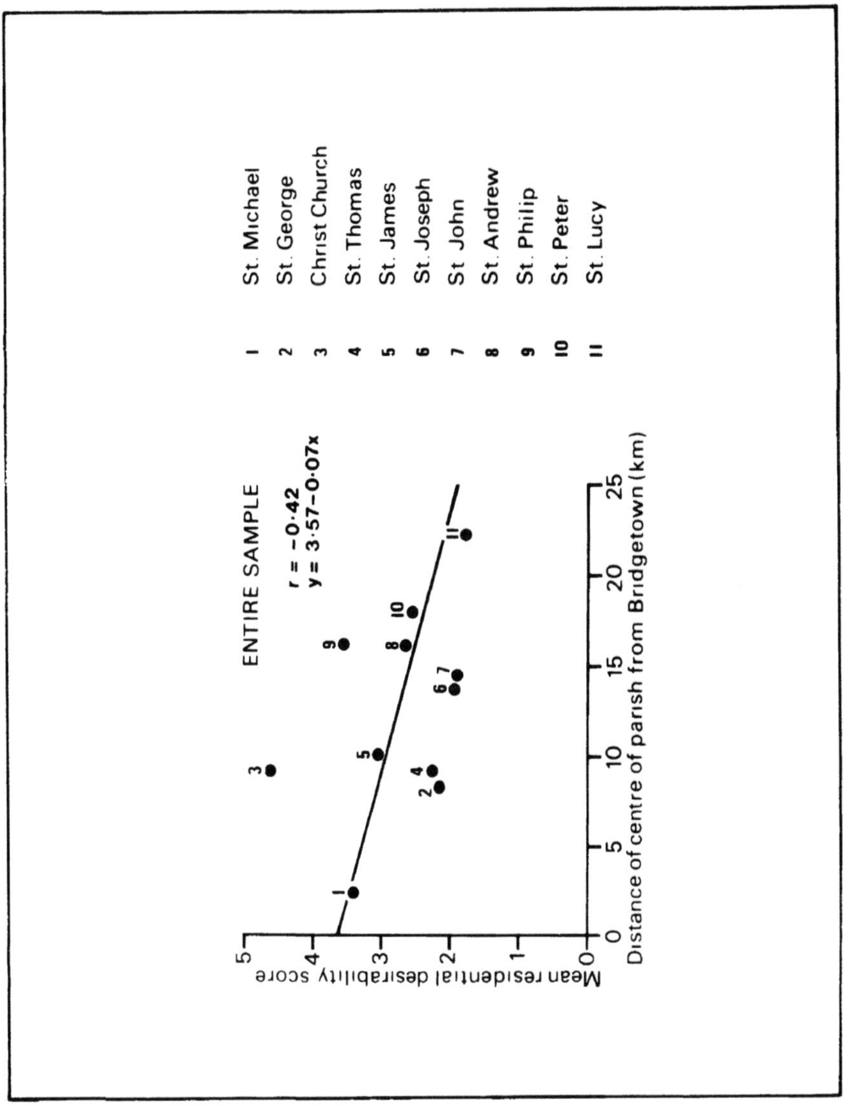

Figure 7.3: Residential desirability against distance from Bridgetown: entire sample

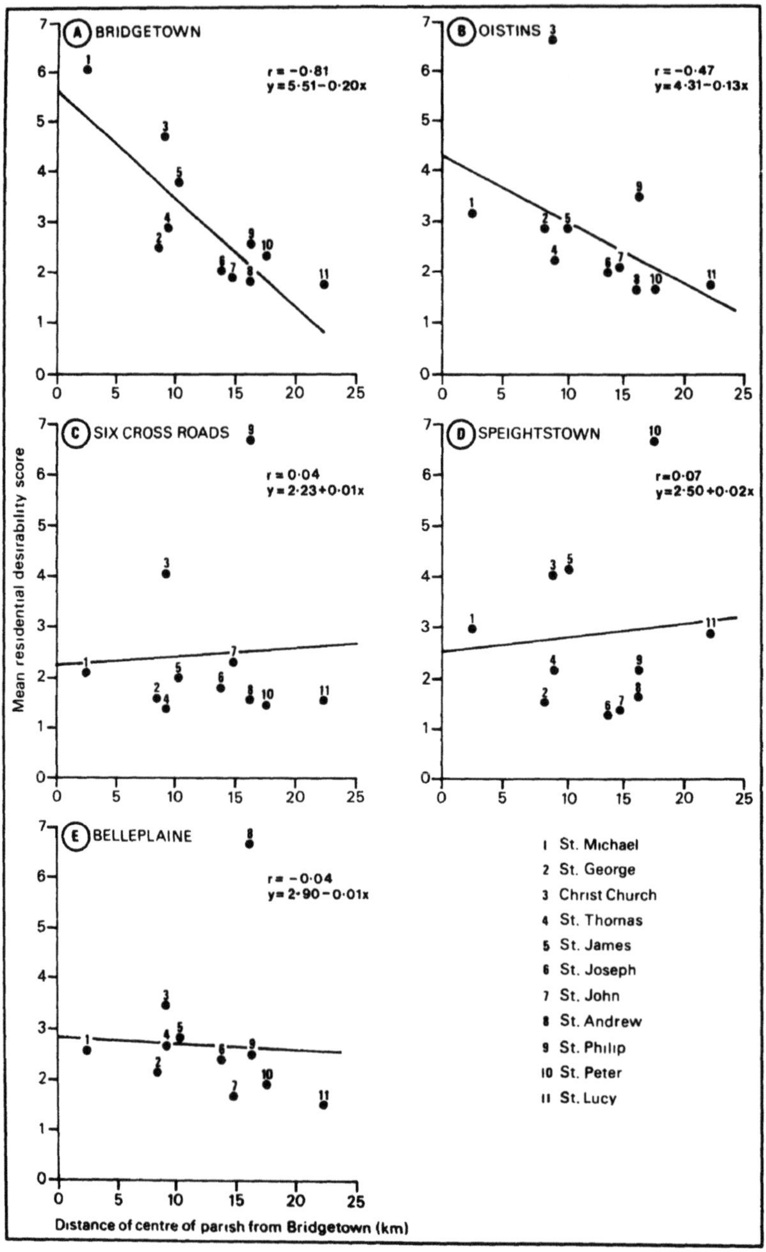

Figure 7.4: Residential desirability against distance from Bridgetown: sub-samples

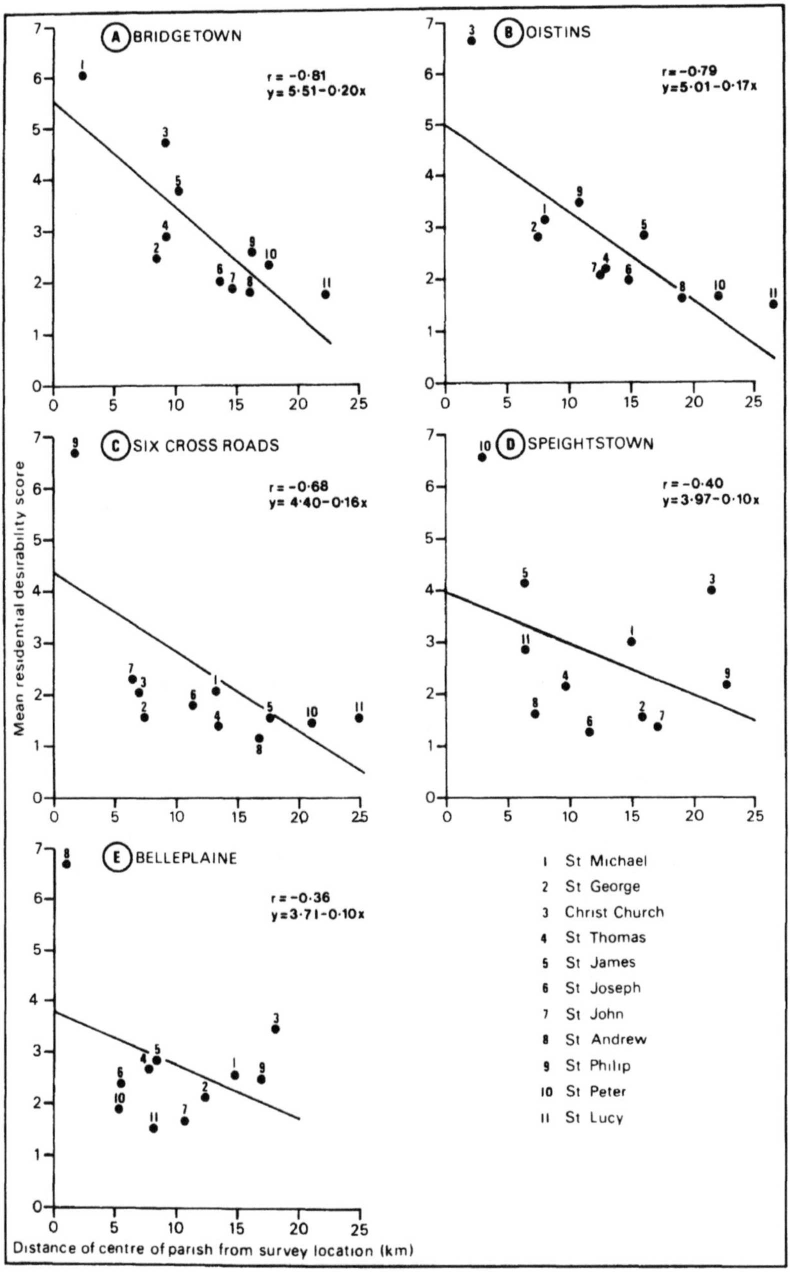

Figure 7.5: Residential desirability against distance from the survey location

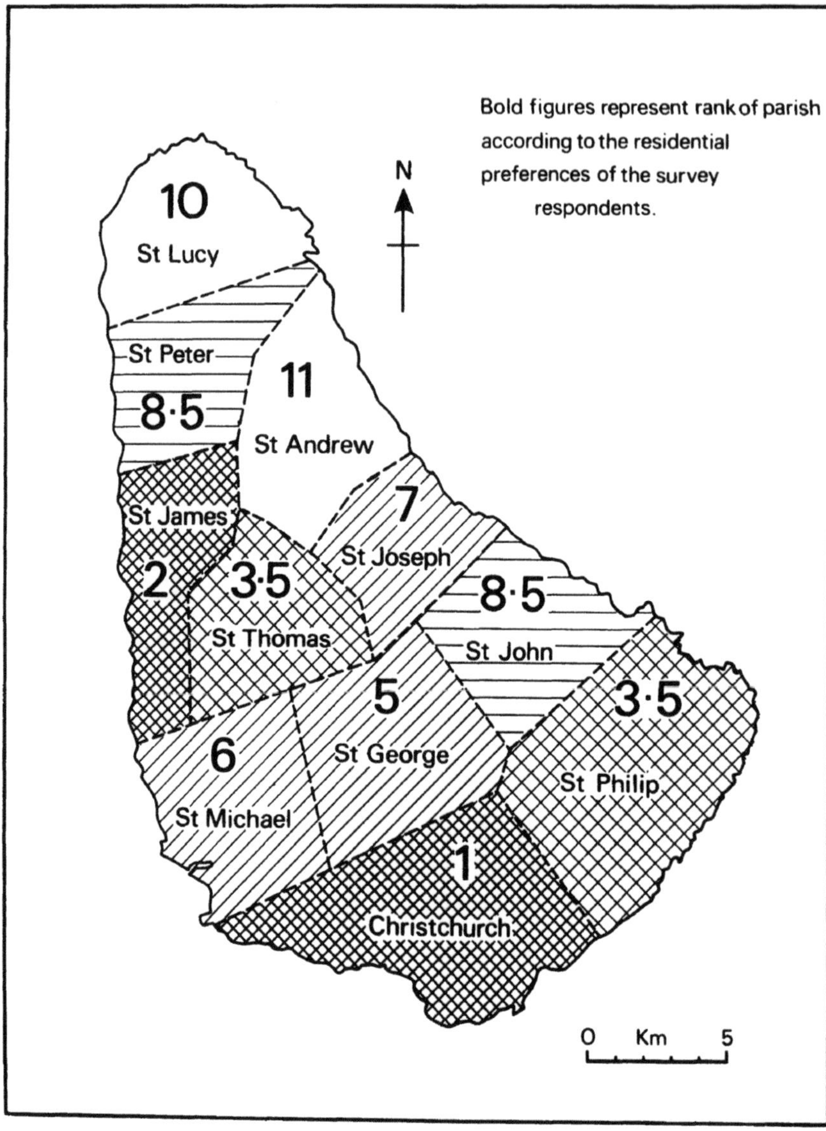

Figure 7.6: Rank of Barbadian parishes according to respondents' residential preferences

Table 7.1: Comparison of the rank order of Barbadian parishes according to residential preferences, socio-economic status and demographic change 1871-1980

Parish	Rank of parishes according to:			
	residential preference	socio-economic index	pop. change 1871-1980	pop. change 1970-1980
Christ Church	1	1	1	2
St. James	2	2	3	1
St. Michael	6	3	2	8
St. Peter	8.5	4	8	7
St. Philip	3.5	5	5	3
St. George	5	6	4	5
St. John	8.5	7	6	9
St. Joseph	7	8	10	11
St. Thomas	3.5	9	7	6
St. Lucy	10	10	9	4
St. Andrew	11	11	11	10

between residential preferences and socio-economic standing is reasonably strong (r_s = 0.70, $p < 0.05$). Thus, Christ Church and St James achieved first and second ranks on both measures, whilst St Lucy and St Andrew occupied the eleventh and twelfth places respectively on both. However, the mis-matches between the rank ordering of parishes according to residential preferences and socio-economic standing are of as much interest. Thus, the metropolitan parish of St Michael, along with St Peter, the parish containing the second largest urban place Speightstown, and St John are all underestimated relative to their socio-economic standing. In contrast, the parishes of St George, St Thomas, St Joseph and St Philip seem to be perceived more favourably than their socio-economic features would suggest. These features provide a good example of the fact that images and preferences and objective characteristics are frequently not the same thing, and therefore it is worth measuring the image as an intervening variable to help explain behavioural decisions such as migrations, and the dynamics of socio-economic and environmental change.

Aspects of perceived residential desirability can also be examined in relation to demographic change. Population change over the 109 year period from 1871 to 1980 is shown for Barbadian parishes in Figure 7.7. During this period, population growth has focused on the urbanised and highly developed west and southern coastal strip made up by St Michael, St James and Christ Church. In contrast, the eastern coastal parishes of St Andrew and St Joseph have experienced absolute population losses during this period. The overall rank correlation existing between present day residential preferences for parishes among the respondents and their long run population change is highly significant (r_s = +0.77, $p < 0.01$). Comparison of the rankings shown in Table 7.1, however, indicates that St Michael, St Peter, St George, St John and St Lucy all record low residential preferences relative to their long term population change, implying that they are less desirable now than they were in the past. As might be expected, therefore, the present day residential desirability expressed for parishes is a little less closely related to recent population change from 1970 to 1980, although the relationship is still significant (r_s = + 0.70, $p < 0.05$). During this period, St James and Christ Church have experienced the major gains in population (Figure 7.8).

Further insights were provided by a set of questions designed to examine respondents' appraisive perceptions of Bridgetown and of planning matters in general (see Appendix A, questions 7 to 9 and 17-26). These simple closed questions indicated that there was a clear consensus (69.09 per cent) among respondents from all areas in Barbados that Bridgetown does not afford a good environment in which to live. Although, saliently, an overwhelming 98.07 per cent did not regard Bridgetown as being too big, the majority felt it to be both too crowded with people (76.32 per cent) and traffic (67.15 per cent), as well as suffering from too much noise (48.79 per cent) and air pollution (60.39 per cent). However, nationals did not consider the pace of life too fast in the capital (66.18 per cent). The answers to these questions tend to suggest that in the public's percep-

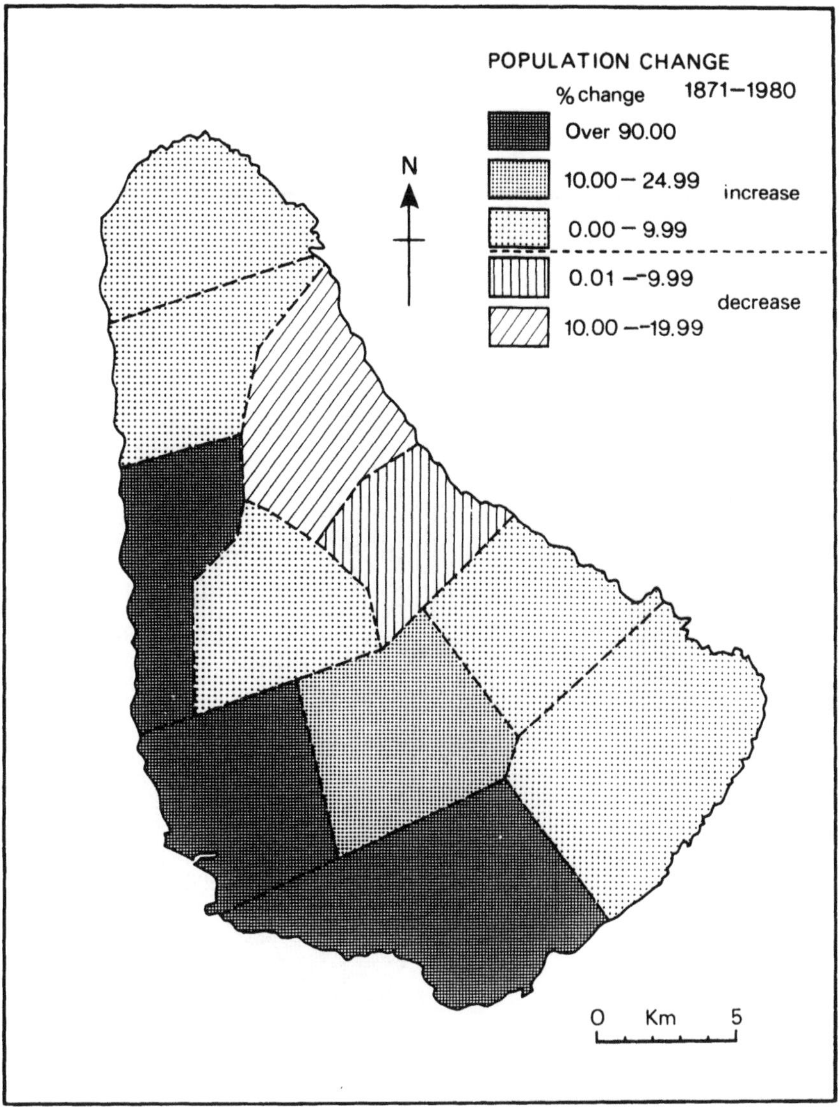

Figure 7.7: Population change 1871-1980 in Barbados

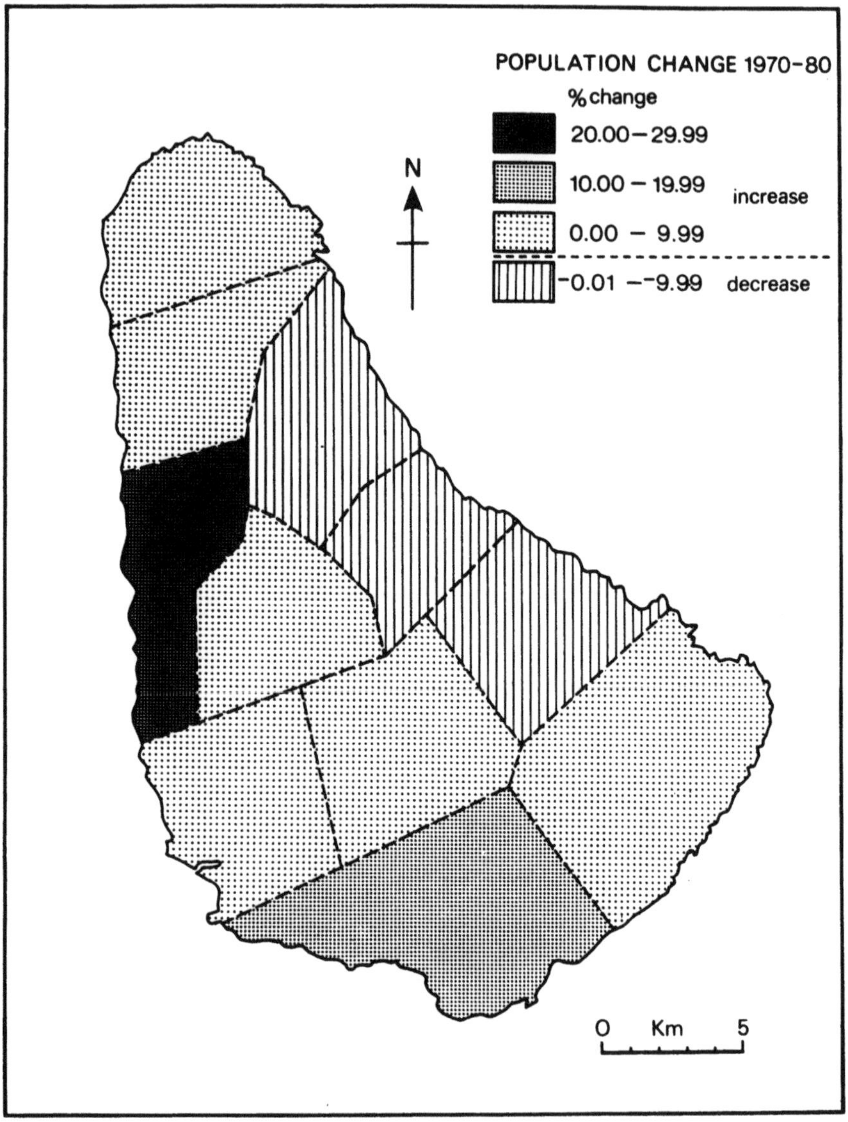

Figure 7.8: Population change 1970-1980 in Barbados by parish

tions, rather than Bridgetown being too large in absolute terms, the real problem is one of inadequate intra-urban design and ensuing congestion.

Questions concerning the respondents' attitudes to planning matters were also instructive. In summary, these revealed that only 28.5 per cent of those interviewed had read or even heard about the Physical Development Plan for Barbados and its proposals. This is perhaps not too surprising given the negligible role of public participation involved in the original plan-making sequence. More encouragingly, however, three-quarters of the sample reported that they were aware of the government's commitment to develop new and expanded towns and as high a proportion as 90.34 per cent considered it to be a good idea to decentralise people and activities away from Bridgetown. Finally, the respondents were asked to state where they felt that new urban developments might best be located in Barbados. The responses that were locatable by parish are shown mapped in Figure 7.9. A clear preference was revealed for the two southeastern parishes of Christ Church and St Philip, along with the St Peter (Speightstown) region in the north (see Potter, 1983f).

All of these findings point to the low desirability of the north and eastern areas of Barbados and the enhanced desirability of the south and west. It has already been mentioned in Chapter 6 that a limited analysis of designative mental maps of Barbados was carried out with four age groups of school children (see Potter and Wilson, 1983). For the maps taken as a whole, settlements predominated, accounting for 32.15 per cent of all features depicted. Taking features shown of all types, the southern coastal region was associated with the largest number of features cited. The overall frequency of depiction of settlements was positively related to settlement size ($r_s = +0.55$). However, interestingly, certain places showed low imageability relative to their size, especially Belleplaine, the principal settlement of St Andrew, and also the main settlements of the parishes of St George, St James and St Lucy. Such findings point to clear weaknesses in the settlement system. Interestingly, all of these places were designated as growth points in the 1970 Physical Development Plan, showing how much has yet to be achieved to enhance their imageability in the public's eyes.

In conclusion, it is argued that taken together results such as these are of some salience to urban and regional planners. First, the analysis of residential preferences indicates quite clearly the marked regional anomalies that exist within even so small a country and the challenges they present to planners and politicians. In Barbados, a strong west coast perceptual bias is clearly apparent which matches the perpetuation of long-established socio-economic differentials. Today, residential preferences clearly lie in favour of the coastal parishes of Christ Church, St James and St Philip, rather than the metropolitan area of St Michael. The Bridgetown-St Michael area is seen as having many environmental problems, although it is not regarded by respondents as being too big in absolute terms. Such cognitions dovetail with demographic changes from 1970 to 1980, a period during which the major gains in population

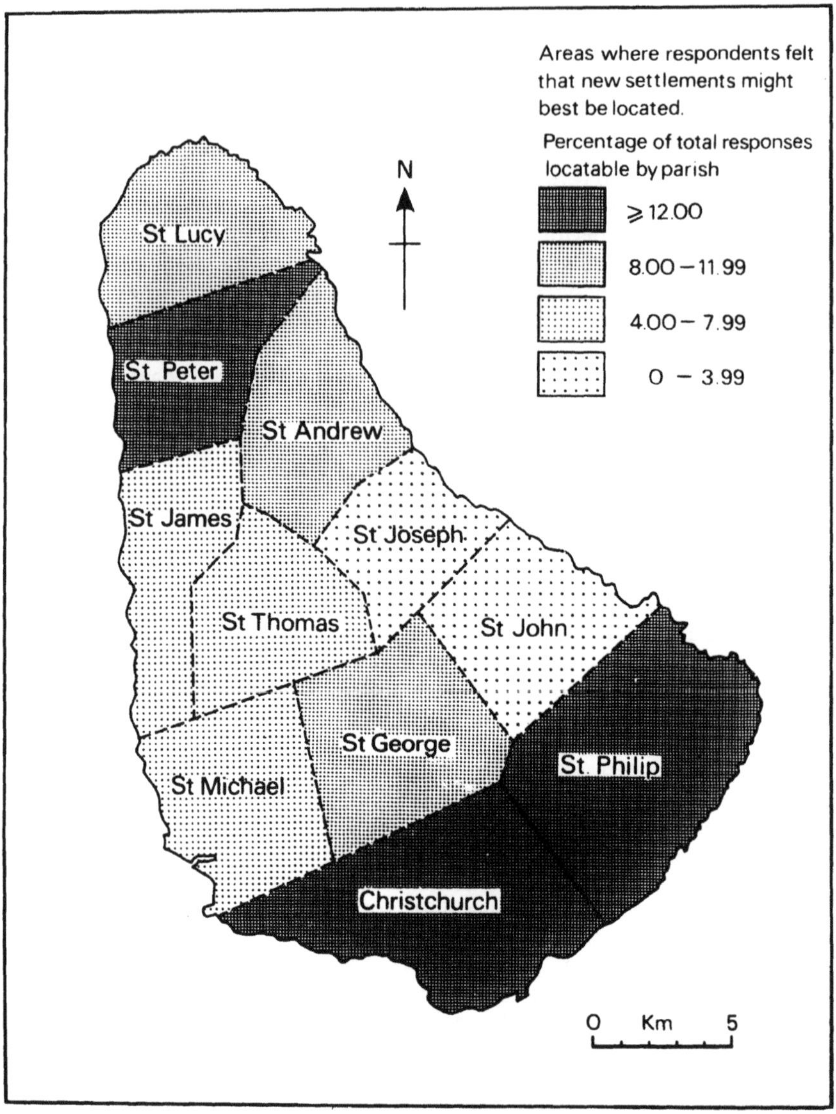

Figure 7.9: The areas where respondents felt that new settlements might best be located in Barbados

occurred in St James, Christ Church and St Philip in that order. In direct contrast, the metropolitan parish of St Michael experienced a relative decline in population (Potter, 1983b).

It is argued that a widening out of population, economic activities and social infrastructure is needed in Barbados and further, the research also showed that the public saw the need for this. However, it is suggested on the basis of the surveys carried out that the 1970 Physical Development Plan was over-zealous in suggesting the adoption of an island-wide central place hierarchical-type settlement pattern. It has already been suggested by the present author that the 1970 Plan is an example of the inappropriate wholesale application of an advanced world model to a Third World problem. In particular, it seemed to take little or no account of the physical geographical realities of the country, especially the distribution of high quality agricultural land and the principal water catchment areas (Potter, 1983b, 1984c). Clearly, however, great efforts will be needed to impart a fresh impetus to the development of the northern and eastern areas of Barbados and a programme of decentralisation into a limited number of key sites such as Oistins, Speightstown, Six Cross Roads and Belleplaine seems appropriate. Such a pattern of concentrated dispersion is justified on social and welfare grounds rather than on purely economic ones.

It is highly desirable for planners to consider how residents feel about different areas of their national space, and it is hoped that the Barbados example presented here has shown that such studies provide a potentially useful data-base for analysis, and also a route for the input of public views into the decision-making process. Saliently, it has been demonstrated once again that although space preferences are closely related to the objective characteristics of areas, they are not synonymous, making the study of perceptions an important avenue of investigation.

SPATIAL COGNITIONS AND PHYSICAL PLANNING ISSUES IN TRINIDAD AND TOBAGO

An essentially similar survey of residents' spatial cognitions of planning-related issues was conducted in Trinidad and Tobago by the present author in 1982. The interview schedule employed is reproduced in full in Appendix B and consisted of a total of 19 major questions. The principal questions posed were very similar to those asked in the Barbados and St Lucia surveys, so that a comparative data set has been derived (see Appendices A to C). The main results of the Trinidad research have been presented elsewhere (Potter, 1983c). In the present account, only a few illustrative examples are given and apart from the space preference survey, emphasis is placed on work of a type not previously described in relation to Barbados.

The background to urbanisation and planning in Trinidad has been set out in Chapters 3 and 4. It may be recalled that Trinidad, the southernmost island of the Caribbean chain, with an area of 5,128 km^2, had a population of 1,114,800 in 1977. It is now very much a middle income developing country, with an economy strongly based on the oil industry. The most notable feature, acknowledged by planners and development experts is the extremely uneven distribution of population and settlement, with a concentration on the northwestern and western coastal regions along the Gulf of Paria (Figure 7.10). The capital, Port of Spain, with a population of around 250,000, occupies a highly constricted site and has developed on a linear basis as far east as Arima. This wider capital region houses a population of some 506,400, that is approximately 44 per cent of the national total. The second ranking town, San Fernando, with a population of 48,950 is also located on the western coast. Thus, great inequalities characterise the country and these have tended to increase with recent economic growth. In Port of Spain there are problems of congestion and poor housing, especially in the eastern environs of the city (Conway, 1976, 1981; Conway and Brown, 1980). The planning system of Trinidad was discussed in Chapter 4. The 1978 Physical Development Plan (Town and Country Planning Division, Trinidad, 1978) specifically acknowledged the need to replan the overall settlement system. Interestingly, four alternative strategies were considered and one of concentrated dispersal was adopted. Thereby, it was decided that decentralisation from the capital should be achieved by channelling new growth to the year 2000 into four main areas: Point Lisas/Couva, Point Fortin/La Brea, Sangre Grande, and Guayaguayare (see Figure 7.10, and also Figure 4.10). In addition, six rural development areas were designated in the plan (Potter, 1983c).

In the same way as the Barbados survey, the interview endeavoured to isolate and investigate the spatial cognitions of Trinidadians that might have some direct bearing on existing environmental conditions and plans. Interviews were carried out at four settlements: Port of Spain, San Fernando, Sangre Grande and Mayaro/Guayaguayare (Figure 7.10). Two interviewers were employed and a day was spent interviewing at each location. A total of 148 interviews were completed.

As an important part of the survey, respondents' appraisive perceptions were examined by means of open-ended questions. For example, in question 2, the respondents were requested to state the first three things they thought of in relation to Trinidad and Tobago. A total of 267 responses were recorded at an average of 1.8 per respondent. Content analysis grouped these into 92 different responses. Within the 20 most frequently cited factors, there was a preponderance of negative reactions. In particular, these related to inadequacies of infrastructure and public service/utility provision, notably housing problems (mentioned by 10.14 per cent of the respondents), transport problems (5.41 per cent), water problems (5.41 per cent), education and schooling difficulties (4.73 per cent). On the other hand, positive responses had much more to do with the perceived character of the population, the favourable natural environment and the easy-going way of life.

Given the primacy of Port of Spain, further open-ended questions were used to probe the respondents' appraisive perceptions of the capital. Thus, in questions 6 and 7 they were asked to cite what they

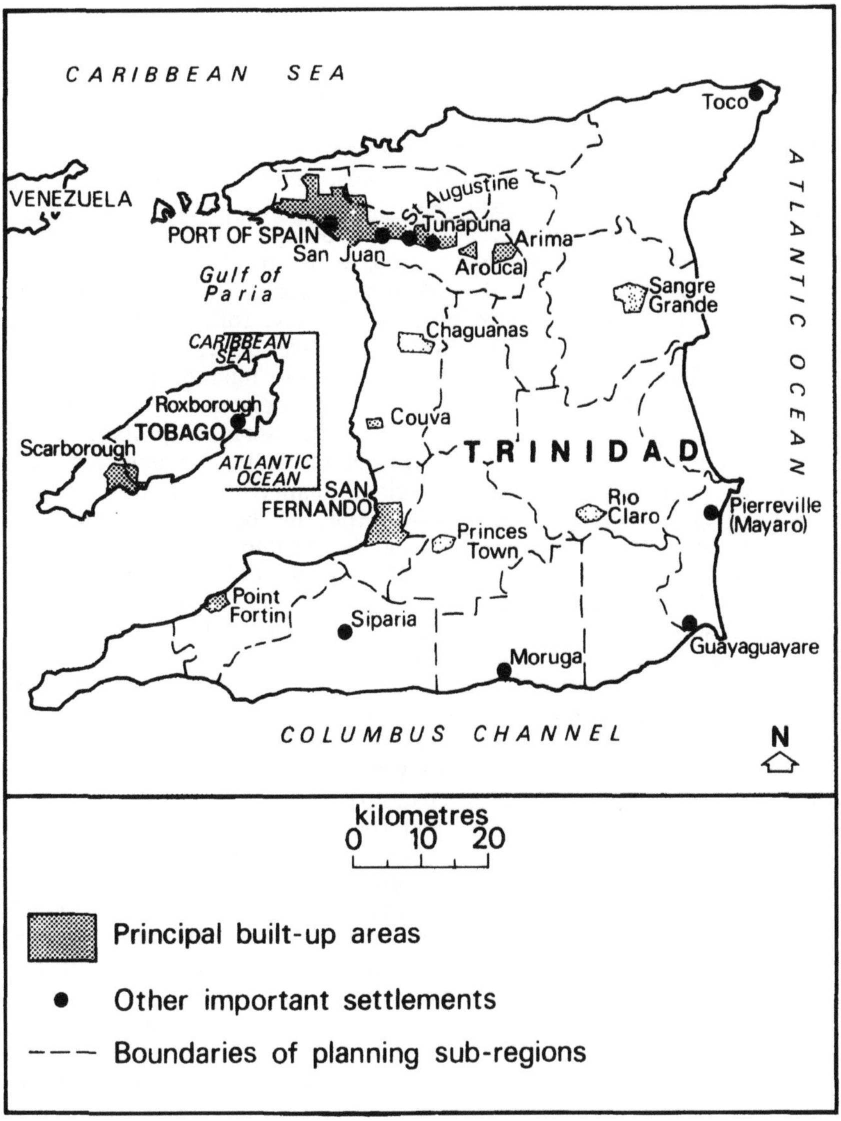

Figure 7.10: Trinidad-principal settlements and survey areas

regarded as the three best and three worst things about Port of Spain. Although naturally enough some highly idiosyncratic responses were elicited, broad areas of consensus also appeared. The 20 most frequently cited views of a favourable and unfavourable nature are listed in Tables 7.2 and 7.3 respectively. Considering first the positive reactions to the capital, its role as a central place and employment node is clearly highlighted by the respondents, with its shopping facilities being mentioned by as high a proportion as 34.46 per cent of them, followed by its commercial importance, accessibility, entertainment facilities, plentiful goods, lively atmosphere and job openings. In all, 46 separate positive features were enumerated at an average of 1.36 per respondent. Perhaps significantly, slightly more negative attributes were cited, a total of 57 different ones at a mean response rate of 1.57 per person. The most frequently mentioned factor was the dirtiness of the city, followed by its traffic and congestion problems and the occurrence of crime, robbery and violence. Other factors mentioned by a large number of respondents included high pollution levels, flooding and beggars. Together these perceptions provide a fair impression of the sorts of problems faced by a developing world city such as Port of Spain, for example, those posed by water supply, overcrowding, shanty towns and squatters and poverty. Thereby, it was shown that in the eyes of the public, Port of Spain is far from being a favourable social and physical environment, but does on the other hand offer a multitude of commercial and business opportunities. This impression was borne out by responses to other appraisive questions asked in the survey concerning Port of Spain (Potter, 1983c).

In order to investigate national space preference patterns the two customary questions were asked. The first gave complete freedom of choice in asking respondents where exactly in Trinidad and Tobago they would most like to live if they were completely free. Secondly, they were asked to assess the desirability of each of the 21 constituent planning sub-regions of Trinidad and Tobago. The analysis of responses to the first question is shown in Figure 7.11. Virtually all of the replies were in terms of named settlements and the aggregate responses for particular places have been mapped. For the Port of Spain sample, a very strong preference is exhibited for the capital itself and the adjacent settlements of Arima and San Juan (Figure 7.11a). In fact, 56.41 per cent of the respondents named the capital as their most preferred settlement. The San Fernando sample's preferences showed a somewhat less strong home bias (55.22 per cent) and locations throughout the rest of the island were mentioned, albeit by only a relatively small number of respondents (Figure 7.11b). The image of the residents of Sangre Grande is exclusively confined to the northern half of Trinidad (Figure 7.11c). Clearly, the view is highly parochial with 60 per cent wanting to remain in the home locality. Outside this area, the respondents are noticeably keen on the eastern tracts of the capital region (St Augustine, Tunapuna, Arima, Arouca), but significantly, not the capital itself. This reflects the pattern of eastward decentralisation that has been occurring in Port of Spain for some time. The Mayaro-Guayaguayare sample also shows a high

Table 7.2 : The respondents' views as to the best features of Port of Spain, Trinidad.

No. Feature	No. of respondents	% respondents
1.Shopping facilities/ centres/stores	51	34.46
2.Commerce and business places	17	11.49
3.Convenient/accessible	15	10.14
4.Entertainment facilities/night life	9	6.08
5.Can get what you want	9	6.08
6.Lively/plenty of activity	7	4.73
7.Job opportunities/near to jobs	6	4.05
8.Parks and recreational facilities	5	3.38
9.Lovely place/beautiful	5	3.38
10.Prices cheaper than elsewhere	5	3.38
11.Savannah	5	3.38
12.Industrial environment	4	2.70
13.National festivals and carnivals held there	4	2.70
14.Few transport problems	4	2.70
15.Many buildings/new buildings	4	2.70
16.Health facilities/ hospitals	3	2.03
17.Meet a lot of people	3	2.03
18.The zoo	3	2.03
19.The Botanical Gardens	3	2.03
20.Main port/shipping area	3	2.03

Table 7.3 : The respondents' views as to the worst features of Port of Spain, Trinidad.

No. Feature	No. of respondents	% respondents
1.Dirty/stinks/too much garbage	44	29.73
2.Traffic/transport problems	28	18.92.
3.Too crowded/ congested	26	17.57
4.Crime/robbery/ violence	21	14.19
5.Pollution/no fresh air	9	6.08
6.Flooding/poor drainage	8	5.41
7.Beggars and madmen in the streets	8	5.41
8.Rastas	6	4.05
9.Sanitation	6	4.05
10.Street traders/vendors	5	3.38
11.Too hot	5	3.38
12.Water shortages/ problems	4	2.70
13.Buildings over-crowded	4	2.70
14.Streets too narrow/ poor repair	3	2.03
15.Shanty towns/squatters	3	2.03
16.Suburbanisation/ decentralisation	3	2.03
17.Lack of security/ safety	2	1.35
18.Poor environment	2	1.35
19.Noise	2	1.35
20.People not friendly/ not loving	2	1.35

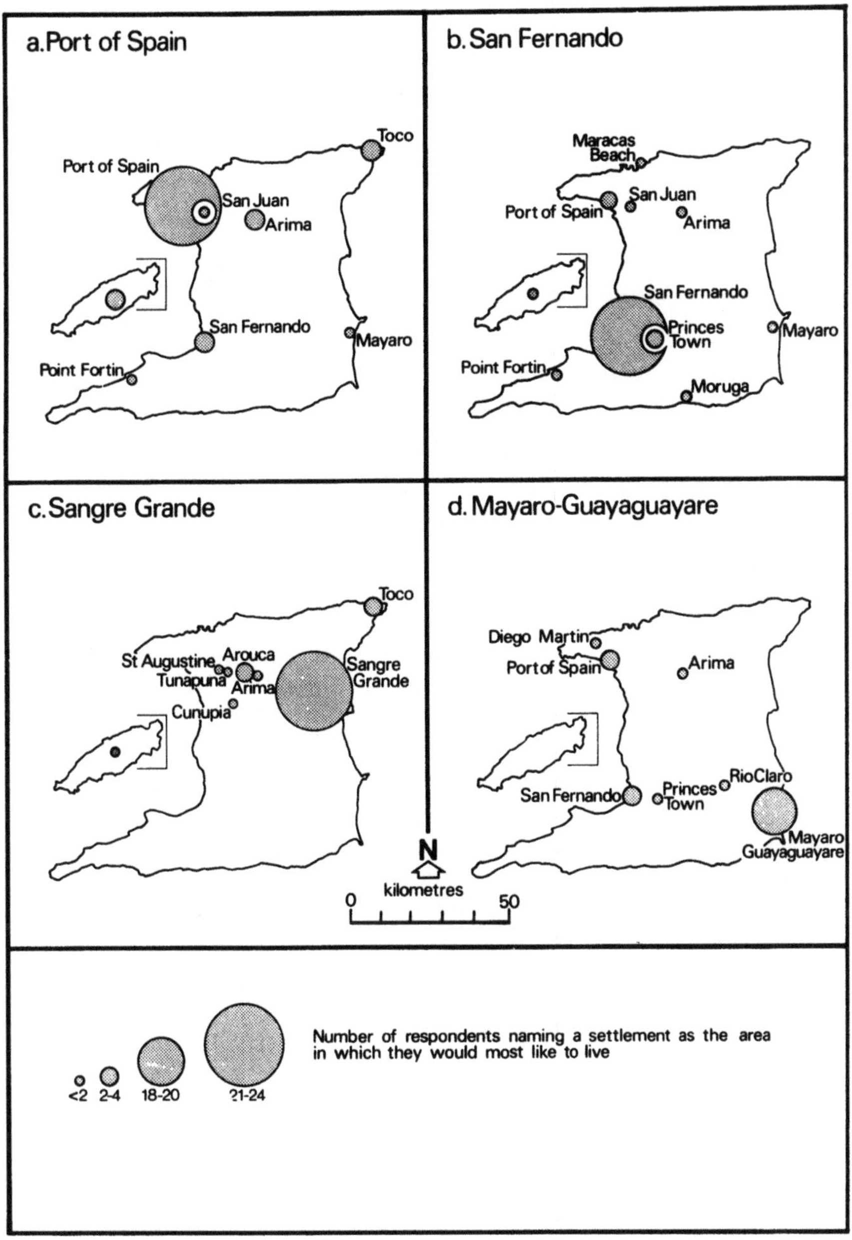

Figure 7.11: Residential preference maps of Trinidad and Tobago

degree of home area bias (62.07 per cent) and a marked corridor patterning of space preferences exists across to the San Fernando urban area via Princes Town and Rio Claro (Figure 7.11a). Port of Spain, Diego Martin and Arima are all mentioned in the capital region. The occurrence of Arima as the only settlement common to all the space preference maps is salient and clearly reflects the popularity of the eastern part of the capital region.

The mean residential desirability scores attributed to the various planning sub-regions by the sample respondents are shown mapped in Figure 7.12. The view from Port of Spain shown in Figure 7.12a stresses the desirability of the broad Capital region extending from Chaguaramas in the west to rural Arima in the east, along with south-west Tobago and the San Fernando urban region in the south. The rural areas of Western and Eastern Caroni and Nariva are very poorly perceived. The residents of San Fernando exhibit a customary strong preference for their home locality and the nearby environs of Princes Town and Siparia (Figure 7.12b). Once more, Tobago is well regarded, whilst Western Caroni, Nariva and Aripo-Toco are held in low esteem. The view of residential desirability from Sangre Grande (Figure 7.12c) shows that a premium is placed on the broad area between the home locality and the Capital region. More distant areas that are held in high regard are again Tobago, and also Mayaro-Guayaguayare in the southeast. Finally, turning to the Mayaro sample, a more complex configuration of residential desirability is revealed (Figure 7.12d). The desirability of the home region is clear, but this is the only example where there is also a strong preference for the Port of Spain/Arima area. There is also a dual corridor effect of enhanced residential desirability linking Mayaro with both the Port of Spain and San Fernando urban areas.

Mean residential desirability scores achieved by the sub-regions are shown plotted against distance from the respective survey locations in Figure 7.13. In each case, the survey sub-region stands out by virtue of its enhanced desirability, although as already observed, this is somewhat less marked in the case of Port of Spain and Mayaro-Guayaguayare. The generalised distance decay or negative relation between perceived desirability and distance is affirmed by the four regression lines and negative correlation coefficients, ranging from $r = -0.15$ for San Fernando to $r = -0.36$ for Mayaro-Guayaguayare. Residuals from the computed regression lines are revealing. The sub-regions of northeast and southwest Tobago (points 20 and 21) record positive residuals in the perception maps of the four sub-samples, whilst Port of Spain (1), Rural Arima (7), San Fernando (10), Western Mayaro (18) and Mayaro-Guayaguayare (19) each record positive residuals on three occasions. In contrast, the North coast region (4), Western Caroni (8), Eastern Caroni (9), Rio Claro (12), Nariva (13), Cedros (14) and Moruga (17) record consistent negative residuals. The implication is that a pronounced urban versus rural dichotomy forms a strong component in people's spatial cognitions of Trinidad. Positive residuals have been mapped in Figure 7.14a-d. In each case the local region and contiguous ones fare well in the eyes of the respondents, thereby showing the operation of a strong

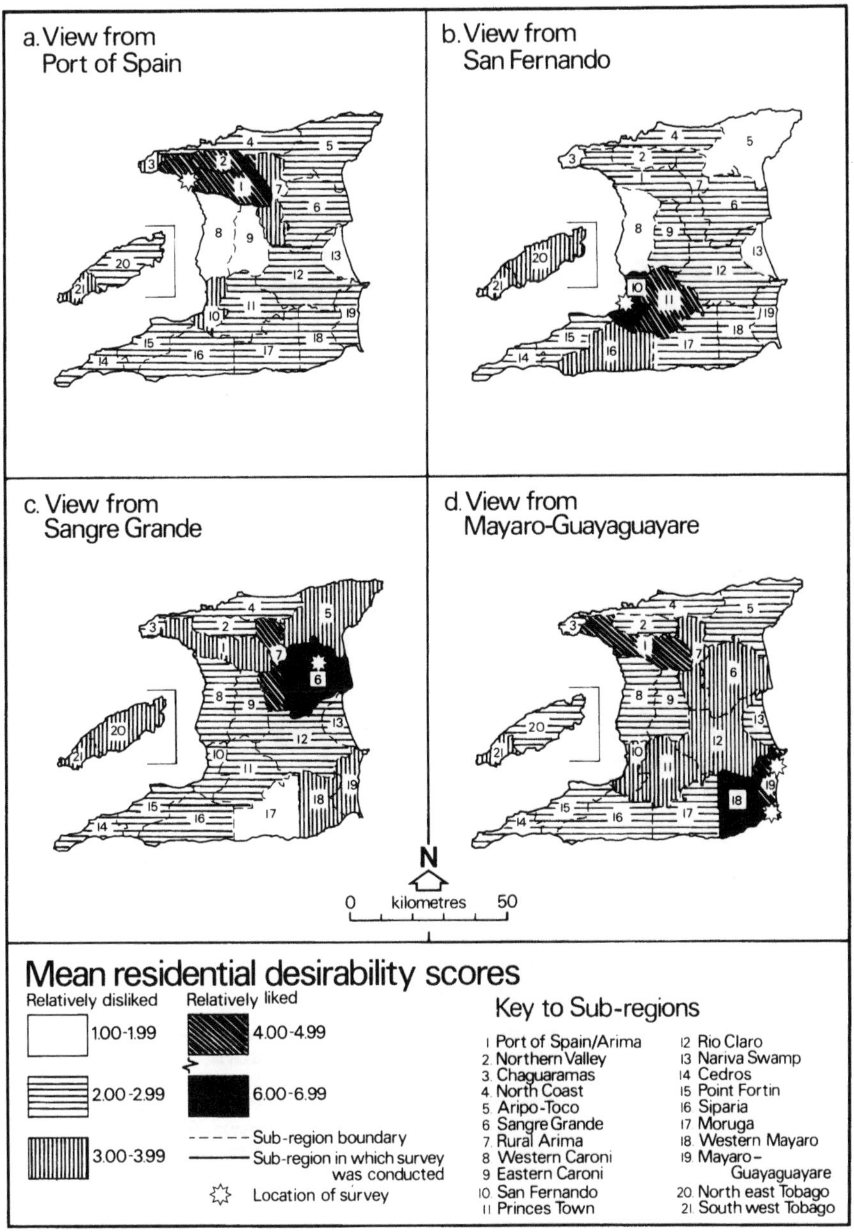

Figure 7.12: Residential desirability maps of Trinidad and Tobago

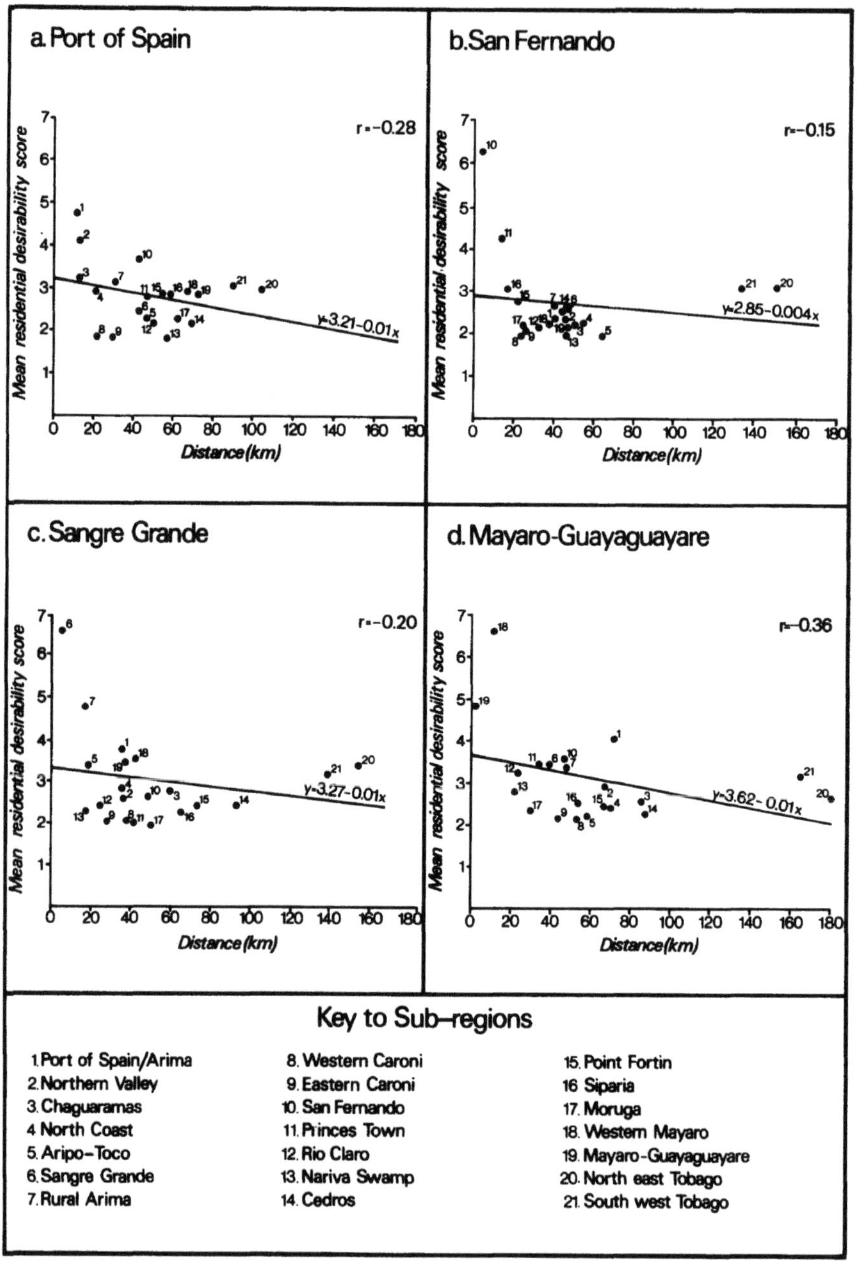

Figure 7.13: Residential desirability against distance from the survey location

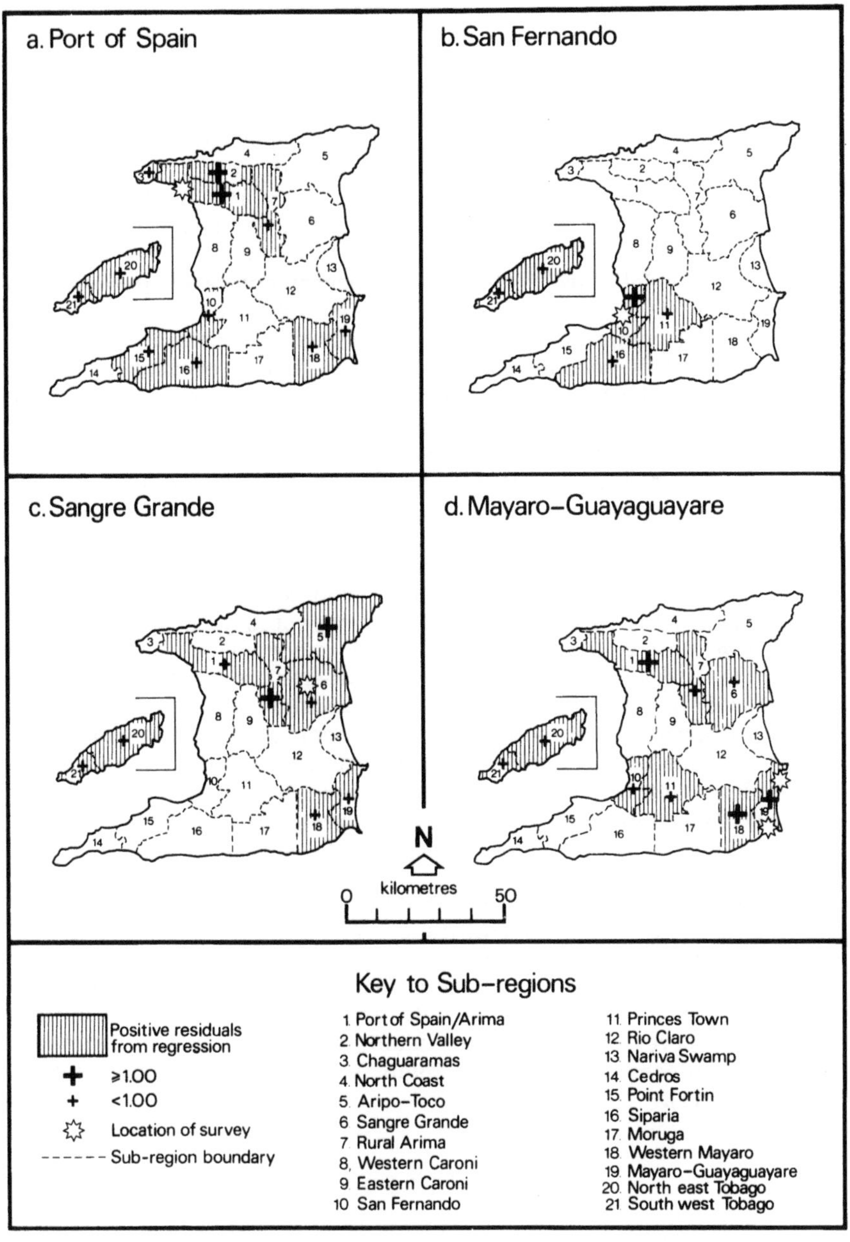

Figure 7.14: Positive residuals from regression of desirability on distance

neighbourhood effect. In each map save for that pertaining to San Fernando, the west-east Capital region stands out clearly as a favourably perceived zone. Further, the two constituent sub-regions of Tobago appear as positive residuals in all four maps, presumably reflecting the island's scenic beauty and tourist connotations. Of similar interest is the fact that the southeastern sub-regions of Trinidad, that is Western Mayaro and Mayaro-Guayaguayare are favourably perceived by all but the residents of San Fernando. This bodes well for efforts to promote a growth centre or new town in this area in the future. Further, the San Fernando area does well in all of the maps, save for that relating to the residents of Sangre Grande. The areas which are consistently negatively evaluated by the interviewees show a close correspondence with four of the six areas designated for comprehensive rural development in the 1978 plan.

Hopefully this second example of the analysis of citizens' spatial cognitions of their national territory will have again exemplified the potential utility of such investigations in environmental policy decision-making. Certainly, the survey indicated that the southwestern region of the country looks a popular one for development, as to a lesser extent do the other designated growth areas of Couva/Pt Lisas and Sangre Grande. In contrast, however, the projected growth centre of Point Fortin-La Brea does not appear to possess very high standing in the public's view. The problems faced by the main rural areas of the country are reflected in their low desirability in the respondents' eyes. The survey essentially shows the importance of being on the map, both literally and figuratively. Clearly, space preferences lie in favour of the principal urban and industrial zones of Trinidad and reflect disparities in incomes, employment opportunities and other facilities. The attraction of the eastern part of the capital region needs close monitoring and careful land use planning. Some of these features are reminiscent of Shiva Naipaul's description of the birthplace of the leading character in his novel The Chip-Chip Gatherers:

> "He had been born in the Settlement, which was not deemed worthy of mention on even the larger maps of the island. The map-makers did not acknowledge its existence and it can hardly have existed in the minds of its unfortunate inhabitants. The eye shied away from focusing on the mean huts and houses clinging despairingly to the curves of the narrow main road which wound its way to distant places like Port-of-Spain and San Fernando" (Naipaul, 1973, p. 11).

CONCLUDING COMMENTS

The employment of spatial cognition and perception surveys, looking at appraisive, designative and prescriptive components of environmental imagery is recommended for planning-related research at both the inter-urban and intra-urban scales. It is hoped that, despite their limited scope, the two illustrative examples presented here have given at least some impression of the utility of such research, both as a form of early reconnaissance work at the survey stage of the planning sequence and also in connection with the subsequent evaluation of alternative plans. At both of these stages, appropriately handled and locally sensitive surveys can be effective in bringing the public's views into the planning process, a particularly important function in many poor countries where, as noted in Chapter 5, other means of stimulating public involvement may well be somewhat less appropriate. Further, the costs of implementing such surveys is potentially extremely low and can be measured in a few person-work days. Additionally, information collected by such surveys can act as a useful surrogate for census data, or at least a very useful complement to it. It has also been shown how the relationships and mis-matches existing between perceptual data, demographic change and socio-economic conditions can reveal much about patterns of migration and urban development. Such information is also highly salient in confronting the professional planner with the views of the public. This attribute, coupled with the relative ease with which such surveys may be implemented and analysed makes them appear as a natural "appropriate planning technology" for many Third World countries.

As is argued elsewhere in this book, when reduced to fundamentals, planning essentially involves the premeditated changing of environmental conditions and images, so that the study of the latter stands as an essential part of the planning sequence (Potter, 1983c, 1984c). This is compounded by the fact that urbanisation is itself closely related to spatial perceptions of structural conditions. Thus, perception-related studies should not be dismissed as being a sophisticated digression, a status quo indulgence or a form of reductionism - they are basic and highly germane to the urban planning process in Third World countries.

Chapter 8

Conclusions

It is reliably estimated that at the present time, in the region of 50,000 people migrate to urban areas within Third World territories every day. It is as a result of this rapid cityward migration and the accompanying high rates of natural increase of urban population that the total number of urban dwellers in Third World countries is projected to double between 1980 and 2000. This absolute urban increase, from 1,045 millions to 2,080 millions will entail the provision of homes, jobs and services for the addition of the equivalent of almost one quarter of the world's existing population. The scale of the problems that this will involve is immense. In late 1983, under the title "An age of nightmare cities: flood tides of humanity will create mammoth urban problems for the Third World", the magazine Newsweek presented the following hypothetical scenario:

> It is a sweltering afternoon in the year 2000, in the biggest city ever seen on earth. Twenty-eight million people swarm about an 80-mile-wide mass of smoky slums, surrounding walled-in, high-rise islands of power and wealth. Half the city's work force is unemployed, most of the rich have fled and many of the poor have never even seen downtown. In a nameless, open-sewer shanty-town, the victims of yet another cholera epidemic are dying slowly, without any medical attention. Across the town, the water truck fails to arrive for the third straight day; police move in with tear gas to quell one more desultory riot. And at a score of gritty plazas around the city, groaning buses from the parched country-side empty a thousand more hungry peasants into what they think is their city of hope (Newsweek, 1983, p. 26).

Despite some of the crude inferences presented here, it is probably fair to argue as the article does, that this could easily be a description of Cairo, Jakarta or Mexico in the year 2000. It is estimated, for example, that by the turn of the century, the latter will be the largest city in the world, with a total population approaching 31 millions. The poor housing, unemployment and lack of water, sanitation and medical services described in this extract are already characteristic of so many Third World cities. Similarly, without radical measures, it seems highly probable that in many countries, the existing marked gross income inequalities will continue to increase, and that urban conditions will worsen; perhaps even resulting in civil upheaval, violence and disorder, despite the traditional conservatism and law-abiding nature of the urban poor.

In this concluding chapter, little in the way of entirely new material is introduced. Rather, an effort is made to draw together some of the principal themes that have been outlined previously in the present volume. Two main approaches are followed. First, a consideration of the scales at which the processes and problems of Third World urbanisation can be considered is provided. From a primarily geographical viewpoint, a basic distinction must be made between aspects of social organisation and those of spatial organisation. It will be argued, as it has already been in several places, that many of the processes giving rise to the character of present-day urbanisation are macro-societal rather than spatial. This type of consideration also involves the fundamental point as to whether emphasis in analysis and policy-making should be placed on the individual and aggregates thereof, or alternatively, on the entire society. This is a complex and important issue and it is argued here that these approaches must increasingly be seen as complementary and not mutually exclusive. In the second part of the chapter, these issues are briefly treated in relation to the four main topics of concern in this book: the process of Third World urbanisation, urban planning responses, and the respective roles of public participation and spatial perceptions.

SOCIETAL ORGANISATION AND SPATIAL ORGANISATION IN THE STUDY OF THIRD WORLD URBANISATION

It is an obvious truism to stress that without the genuine will to improve the lot of the urban poor, little or no progress is likely to be made in particular countries. Thus, no amount of potentially useful and appropriate techniques and theories can possibly compensate for a basic unwillingness or inability to tackle the conditions which perpetuate poverty and inequality. Thus, it is undoubtedly true that a cautionary note of realism must be borne in mind when considering some of the highly specific recommendations made in the second half of this book, concerning the use of techniques for examining spatial cognitions in the Third World urban planning and development process.

This issue really involves accepting that a distinction must be drawn between spatial patterns on the one hand and the processes that are responsible for generating them on the other. For example, many investment and sectoral allocation decisions are made at the aggregate national level, perhaps without any direct consideration being given to their spatial implications. Afterward, however, these decisions are played out in space, sometimes almost incidentally. This point is well illustrated in Figure 8.1, which has been adapted from a diagram contained in Yeates and Garner (1980). Thus, it must be appreciated that the spatial organisation of human activity

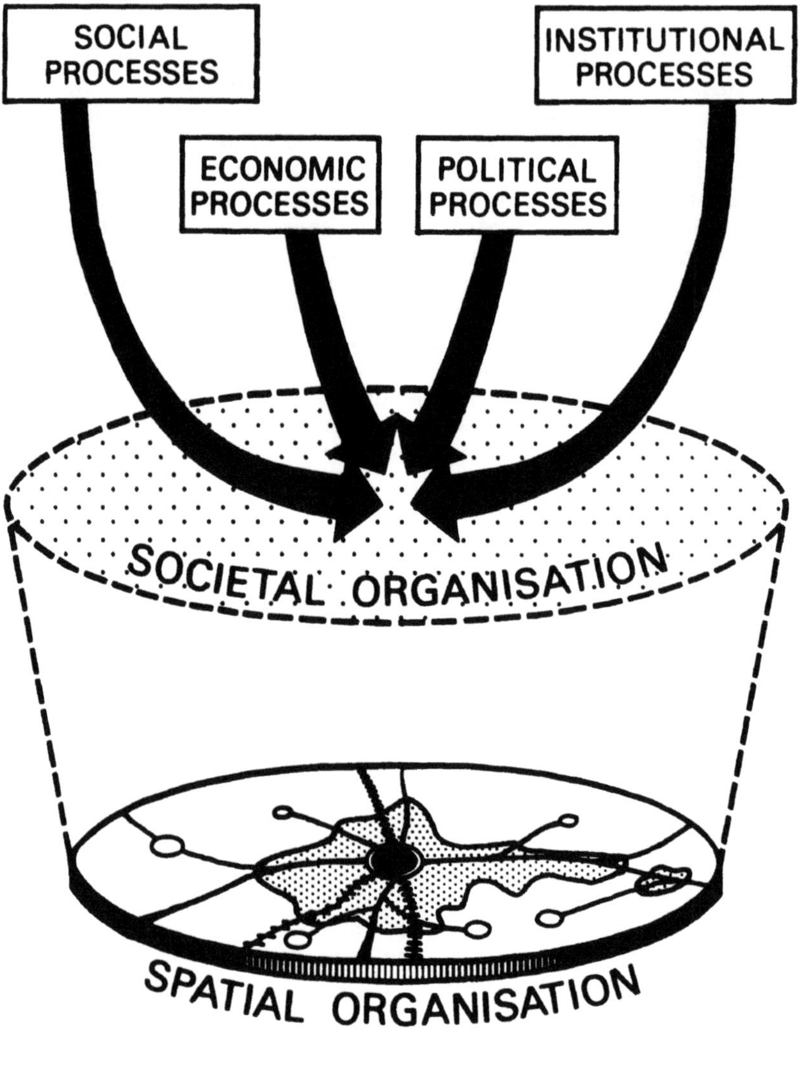

Figure 8.1: Societal organisation and spatial organisation

is mainly a reflection of the non-spatial organisation of society itself. Hence, urban and regional spatial organisation is the direct outcome of social, economic, political and institutional processes of change. These processes operate in space, but are not spatial processes *per se*.

This distinction carries important implications for physical, social, economic and other types of development planning. Most particularly, it suggests that entirely spatial or area-based policies for urban and regional change are unlikely to have great impact in the absence of more fundamental social, economic and institutional reforms. However, it should be stressed that this observation is not intended to decry the importance of policies which deal with activities in space, but rather to argue that these considerations are of little importance on their own. Similarly, purely structural policies without any spatial content would inevitably lead to distributional problems. For example, clearing a group of irregular settlements within a city does nothing of itself to remove problems of poverty and poor housing. Clearly, some form of structural change is required to increase employment, enhance incomes and to improve security of tenure, so as to facilitate housing consolidation and improvement. However, if structural reform does occur, spatial planning becomes vitally important: where are new jobs to be encouraged, where should new housing or upgraded areas be developed? The basic point is that spatial planning on its own will be little more than cosmetic, serving merely to paper over the proverbial socio-economic cracks. The real problem is that generally there has been too much spatial planning and too little accompanying structural reform. Both sets of changes are needed in unison for the promotion of genuine planning and development.

This argument is particularly important for policy-oriented geographical work, especially given the subject's emphasis on the theme of spatial organisation since the 1960s. This latter approach has been based on the positivistic philosophy and associated with the scientific method. As a result, whilst our understanding of spatial patterns may have been advanced, such work has had much less to say about societal processes. Some human geographers and other social scientists would argue that this work has been *status quo* in that it has served to divert attention away from the crucial historical and contemporary processes that have promoted inequalities. Recently, a growing band of human geographers have moved toward structuralist perspectives, that is examining the wider structures that contain and mould the phenomena under study (Johnston, 1983). Viewed with respect to Figure 8.1, this paradigm shift has represented a progressive redirection of attention away from the sub-structure of spatial organisation to the superstructure of societal organisation. A parallel argument of the structuralist school is that the focus of study should not be placed on the individual, but rather on the organisation and functioning of society as a whole. This holistic societal approach would, for example, be particularly typical of Marxist and neo-Marxist writers.

These arguments have many important implications regarding the subject matter covered in the present book and indeed they have already been discussed at length in several places in the text. It is argued below in conclusion that in future urban geographical work set in less developed countries, both aggregate-structuralist and individual-positivist oriented work is required in order to aid effective urban planning and development.

URBANISATION, PLANNING, PARTICIPATION AND PERCEPTION

In the first half of this book, it was stressed that urbanisation and the form of cities in contemporary Third World countries can only be understood in the light of global historical economic, social and political patterns of development. In particular, the rise of mercantile trade and the spread of capitalism as a dominant mode of economic production have influenced significantly the form of world urbanisation. Thus, Third World urbanisation cannot be explained in purely spatial terms, nor with respect to current processes of demographic change, migration and economic development alone. Quite simply, the Third World city is not an independent phenomenon, but rather one that has been fashioned by the historical integration of less developed countries into the world economic system. This interdependence will undoubtedly remain the case, suggesting that the most pressing problems of Third World urbanisation and development must be tackled at the national and international scales. The need is for a growing awareness on the part of governments and peoples of the realities of twentieth century global interdependence, along with the inhumanity of, and threats posed by, the ever-widening gap between the rich and the poor. The level of total world population increase may well have started to decline, from 1.9 per cent per annum in 1970/75 to a current level of 1.7 per cent per annum, but we are still facing a world population of some 6000 millions, half of which will be urban, by the year 2000 (see, for example, Hall, 1984). There is little room for complacency and fresh structural perspectives and reforms are undoubtedly required urgently.

As previously suggested, therefore, regardless of the scale of consideration, spatial planning is not likely to be enough on its own. But this is not to argue that spatial or physical planning is of no relevance. Some analysts argue that if there were social equality, then there would be no spatial inequalities. Such an argument begs the questions: "Is it only political changes and modifications in development style that can cure these imbalances? Is spatial policy necessary if such reforms take place?" (Gilbert and Gugler, 1982, p. 162). The view is taken here that this thesis of the 'irrelevance of the spatial problematic' (Gilbert and Gugler, 1982) seems to ignore the need for sound locational decision-making and spatial planning frameworks in association with fundamental structural reforms. If we are to be realistic, it has to be accepted that such frameworks may also be required to do an ameliorative job in the absence of more radical structural reforms. Thus, just as spatial-allocative planning must dovetail with major structural reforms, so research must proceed along these two distinct but related paths.

The real need is for the development of planning systems that are appropriate to the needs and circumstances of particular countries. But, as stressed in this volume, unfortunately, there is no one plan, no one miracle cure or universal panacea. Thus, with regard to housing, politicians and planners may increasingly agree that the need is for some form of government-sponsored sites and services or core housing scheme, or for aided self-help and in situ upgrading. But the precise balance and possible contribution of these and other housing policies will depend entirely on local circumstances. Similarly, with regard to the national urban system, the spread of development to peripheral regions, perhaps along the lines of agropolitan development (Friedmann and Douglas, 1975; Friedmann and Weaver, 1979), or by the promotion of planned intermediate cities (Rondinelli, 1982, 1983a, 1983b) also needs to be interpreted afresh in each and every socio-cultural and environmental context.

The general framework of planning and its specific imperatives must be adapted and customised to suit local socio-economic, cultural, institutional and other circumstances. What is required first and foremost is the development of appropriate planning systems and policies that afford the opportunity for people to become involved in the decision-making processes affecting their lives, and which thereby encourage genuine planner-public interaction. Hence, the call in this volume and by many other writers for the promotion of public participation in planning at all levels, from that of the community to that of the nation. As was argued in Chapter 5, it is clearly not possible for all of the people to participate in environmental decision-making, so that participation must be optimised, not maximised in purely quantitative terms. The need is to ensure that planning in Third World countries is not unnecessarily technocratic, elitist and based on inappropriate Westernised values and perspectives. The establishment of such people-based planning systems would itself represent a fundamental structural change in many Third World countries. Thus, in the context of a social-anthropological view of Third World urbanisation, Lloyd (1979, pp. 213-14) has observed that:

> ... man's actions are to be understood with reference to his view of the world. Man is a strategist, continually making choices between alternative courses of action. He seeks to understand the world in order to control it, and in so doing he both enhances his understanding and changes the world. His understanding derives from the concepts and propositions which he has learned, his past and present experiences.

Such views have clear and unmistakable links with cognitive-behavioural research perspectives, especially Kelly's personal construct theory reviewed in Chapter 6. Hence the argument that what is desperately needed is an approach to urban environmental planning which seeks to identify and take account of the perceptions, cognitions and aspirations of individuals. The genuine desire to achieve such aims will do much to eradicate the problems of elitist and Westernised biases in Third World planning. Lloyd (1979, p. 214) continues with an argument that all planners, public servants and academics involved in planning and development studies would do well to bear in mind at all times:

> As observers we may bring to the situation a different set of concepts. Our experiences are unlike those of our actors. We have access to other sources of information. But the actor's view of his world should be an essential ingredient in our own picture. Our understanding and predication of his actions will be much more sophisticated with this ingredient than without it.

This is the crux of the matter, for behavioural approaches give us the opportunity, however imperfectly, to try and see the world as others do. The need for such empathy with the individual in the street forms the cornerstone of perception studies. Despite its initial humanistic stance, the fact that much behavioural geography has been strongly grounded in positivism has probably led some to question its potential in this context. But as Couclelis and Golledge (1983) have recently stressed, the field now displays signs that it is shedding some of its tenets and is adopting a more subjective orientation. The claim would still be made by many radical and structuralist geographers, however, that perception studies have "actively supported the status quo, i.e. ruling class view of society that all its problems [are] solvable through gradual reform" (Rieser, 1973, p. 53). The same author continued by arguing that methods of behavioural geographical research have a "ready application in the planning field as a sop to arguments for participation as a substitute for democracy" (Rieser, 1973, p. 53). Certainly this must be recognised as an ever-present danger, but the overall assessment seems rather harsh. Structural approaches are desperately needed, but they must not be seen as incompatible with an individualist focus of study. Just as we need to study how societies function within the world economy, so it is equally important for us to strive to understand how these macro-circumstances affect people, their hopes and aspirations at the local level.

FINAL COMMENTS

Continuing rapid urbanisation is an inevitable process at the global scale. Recently, Castells (1977, p.1) has observed that "urban problems are becoming an essential element in the policies of governments, in the concerns of the mass media and, consequently, in the everyday life of a large section of the population". This comment stands whether we are talking of the First, Second or Third Worlds, and individuals, communities and governments are all locked in the struggle of finding ways of making existing levels of urbanisation acceptable by means of careful redevelopment and planning. However, dealing with the problems afforded by the processes of urbanisation and urban growth in the last quarter of the twentieth century and beyond will involve more than piecemeal and *ad hoc* responses to past processes. It will necessitate an ideological commitment to promoting more balanced and more equitable processes and patterns of development. If this is to amount to anything more than a pious hope, the international character of the problems involved in the process of world urbanisation must come to be more fully appreciated. However daunting this task may sound, tackling it stands as a priority of the highest order.

It is argued here that recognition of the aspirations and perceptions of the general populace is urgently required, along with an increasing emphasis on individual and community participation in the planning process. Planning is obviously enmeshed in the political milieux and it is hoped that efforts to foster participatory democracy will help to eradicate *status quo* systems which because they work to the advantage of particular groups and individuals in society, provide little impetus for change and development. At the same time, no matter how utopian or idealistic it may sound, structural changes are urgently needed at the national and international scales, in order to facilitate a more realistic and genuine process of development. But it is just as important to recognise that an essential step toward achieving this goal will be the development of a greater commitment to equality of opportunity and the recognition of the rights of individuals and the local community.

Appendix A
Barbados Interview Schedule

Bedford College
(University of London)
Department of Geography,
REGENT'S PARK LONDON NW1 4NS
Telephone: 01-486 4400
Telegrams: Edforcoll London NW1

DEPARTMENT OF GEOGRAPHY

1		
2		
3		
4		

Leave Blank

Interviewer.....................
Date...........Location.........
........Interview No. ....

Barbados Survey

Good morning/Good afternoon. We are working on a project that is being directed by Dr Potter, who is a geographer. We are interested in hearing how you feel about living and working in Barbados. Do you have the time to help us by answering a few simple questions? It will not take long and you should find the questions quite interesting. None of them are of a personal nature and your answers will be treated in the strictest confidence. You do live in the..........................area don't you? (ADAPT AS NECESSARY, IF THE RESPONDENT REQUESTS FURTHER INFORMATION BE PREPARED TO GIVE IT IF YOU CAN)

*****START OF INTERVIEW PROPER*****

1. How long have you lived in..........................?______yrs._____mths.
2. If you think about Barbados, what are the first THREE things that come into your mind?
   i)____________________
   ii)____________________
   iii)____________________
3. What are the THREE things that you most like about living in Barbados?
   i)____________________
   ii)____________________
   iii)____________________
4. What are the THREE things that you most dislike about living in Barbados?
   i)____________________
   ii)____________________
   iii)____________________
5. Think about Barbados for a moment; now, if you were free to live in any part of the country, where exactly would you most like to live? __________
6. (GIVE PARISH MAP OF BARBADOS WITH LIST OF RESPONSES TO INTERVIEWEE) Look at the map I have just given you. As you will see, it shows the island of Barbados divided into the 11 parishes. I would like you to think about each of the parishes as I read their names out. For each one, think how much you would like to live in that parish. Please read out from the bottom of the map the phrase which best describes your feelings about the area as a possible place to live. (MAKE SURE THE INTERVIEWEE UNDERSTANDS WHAT IS REQUIRED. READ OUT PARISH NAMES AND NOTE RESPONSE IN TERMS OF SCALE FROM 1 to 7.)

St. Lucy		St. Thomas		St. Michael	
St. Peter		St. Joseph		Christ Church	
St. Andrew		St. John		St. Philip	
St. James		St. George			

-1-

7. Do you think that the Bridgetown area is a good place to live? Yes/No/Don't know

8. What do you think are the THREE best things about Bridgetown?
   i) ______
   ii) ______
   iii) ______

9. What do you think are the THREE worst things about Bridgetown?
   i) ______
   ii) ______
   iii) ______

10. Do you have a job? Yes/No

11. (IF ANSWER TO 10 IS 'YES') In which Parish do you work? ______

12. How often do you go into Bridgetown? ______ visits per ______

13. What do you go there for? ______

14. Where do you normally buy your bread? ______

15. Where do you normally buy your meat? ______

16. Where do you normally buy your clothes? ______

17. Do you think that Bridgetown is too big? Yes/No/Don't know

18. Do you think that Bridgetown is too crowded with people? Yes/No/Don't know

19. Do you think that Bridgetown is too crowded with cars and buses? Yes/No/Don't know

20. Do you think that the pace of life in Bridgetown is too fast? Yes/No/Don't know

21. Do you think that Bridgetown suffers from too much noise? Yes/No/Don't know

22. Do you think that Bridgetown suffers from too much pollution? Yes/No/Don't know

23. Have you read or heard about the Physical Development Plan for Barbados, which was published in 1970? Yes/No

24. Did you know that the present Government wants to develop new urban areas in Barbados? Yes/No

25. Do you think that this is a good idea? Yes/No/Don't know

26. Where do you think the Government might best locate new settlements?
   ______

THE INTERVIEW IS NOW FINISHED. THANK YOU VERY MUCH FOR YOUR HELP.

(TO BE FILLED IN BY THE INTERVIEWER ; PLEASE TICK APPROPRIATE BOXES:-

AGE:	Under 30	30-50	Over 50

SEX:	Male	Female

-2-

MAP AND RESPONSES FOR QU. 6

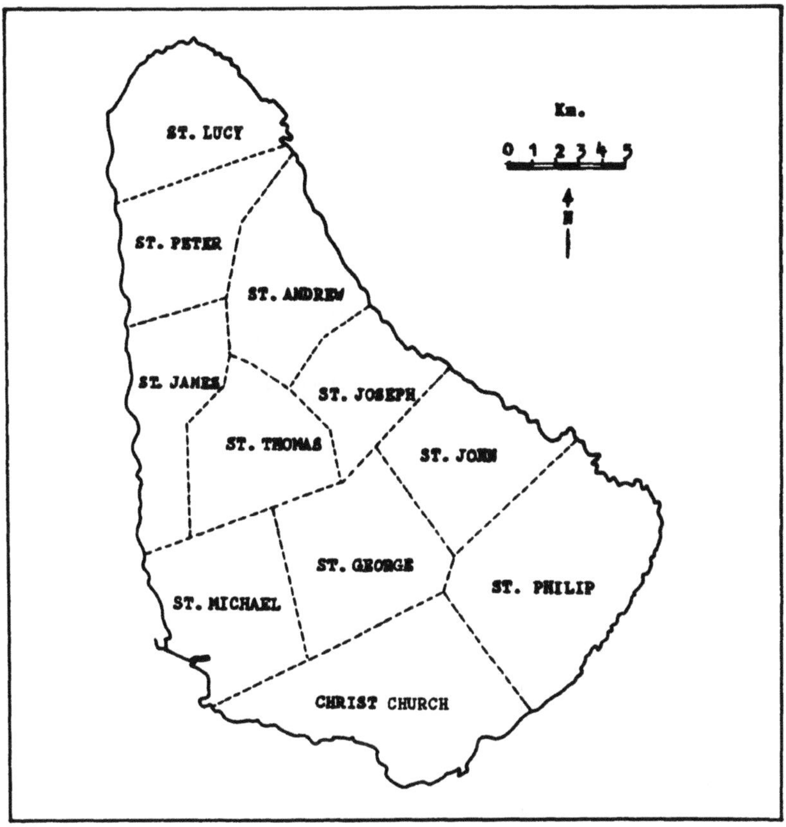

1 WOULD NOT LIKE TO LIVE THERE AT ALL
2 WOULD NOT LIKE TO LIVE THERE
3 WOULD NOT LIKE TO LIVE THERE MUCH
4 UNDECIDED / DO NOT KNOW
5 WOULD QUITE LIKE TO LIVE THERE
6 WOULD LIKE TO LIVE THERE
7 WOULD LIKE TO LIVE THERE VERY MUCH

Appendix B
Trinidad and Tobago
Interview Schedule

Interviewer:	
Date:	
Location:	
Interview No.:	

Bedford College
(University of London)
Department of Geography,
REGENT'S PARK LONDON NW1 4NS
Telephone: 01-486 4400
Telegrams: Edforcoll London NW1
September 1982

Leave Blank

1	
2	
3	
4	

Trinidad & Tobago Survey

Good morning/Good afternoon. We are working on a project that is being directed by Dr Potter, who is a geographer. We are interested in hearing how you feel about living and working in Trinidad and Tobago. Do you have the time to help us by answering a few simple questions? It will not take long, and you should find the questions quite interesting. None of them are of a personal nature and your answers will be treated in the strictest confidence. You do live in the ______________ area don't you? (Adapt as necessary; if the respondent requests further information be prepared to give it if you can).

*****Start of Interview*****

1. How long have you lived in __________________? ___yrs. ___mths.
2. If you think about Trinidad and Tobago, what are the first THREE things that come into your mind?

   (i) ________________________

   (ii) ________________________

   (iii) ________________________
3. Think about Trinidad and Tobago for a moment; now, if you were free to live in any part of the country, where exactly would you most like to live?

   ____
4. (Give map of Trinidad and Tobago showing sub-regions and with list of responses to interviewees) Look at the map which I have just given you. It shows Trinidad and Tobago divided into 21 sub-regions. I would like you to think about each of these as I read their names out. For each one, think how much you would like to live in that area. Please pick out from the list at the bottom of the map, the phrase which best describes your feelings about the area as a possible place to live. (Make sure the interviewee understands what is required. Read out the names of the sub-regions and note response in terms of 1-7 scale).

1. Port of Spain/Arima	
2. Northern Valley	
3. Chaguaramas	
4. North Coast	
5. Aripo-Toco	
6. Sangre Grande	
7. Rural Arima	
8. Western Caroni	
9. Eastern Caroni	
10. San Fernando	
11. Princes Town	

12. Rio Claro	
13. Nariva Swamp	
14. Cedros	
15. Point Fortin	
16. Siparia	
17. Moruga	
18. Western Mayaro	
19. Mayaro-Guayaguayare	

20. North east Tobago	
21. South west Tobago	

-1-

5. Do you think that the Port of Spain urban area is a good place to live?........................Yes/No/Don't know

6. What do you think are the THREE best things about Port of Spain?
(i)__________
(ii)__________
(iii)__________

7. What do you think are the THREE worst things about Port of Spain?
(i)__________
(ii)__________
(iii)__________

8. Do you think that Port of Spain is too big?.....Yes/No/Don't know

9. Do you think that Port of Spain is too crowded with people?............................Yes/No/Don't know

10. Do you think that Port of Spain is too crowded with cars and buses?...................Yes/No/Don't know

11. Do you think that the pace of life in Port of Spain is too fast?......................Yes/No/Don't know

12. Do you think that Port of Spain suffers from too much noise?............................Yes/No/Don't know

13. Do you think that Port of Spain suffers from too much pollution?........................Yes/No/Don't know

Finally, we would like to ask you one or two questions to see how you feel about planning and development issues in Trinidad and Tobago.

14. What do you feel are the main problems faced by a country like Trinidad and Tobago?__________

15. Have you read or heard about the National Physical Development Plan for Trinidad and Tobago, published in 1978?...........Yes/No

16. Did you know that a main aim of the planners is to develop new and expanded urban areas?......................Yes/No

17. Do you feel that in general this is a good idea?...........Yes/No

18. Do you personally feel that it would be a good idea to expand the settlement at:-
a) Valencia?..........................Yes/No
b) Sangre Grande?.....................Yes/No
c) Chaguanas?.........................Yes/No
d) Couva?.............................Yes/No
e) Point Fortin?......................Yes/No
f) Guayaguayare?......................Yes/No

19. Are there any other places where you feel that new urban areas might be located?__________

THE INTERVIEW IS NOW FINISHED -THANK YOU

To be filled in by the interviewer; please tick appropriate boxes

AGE:	Under 30	30-50	Over 50		SEX:	M	F

-2-

MAP AND RESPONSES FOR QU. 4

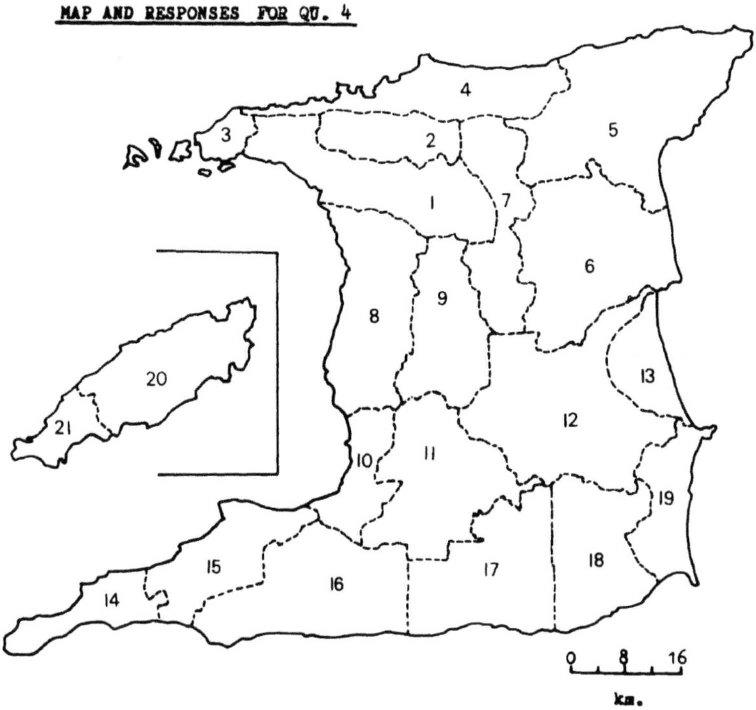

KEY TO SUB-REGIONS:

1. Port of Spain/Arima
2. Northern Valley
3. Chaguaramas
4. North Coast
5. Aripo-Toco
6. Sangre Grande
7. Rural Arima
8. Western Caroni
9. Eastern Caroni
10. San Fernando
11. Princes Town
12. Rio Claro
13. Nariva Swamp
14. Cedros
15. Point Fortin
16. Siparia
17. Moruga
18. Western Mayaro
19. Mayaro-Guayaguayare
20. North east Tobago
21. South west Tobago

RESPONSES:

1 WOULD NOT LIKE TO LIVE THERE AT ALL
2 WOULD NOT LIKE TO LIVE THERE
3 WOULD NOT LIKE TO LIVE THERE MUCH
4 -------- UNDECIDED/DO NOT KNOW ------
5 WOULD QUITE LIKE TO LIVE THERE
6 WOULD LIKE TO LIVE THERE
7 WOULD LIKE TO LIVE THERE VERY MUCH

Appendix C
St. Lucia Interview Schedule

Interviewer
Date
Location
Interview No.

A	
B	
C	
D	

(Leave Blank)

Bedford College
(University of London)
Department of Geography,
REGENT'S PARK LONDON NW1 4NS
Telephone: 01-486 4400
Telegrams: Edforcoll London NW1
August 1982

St.Lucia Survey

Good morning/Good afternoon. We are working on a project that is being directed by Dr Potter, who is a geographer. We are interested in hearing how you feel about living and working in St.Lucia. Do you have the time to help us by answering a few simple questions? It will not take very long, and you should find the questions quite interesting. None of them are of a personal nature and your answers will be treated in the strictest confidence. You do live in the____________ area don't you? (Adapt as necessary, if the respondent requests further information be prepared to give it if you can).

****START OF INTERVIEW****

1. How long have you lived in____________________?___yrs.___mths.
2. If you think about St.Lucia, what are the first THREE things that come into your mind?
   (i)____________________
   (ii)____________________
   (iii)____________________
3. Think about St.Lucia for a moment; now, if you were free to live in any part of the country, where exactly would you most like to live?
   ____
4. (Give Quarter map of St.Lucia with list of responses to interviewee) Look at the map I have just given you. As you will see, it shows St.Lucia divided into the 11 Quarters. I would like you to think about each of these as I read their names out. For each one, think how much you would like to live in that Quarter. Please read out from the bottom of the map the phrase which best describes your feelings about the area as a possible place to live. (Make sure the interviewee understand what is required. Read out Quarter names and note response in terms of scale from 1 to 7).

Castries________		Anse-La-Raye________		Vieux Fort________	
Gros-Islet________		Soufriere________		Micoud________	
Dauphin________		Choiseul________		Praslin________	
Dennery________		Laborie________			

5. Do you think that the Castries urban area is a good place to live?
   Yes/No/Don't know
6. What do you think are the THREE best things about Castries?
   (i)____________________
   (ii)____________________
   (iii)____________________

   What do you think are the THREE worst things about Castries?
   (i)____________________
   (ii)____________________
   (iii)____________________

-1-

7. How often do you visit Castries? ________visits per________

8. What do you go there for?________________________________

9. In which town do you normally buy:-

your bread?________________
your meat?________________
your clothes?________________

10. Do you think that Castries is too big? ....Yes/No/Don't know

11. Do you think that Castries is too crowded with people?......................................Yes/No/Don't know

12. Do you think that Castries is too crowded with cars and buses?..............................Yes/No/Don't know

13. Do you think that the pace of life in Castries is too fast?.......................................Yes/No/Don't know

14. Do you think that Castries suffers from too much noise?......................................Yes/No/Don't know

15. Do you think that Castries suffers from too much pollution?...................................Yes/No/Don't know

16. Do you think that it would be a good idea to develop new urban areas in St.Lucia?.............Yes/No/Don't know

17. Where do you think that these might best be located?

____

Finally, we would like to ask you one or two questions to see how you feel about the importance of tourism to St.Lucia.

18. Taking all things into consideration, do you feel that tourism is a good thing in a country like St.Lucia?.............................Yes/No/Don't know

19. What do you feel are the good things about tourism for a country like St.Lucia? (see if interviewee can give three factors):

____

____

20. What do you feel are the bad things about tourism for a country like St.Lucia? (again, see if interviewee can provide three factors):

____

____

21. Which area do you think is best for tourists to stay in?

____

22. Are you in any way involved in the tourist industry?...........Yes/No
(if 'yes', in what way?________________________________)

THE INTERVIEW IS NOW FINISHED - THANK YOU FOR YOUR HELP

To be filled in by the interviewer; please tick appropriate boxes :-

AGE:	Under 30	30-50	Over 50

SEX:	M	F

____

____

____

-2-

MAP AND RESPONSES FOR QU. 4

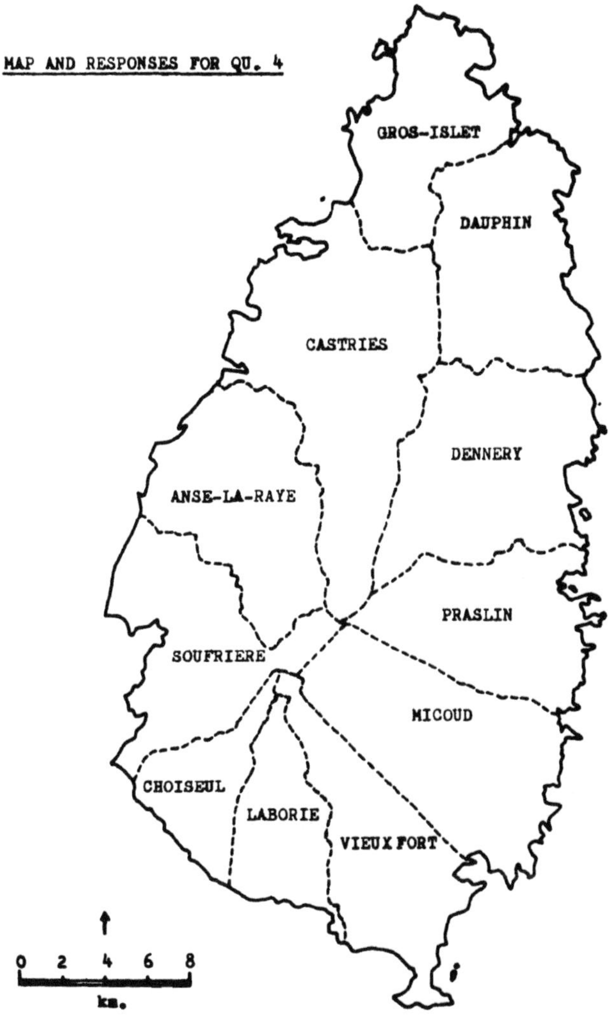

1 WOULD NOT LIKE TO LIVE THERE AT ALL
2 WOULD NOT LIKE TO LIVE THERE
3 WOULD NOT LIKE TO LIVE THERE MUCH
4 -------- UNDECIDED/DO NOT KNOW ------
5 WOULD QUITE LIKE TO LIVE THERE
6 WOULD LIKE TO LIVE THERE
7 WOULD LIKE TO LIVE THERE VERY MUCH

Bibliography

Abrams, C. (1963): Report to the Barbados Government and the Barbados Housing Authority on Land Tenure, Housing Policy and Home Finance. New York.

--- (1964): Housing in the Modern World: Man's Struggle for Shelter in an Urbanizing World. Faber and Faber: London.

Abu-Lughod, J. (1971): Cairo: 1001 Years of the 'City Victorious'. Princetown University Press.

Abu-Lughod, J. and Hay, R. (Eds.) (1977): Third World Urbanization. Methuen: New York.

Adams, R. M. (1960): "The origin of cities", Scientific American Offprint, No. 606.

--- (1966): The Evolution of Urban Society. Weidenfeld and Nicolson: London.

Aitken, S. R. (1981): "Squatters and squatter settlements in Kuala Lumpur", Geographical Review, 71, 158-175.

Alonso, W. (1968): "Urban and regional imbalances in economic development", Economic Development and Cultural Change, 17, 1-14.

--- (1971): "The economics of urban size", Papers of the Regional Science Association, 26, 67-83.

--- (1980): "Five bell shapes in development", Papers of the Regional Science Association, 45, 5-16.

Amato, P. (1970): "Elitism and settlement patterns in the Latin American city", Journal of the American Institute of Planners, 36, 96-105.

Andrews, F. M. and Phillips, G. (1970): "The squatters of Lima: who are they and what do they want?", Journal of Developing Areas, 4, 211-24.

Angel, S. (1983): "Upgrading slum infrastructure: divergent objectives in search of a consensus", Third World Planning Review, 5, 5-22.

Appleyard, D. (1969): "City designers and the pluralistic city", in: Rodwin, L. (Ed.): Planning Urban Growth and Regional Development: the Experience of the Guayana Program of Venezuela, M.I.T. Press, pp. 422-52.

--- (1970): "Styles and methods of structuring a city", Environment and Behavior, 2, 100-16.

Arnstein, S. R. (1969): "A ladder of citizen participation", Journal of the American Institute of Planners, 35, 216-24.

Auerbach, F. (1913): "Das Gesetz der Bevölkerungskonzentration", Petermann's Mitteilungen, 59, 74-6.

Auty, R. M. (1979): "Worlds within the Third World", Area, 11,232-5.

Bairoch, P. (1975): The Economic Development of the Third World since 1900. Methuen: London.

Bannister, D. and Mair, J. M. M. (1968): The Evaluation of Personal Constructs. Academic Press: London.

Barker, D. (1977): "Some methodological issues in the measurement, analysis and evaluation of peasant farmers' knowledge of their environment", M.A.R.C. report No. 9, Chelsea College, London.

Barker, D. and Ferguson, A. G. (1980): "Employment aspirations and the perception of rural and urban environments among village polytechnic trainees, Central Kenya", Bedford College, University of London, Papers in Geography, 11, 22pp.

— (1983): "A goldmine in the sky faraway: rural-urban images in Kenya", Area 15, 185-91.

Barker, D. and Richards, P. (1978): "Repertory grid methods and environmental images in rural Africa", Paper presented to the joint meeting of the I.B.G. Quantitative Methods Study Group and Developing Areas Study Group.

Berry, B. J. L. (1961): "City size distributions and economic development", Economic Development and Cultural Change, 9, 573-87.

— (1972): "Hierarchical diffusion: the basis of development filtering and spread in a system of growth centres", in Hanson, N. M. (Ed.): Growth Centres in Regional Economic Development. The Free Press.

— (1973): The Human Consequences of Urbanisation: Divergent Paths in the Urban Experience of the Twentieth Century. Methuen: London.

— (Ed.) (1976): Urbanization and Counterurbanization. Sage: Beverly Hills.

Berry, B. J. L. and Barnum, H. G. (1962): "Aggregate relations and elemental components of central place systems", Journal of Regional Science, 4, 35-68.

Bird, J. (1977): Centrality and Cities. Routledge and Kegan Paul: London.

Blume, H. (1974): The Caribbean Islands. Longman: London.

Blunt, A. (1982): "Community participation - a Philippine experience", Habitat International, 6, 179-87.

Borchert, J. R. (1967): "American metropolitan evolution", Geographical Review, 57, 301-23.

Boudeville, J. R. (1966): Problems of Regional Economic Planning. Edinburgh University Press.

Boulding, K. E. (1956): "The Image: Knowledge in Life and Society", University of Michigan Press: Ann Arbor.

Bourne, L. S. (1975): Urban Systems: Strategies for Regulation. Oxford University Press: London.

von Boventer, E. (1975): "Regional growth theory", Urban Studies, 12, 1-29.
Brandt, W. (1980): North-South: A Programme for Survival. Pan Books: London.
Briggs, J. (1983): "Rural development in Tanzania: end of an era?", Geography, 68, 66-8.
Brookfield, H. C. (1969): "On the environment as perceived", Progress in Geography, 1, 51-80.
— (1973): "On one geography and a Third World", Transactions of the Institution of British Geographers, 58, 1-20.
— (1975): Interdependent Development. Methuen: London.
Broom, L. (1953): "Urban research in the British Caribbean: a prospectus", Social and Economic Studies, 1, 113-19.
Brunn, S. D. and Williams, J. F. (1983): Cities of the World: World Regional Urban Development. Harper and Row: New York.
Buksmann, P. and Rowley, G. (1978): "Squatter settlements in Medellín, Colombia: a rejoinder", Area, 10, 15-19.
Bunting, T. E. and Guelke, L. (1979): "Behavioral and perception geography: a critical appraisal", Annals of the Association of American Geographers, 69, 448-62.
Burgess, E. W. (1925): "The growth of the city", in: Park, R. E., Burgess, E. W. and McKenzie, R. D. (Eds.): The City. University of Chicago Press.
Burgess, J. A. (1982): "Selling places: environmental images for the executive", Regional Studies, 16, 1-17.
Burgess, R. (1977): "Self-help housing: a new imperialist strategy? A critique of the Turner School", Antipode, 9, 50-9.
— (1978): "Petty commodity housing or dweller control? A critique of John Turner's views on housing policy", World Development, 6, 1105-33.
— (1981): "Ideology and urban residential theory in Latin America", in Herbert, D. T. and Johnston, R. J. (Eds.): Geography and the Urban Environment: Progress in Research and Applications, Volume IV: John Wiley, pp. 57-114.
— (1982): "Self-help housing advocacy: a curious form of radicalism: a critique of the work of John F. C. Turner", in Ward, P. W. (Ed.): Self-Help Housing: A Critique. Mansell Publishing, pp. 56-97.
Burns, L. S. (1983): "Self-help housing: an evaluation of outcomes", Urban Studies, 20, 299-309.
Buttler, F. A. (1975): Growth Pole Theory and Economic Development. Saxon House: Farnborough.

Cannon, T. G. (1975): "Geography and underdevelopment", Area, 7, 212-6.
Canter, D. and Tagg, S. K. (1975): "Distance estimation in cities", Environment and Behavior, 7, 59-80.
Carol, H. (1964): "Stages of technology and their impact upon the physical environment: a basic problem in cultural geography", Canadian Geography, 8, 1-9.
Carter, H. (1977): "Urban origins: a review", Progress in Human Geography, 1, 12-32.

Carter, H. (1981): The Study of Urban Geography (Third Edition). Arnold: London.
— (1983): An Introduction to Urban Historical Geography. Arnold: London.
Castells, M. (1977): The Urban Question: A Marxist Approach. Arnold: London.
Chang, S-D. (1982): Modernization and urbanization problems in China. Occasional Paper 26, The Chinese University of Hong Kong.
Childe, V. G. (1950): "The urban revolution", Town Planning Review, 21, 3-17.
— (1951): Man Makes Himself. Merton: New York.
Christaller, W. (1933): Die zentralen Orte in Suddeutschland. Verlag.
Clark, D. (1982): Urban Geography: An Introductory Guide. Croom Helm: London.
Clark, J. A. (1982): "The role of the State in regional development", Chapter 6 in: Flowerdew, R. (Ed.): Institutions and Geographical Patterns. Croom Helm.
Clarke, C. G. (1966): "Population pressure in Kingston, Jamaica: a study of unemployment and overcrowding", Transactions of the Institute of British Geographers, 38, 165-87.
— (1974): "Urbanization in the Caribbean", Geography, 59, 223-32.
— (1975): Kingston, Jamaica: Urban Growth and Social Change, 1692-1962. University of California Press: Berkeley, Los Angeles.
— (1975): "The Commonwealth Caribbean", in Jones, R. (Ed.): Essays on World Urbanization. George Philip & Son Ltd: London, pp. 341-51.
Clarke, C. G. and Ward, P. (1976): "Stasis in make-shift housing: perspectives from Mexico and the Caribbean", Actes du XLII Congrés International des Americanistes, 10, 351-8.
Conway, D. (1976): "Residential area change and residential relocation in Port of Spain, Trinidad", Unpublished Ph.D. thesis, University of Texas at Austin.
— (1980): "Step-wise migration: toward a clarification of the mechanism", International Migration Research, 14, 3-14.
— (1981): "Fact or opinion on uncontrolled peripheral settlement in Trinidad: or how different conclusions arise from the same data", Ekistics, 286, 37-43.
— (1982): "Self-help housing, the commodity nature of housing and amelioration of the housing deficit: continuing the Turner-Burgess debate", Antipode, 14, 40-6.
— (1983): "The commuter zone as a relocation choice of low income migrants moving in a step-wise pattern to Port of Spain, Trinidad", Caribbean Geography, 1, 89-106.
Conway, D. and Brown, J. (1980): "Intraurban relocation and structure: low-income migrants in Latin America and the Caribbean", Latin American Research Review, 15, 95-125.
Conyers, D. (1982): An Introduction to Social Planning in the Third World. Wiley: Chichester.
Couclelis, H. and Golledge, R. (1983): "Analytic research, positivism, and behavioral geography", Annals of the Association of American Geographers, 73, 331-9.

Cox, R. (1981): "Bourgeois thought and the behavioral geography debate", Ch. 11 in: Cox, K. R. and Golledge, R. G. (Eds.): Behavioural Geography Revisited. Methuen: London.

Cross, M. (1979): Urbanization and Urban Growth in the Caribbean: an essay on Social Change in Dependent Societies. Cambridge University Press.

Dann, G. (1984): The Quality of Life in Barbados. Macmillan: London.

Dansereau, P. (1978): "An ecological grading of human settlements", Geoforum, 9, 161-210.

Darwent, D. F. (1969): "Growth poles and growth centers in regional planning - a review", Environment and Planning, 1, 5-31.

Davis, K. (1955): "The origin and growth of urbanization in the world", American Journal of Sociology, 60, 429-37.

— (1965): "The urbanization of the human population", Scientific American, 213, 40-53.

— (1969): World Urbanisation 1950-1970, Volume I: Basic Data for Cities, Countries and Regions. Institute for International Studies, University of California: Berkeley.

— (1972): World Urbanisation 1950-1970, Volume II: Analysis of Trends, Relationships, and Development. Institute for International Studies, University of California: Berkeley.

— (Ed.) (1973): Cities: Their Origin, Growth, and Human Impact. Freedman: San Francisco.

Devas, N. (1983): "Financing urban land development for low income housing: an analysis with particular reference to Jakarta, Indonesia", Third World Planning Review, 5, 209-25.

Dickenson, J.P.et al.(1983):A Geography of the Third World. Methuen: London.

Dijkink, D. and Elbers, E. (1981): "The development of geographic representation in children: cognitive and affective aspects of model-building behaviour", Tijdschrift voor Economische en Sociale Geografie, 72, 2-16.

Downs, R. M. (1981): "Maps and metaphors", Professional Geographer, 33, 287-93.

Downs, R. M. and Stea, D. (1973): Image and Environment. Aldine: Chicago.

Doxiadis, C. A. (1967): "Developments toward Ecumenopolis - the Great Lakes Megalopolis", Ekistics, 22, 14-31.

Doxiadis, C. A. and Papaioannou, J. G. (1974): Ecumenopolis: the Inevitable City of the Future. W. W. Norton: New York.

Drakakis-Smith, D. W. (1981): Urbanisation, Housing and the Development Process. Croom Helm: London.

Dunn, P. D. (1978): Appropriate Technology: Technology with a Human Face. Macmillan: London.

Dwyer, D. J. (Ed.) (1974): The City in the Third World. Macmillan: London.

— (1975): People and Housing in Third World Cities. Longman: London.

Dyson, A. (1967): "Population trends in the Eastern Caribbean", in: Steel, R. W. and Lawton, R.(Eds.): Liverpool Essays in Geography, Longman, pp. 381-405.

Eastwood, D. A. (1983): "Venezuela: the 1980 Census shows continued rapid population growth", Geography, 68, 345-7.

Ebong, M. O. (1983): "The perception of residential quality", Third World Planning Review, 5, 273-85.

El-Shakhs, S. (1971): "National factors in the development of Cairo", Town Planning Review, 42, 235-49.

--- (1972): "Development, primacy and systems of cities", Journal of Developing Areas, 7, 11-36.

England, L. (1974): "The public's view", New Society, 28, 315-16.

Eyre, L. A. (1972): "The shantytowns of Montego Bay, Jamaica", Geographical Review, 62, 394-412.

Fagence, M. (1977): Citizen Participation in Planning. Pergamon: Oxford.

Farmer, B. H. (1983): "British geographers overseas, 1933 -1983", Transactions of the Institute of British Geographers, New Series, 8, 70-9.

Ferguson, A. G. (1979): "Some aspects of urban spatial cognition in an African student community", Transactions of the Institute of British Geographers, New Series, 4, 77-93.

Fox, H. S. A. (1970): "Going to market in 13th century England", Geographical Magazine, 42, 658-67.

Frank, A. G. (1969): Capitalism and Underdevelopment in Latin America. Monthly Review Press: New York.

Franklin, E. H. M. (1979): "The housing problem in Caracas, Venezuela", Unpublished Ph.D. thesis, University of Liverpool.

Franklin, G. H. (1979): "Physical development planning and the Third World", Third World Planning Review, 1, 7-22.

Fransella, F. and Bannister, P. (1977): A Manual for Repertory Grid Technique. Academic Press: London.

Friedmann, J. (1966): Regional Development Policy: A Case Study of Venezuela. M.I.T. Press.

Friedmann, J. and Douglas, M. (1975): "Agropolitan development: towards a new strategy for regional development in Asia", United Nations Centre for Regional Development, 333-87.

Friedmann, J. and Weaver, C. (1979): Territory and Function: the Evolution of Regional Planning. Arnold: London.

Friedmann, J. and Wulff, R. (1976): The Urban Transition: Comparative Studies of Newly Industrializing Societies. Arnold: London.

Fuller, G. and Chapman, M. (1974): "On the role of mental maps in migration research", International Migration Review, 8, 491-506.

Funnell, D. C. (1973): "Rural business centres in a low income economy: some theoretical problems", Tijdschrift voor Economische en Sociale Geografie, 64, 86-92.

Gilbert, A. G. (1976): "The arguments for very large cities reconsidered", Urban Studies, 13, 27-34.

--- (1977): "The argument for very large cities reconsidered: a reply", Urban Studies, 14, 225-7.

--- (1981): "Pirates and invaders: land acquisition in urban Colombia and Venezuela", World Development, 9, 657-78.

Gilbert, A. G. and Goodman, D. E. (1976): "Regional income disparities and economic development", in Gilbert, A. G. (Ed.): Development Planning and Spatial Structure, Wiley: Chichester.

Gilbert, A. G. and Gugler, J. (1982): Cities, Poverty and Development: Urbanization in the Third World. Oxford University Press.

Gilbert, A. G. and Sollis, P. J. (1979): "Migration to small Latin American cities: a critique of the concept of 'fill-in' migration", Tijdschrift voor Economische en Sociale Geografie, 70, 110-13.

Gilbert, A. and Ward, P. (1981): "Public intervention, housing and land use in Latin American cities", Bulletin of Latin American Research, 1, 97-104.

Gilbert, A. G. and Ward, P. M. (1982a): "Residential movement among the poor: the constraints on housing choice in Latin American cities", Transactions of the Institute of British Geographers, New Series, 7, 129-49.

— (1982b): "Low-income housing and the State", Chapter 5 in: Gilbert, A., Hardoy, J. E. and Ramirez, R. (Eds.): Urbanization in Contemporary Latin America. John Wiley, pp. 79-127.

Glass, R. (1976): "Urban images", Chapter 18 in: Harrison, G. A. and Gibson, J. B. (Eds.): Man in Urban Environment. Oxford University Press, pp. 349-67.

Gold, J. R. (1980): An Introduction to Behavioural Geography. Oxford University Press.

Goodey, B. (1973): "Perception of the Environment: an introduction to the literature", Centre for Urban and Regional Studies, University of Birmingham, Occasional Paper No. 17.

Gottmann, J. (1957): "Megalopolis, or the urbanization of the Northeastern seaboard", Economic Geography, 33, 189-200.

— (1978): "Megalopolitan systems around the world", in Bourne, L. S. and Simmons, J. W. (Eds.): Systems of Cities. Oxford University Press, pp. 53-60.

— (1983): "Third World cities in perspective", Area, 15, 311-13.

Gould, P. R. (1966): "On mental maps", Michigan Inter-University Community of Mathematical Geographers Discussion Paper 9, Reprinted in Downs, R. M. and Stea, D. (1973): Image and Environment. Aldine Press: Chicago, pp. 182-220.

Gould, P. R. (1967): "Structuring information on spacio-temporal preferences", Journal of Regional Science, 7, 259-74.

— (1969a): "The structure of space preferences in Tanzania", Area, 1, 29-35.

— (1969b): "Problems of space preference measures and relationships", Geographical Analysis, 1, 31-44.

— (1970): "Tanzania, 1920-63: the spatial impress of the modernisation process", World Politics, 22, 149-70.

Gould, P. R. and Lyew-Ayee, A. (1983): "Jamaican television: images and geographic connections", Caribbean Geography, 1, 36-50.

Gould, P. and Ola, D. (1970): "The perception of residential desirability in the Western Region of Nigeria", Environment and Planning, 2, 73-87.

Gould, P. R. and White, R. (1968): "The mental maps of British school leavers", Regional Studies, 2, 161-82.

Gould,P.R.and White, R.(1974): Mental Maps. Penguin: Harmondsworth.
Graham, E. (1976): "What is a mental map?", Area, 8, 259-62.
— (1982): "Maps, metaphors and muddles", Professional Geographer, 34, 251-60.
Gray, F. (1975): "Non-explanation in urban geography", Area, 7, 228-34.
Griffin, E. and Ford, L. (1980): "A model of Latin American city structure", Geographical Review, 70, 397-422.
Grimshaw, P. N. and Briggs, K. (1970): "Geography and citizenship: pupil participation in town and country planning", Geography, 55, 307-13.
Gulick, J. (1963): "Images of an Arab City", Journal of the American Institute of Planners, 29, 179-97.
Gwynne, R. N. (1978): "City size and retail prices in less-developed countries: an insight into primacy", Area, 10, 136-40.

Hagerstrand, T. (1953): Innovationsforloppet ur Korologisk Synpunkt. Lund.
Haggett, P. (1965): Locational Analysis in Human Geography. Arnold: London.
Haggett, P., Cliff, A. D. and Frey, A. (1977): Locational Analysis in Human Geography: Locational Models. Volume 1. Arnold: London.
Hall, D. R. (1981a): "Town and country planning in Cuba", Town and Country Planning, 50, 81-3.
— (1981b): "External relations and current development patterns in Cuba", Geography, 66, 237-40.
Hall, P. (1974): Urban and Regional Planning. Penguin: Harmondsworth.
Hall, R. (1984): "Changing fertility in the developing world and its impact on global population growth", Geography, 69, 19-27.
Hamnett, C. (1977): "Non-explanation in urban geography: throwing the baby out with the bath water?", Area, 9, 143-5.
Hampton, W. (1977): "Research into public participation in structure planning", Chapter 3 in: Sewell, W. R. D. and Coppock, J. T. (Eds.): Public Participation in Planning. John Wiley: Chichester.
Hansen, N. M. (1967): "Development pole theory in a regional context", Kyklos, 20, 709-25.
— (Ed.) (1972): Growth Centers in Regional Economic Development. Free Press: New York.
— (1981): "Development from above: the centre-down development paradigm", in Stohr,W.B.and Taylor, D. R. F. (Eds.): Development from Above or Below?: The Dialectics of Regional Development in Developing Countries. Wiley: Chichester.
Harms, H. (1982): "Historical perspectives on the practice and purpose of self-help housing", Chapter 2 in: Ward, P. M. (Ed.): Self-help Housing: A Critique. Mansell: London.
Hartshorn, T. A. (1980): Interpreting the City: An Urban Geography. Wiley: New York.
Harvey, D. (1973): Social Justice and the City. Arnold: London.
Helson, H. (1964): Adaptation Level Theory. Harper and Row: New York.

Herbert, D. T. and Raine, J. W. (1976): "Defining communities within urban areas: an analysis of alternative approaches", Town Planning Review, 47, 325-38.

Herbert, D. T. and Thomas, C. J. (1982): Urban Geography: a First Approach. Wiley: Chichester.

Hirschman, A. O. (1958): The Strategy of Economic Development. Yale University Press: New Haven.

Hirst, M. (1978): "Recent villagization in Tanzania", Geography, 63, 122-5.

Hoa, L. (1984): "Rural industrialisation and stress on small towns: China's unchanged policy", Third World Planning Review, 6, 27-36.

Hoare, A. C. (1981): "Why they go where they go: the political imagery of industrial location", Transactions of the Institute of British Geographers, New Series, 6, 152-75.

Hoch, I. (1972): 'Income and city size', Urban Studies, 9, 299-328.

Hollnsteiner, M. R. (1977): "The case of 'the people versus Mr. Urbano Planner y administrador'", Chapter 25 in Abu-Lughod, J. and Hay, R. (Eds.): Third World Urbanization. Maaroufa Press.

Honour, L. (1979): 'Jamaican development: policies and practice", Bedford College, University of London, Papers in Geography, 7, 42 pp.

Hope, K. R. (1983): "Urban population growth in the Caribbean", Cities, 167-74.

Hope, K. R. and Ruefli, T. (1981): "Rural-urban migration and the development process: a Caribbean case study", Labour and Society, 6, 145-6.

Horvath, R. J. (1969): "In search of a theory of urbanisation: notes on the colonial city", East Lakes Geographer, 5, 69-82.

— (1972): "A definition of colonialism", Current Anthropology, 13, 45-57.

Hoselitz, B. F. (1955): "Generative and parasitic cities", Economic Development and Cultural Change, 3, 278-94.

Hoyle, B. S. (1979): "African socialism and urban development: the relocation of the Tanzanian capital", Tijdschrift voor Economische en Sociale Geografie, 70, 207-16.

— (1983): Seaports and Development: the Experience of Kenya and Tanzania. Gordon and Breach: London.

Hoyos, F. A. (1979): Barbados: Our Island Home (Third Edition). Macmillan: London.

Hoyt, H. (1939): The Structure and Growth of Residential Neighbourhoods in American Cities". Federal Housing Administration: Washington D.C.

Hudson, J. C. (1969): "Diffusion in a central place system", Geographical Analysis, 1, 45-58.

Isard, W. (1956): Location and Space-Economy. M.I.T. Press.

Jackson, J. C. (1974): "Urban squatters in south-east Asia", Geography, 59, 24-30.

Jackson, R. T. (1975): "In praise of Burundi, Paraguay et al"Area,7, 83-6.

Jacobs, J. (1969): The Economy of Cities. Random House: New York.
Jefferson, M. (1939): "The law of the primate city", Geographical Review, 20, 226-32.
Johnson, B. L. C. (1983): India: Resources and Development (Second Edition). Heinemann: London.
Johnson, E. A. J. (1965): Market Towns and Spatial Development in India. New Delhi.
— (1970): The Organization of Space in Developing Countries. Harvard University Press.
Johnson, J. H. (1967): Urban Geography: An Introductory Analysis. Pergamon: Oxford.
Johnston, R. J. (1971): "On the progression from primacy to rank-size in an urban system: the deviant case of New Zealand", Area, 3, 180-4.
— (1977): "Regarding urban origins, urbanization and urban patterns", Geography, 62, 1-8.
— (1980): City and Society: An Outline for Urban Geography. Penguin: Harmondsworth.
— (1983): Philosophy and Human Geography: an Introduction to Contemporary Approaches. Arnold: London.
Jones, E. (1964): "Aspects of urbanisation in Venezuela", Ekistics, 18, 420-25.
— (1966): Towns and Cities. Oxford University Press.
Jones, E. and Eyles, J. (1977): An Introduction to Social Geography. Oxford University Press.
Jones, H. R. (1977): "Metropolitan dominance and family planning in Barbados", Social and Economic Studies, 26, 327-38.
— (1981): A Population Geography. Harper and Row: London.
Jones, R. C. (1975): "Latent migration potential between a depressed region and alternative destinations: a Venezuelan case study", Proceedings of the Association of American Geographers, 7, 104-9.
— (1978): "Myth maps and migration in Venezuela", Economic Geography, 54, 75-91.
— (1980): "The role of perception in urban in-migration: a path analytic model", Geographical Analysis, 12, 98-108.
Jones, R. C. and Zannaras, G. (1978): "The role of awareness space in urban residential preferences: a case study of Venezuelan youth", Annals of Regional Science, 12, 36-52.

Kabagambe, D. and Moughtin, C. (1983): "Housing the poor: a case study in Nairobi", Third World Planning Review, 5, 227-48.
Karan, P. P. and Bladen, W. A. (1982): "Perception of the urban environment in a Third-World country", Geographical Review, 72, 228-32.
Karch, C. A. and Dann, G. H. S. (1981): "Close encounters of the Third World", Human Relations, 34, 249-68.
Kelly, G. A. (1955): The Psychology of Personal Constructs, Vols. 1 and 2. Norton: New York.
Kent, G. (1981): "Community-based development planning", Third World Planning Review, 3, 313-26.

Keyfitz, N. (1980): "Do cities grow by natural increase or migration?", Geographical Analysis, 12, 142-56.

Kirk, W. (1963): "Problems of geography", Geography, 48, 357-71.

Knox, P. (1982): Urban Social Geography: An Introduction. Longman: London.

Koenigsberger, O. H. (1983): "The role of the planner in a poor (and in a not quite so poor) country", Habitat International, 7, 49-55.

Kolars, J. P. and Nystuen, J. D. (1974): Geography: the Study of Location, Culture and Environment. McGraw Hill: New York.

Kosinski, L. A. (1976): "Demographic aspects of urbanization", Geoforum, 7, 313-25.

Kuipers, B. (1978): "Modeling spatial knowledge", Cognitive Science, 2, 129-53.

— (1981): "The cognitive map: could it have been any other way?", Tufts University Working Papers in Cognitive Science, 16.

— (1982): "The 'map in the head' metaphor", Environment and Behavior, 14, 202-20.

Kuklinski, A. (Ed.) (1972): Growth Poles and Growth Centres in Regional Planning. Mouton.

Kulaba, S. M. (1982): "Rural settlement policies in Tanzania", Habitat International, 6, 15-29.

Kuznets, S. (1955): "Economic growth and income inequality", American Economic Review, 45, 1-28.

Lampard, E. E. (1955): "The history of cities in the economically advanced areas", Economic Development and Cultural Change, 3, 81-102.

— (1965): "Historical aspects of urbanization", Chapter 14 in: Hauser, P. M. and Schnore, L. F. (Eds.): The Study of Urbanization. John Wiley.

Langton, J. (1975): "Residential patterns in pre-industrial cities: some case studies of the seventeenth century", Transactions of the Institute of British Geographers, 65, 1-27.

— (1978): "Industry and towns, 1500 to 1730", in: Dodgson, R. A.and Butlin, R. A. (Eds.): An Historical Geography of England and Wales. Academic Press: London, pp. 173-98.

Laquian, A. A. (1977): "Whither site and services?", Habitat, 2, 291-301.

— (1983): "Sites, services and shelter - an evaluation", Habitat International, 7, 211-25.

Lasuen, J. R. (1969): "On growth poles", Urban Studies, 6, 137-61.

— (1973): "Urbanisation and development - the temporal interaction between geographical and sectoral clusters", Urban Studies, 10, 163-88.

Lee, E. (1966): "A theory of migration", Demography, 3, 47-57.

Lee, T. R. (1963): "Psychology and living space", Transactions of the Bartlett Society, 2, 9-36.

— (1970): "Perceived distance as a function of distance in a city", Environment and Behaviour, 2, 40-51.

— (1976): Psychology and the Environment. Methuen: London.

Lee Fong, M. K. (1980): "Some aspects of perception of the residents of Tai Po on their living environment", Department of Geography and Geographical Research Centre, The Chinese University of Hong Kong, Occasional Paper No. 3.

Lehmann, D. (1982): "Agrarian structure, migration and the state in Cuba", in: Peek, P. and Standing, G. (Eds.): State Policies and Migration: Studies in Latin America and the Caribbean. Croom Helm, pp. 321-90.

Lewis, G. J. and Maund, D. J. (1976): "The urbanization of the countryside: a framework for analysis", Geografiska Annaler, 58B, 17-27.

Lewis, L. A., Degani, A. and Hudson, T. (1980): "The measurement and explanation of the spatial perception of Africa: a Nigerian viewpoint", Geografiska Annaler, 62B, 33-8.

Lewis, O. (1959): Five Families: Mexican case studies in the culture of poverty, Basic Books.

--- (1966): "The Culture of Poverty", Scientific American, 215, 19-25.

Ley, D. (1981): "Behavioral geography and the philosophies of meaning", Chapter 9 in: Cox, K. R. and Golledge, R. G. (Eds.): Behavioral Geography Revisited. Methuen: London.

Lightfoot, P. and Fuller, T. (1983): "Circular rural-urban movement and development planning in northeast Thailand", Geoforum, 14, 277-87.

Lightfoot, P., Fuller, T. and Kamnuansilpa, P. (1981): "Impact and image of the city in the Northeast Thai countryside", Cultures et développement, 13, 97-122.

Linsky, A. S. (1965): "Some generalizations concerning primate cities", Annals of the Association of American Geographers, 55, 506-13.

Lipton, M. (1977): Why Poor People Stay Poor: Urban Bias in World Development. Temple Smith: London.

--- (1980): "Migration from rural areas of poor countries: the impact on rural productivity and income distribution", World Development, 8, 1-24.

Lloyd, P. (1979): Slums of Hope? Shanty towns of the Third World. Penguin: Harmondsworth.

--- (1980): The 'Young Towns' of Lima: Aspects of Urbanization in Peru. Cambridge University Press.

Lloyd, R. (1982): "A look at images", Annals of the Association of American Geographers, 72, 532-48.

Logan, M. I. (1972): "The spatial system and planning strategies in developing countries", Geographical Review, 62, 229-44.

Losch, A. (1940): Die räumliche Ordnung der Wirtschaft, Jena. Translated by Woglom, W. H. and Stolper, W. F. (1954): The Economics of Location. Yale University Press.

Lowenthal, D. (1957): "The population of Barbados", Social and Economic Studies, 6, 445-501.

--- (1972): West Indian Societies. Oxford University Press.

Lundqvist, J. (1981): "Tanzania: Socialist ideology, bureaucratic reality, and development from below", Chapter 13 in: Stohr, W. B. and Fraser Taylor, D. R. (Eds.): Development from Above or Below?. Wiley: Chichester, pp. 329-49.

Lynch, K. (1960): The Image of the City. M.I.T.: Cambridge, Mass.

Ma, J. C. (1976): "Anti-urbanism in China", Proceedings of the Association of American Geographers, 8, 114-18.

Mabogunje, A. L. (1968): Urbanization in Nigeria. University of London Press.

— (1980): The Development Process: A Spatial Perspective. Hutchinson: London.

MacGregor, M. T. G. and Valverde, V. C. (1975): "Evolution of the urban population in the arid zones of Mexico, 1900-1970", Geographical Review, 65, 214-28.

Mangin, W. (1967): "Latin American squatter settlements: a problem and a solution", Latin American Research Review, 2, 65-98.

Mangin, W. and Turner, J. (1968): "The Barriada movement", Progressive Architecture, 5, 154-62.

McGee, T. G. (1971): The Urbanization Process in the Third World: Exploration in Search of a Theory. Bell: London.

McLoughlin, J. B. (1973): Urban and Regional Planning: A Systems Approach. Faber: London.

Mehta, S. K. (1964): "Some demographic and economic correlates of primate cities: a case for revaluation", Demography, 1, 136-47.

Meier, G. M. and Baldwin, R. E. (1957): Economic Development: Theory, History, Policy. John Wiley: New York.

Mellaart, J. (1964): "A Neolithic city in Turkey", Scientific American Reprint, No. 620, 10 pp.

— (1967): Catal Huyuk: a Neolithic City in Anatolia. Oxford University Press.

Mera, K. (1973): "On the urban agglomeration and economic efficiency", Economic Development and Cultural Change, 21, 309-24.

— (1975): Income Distribution and Regional Development. University of Tokyo Press.

— (1978): "The changing pattern of population distribution in Japan and its implications for developing countries", in: Lo, F. C. and Salih, K. (Eds.): Growth Pole Strategies and Regional Development Policy. Pergamon Press.

Misra, R. P. and Natraj, V. K. (1981): "India: blending central and grass-roots planning", Chapter 10 in: Stohr, W. B. and Taylor, D. R. F. (Eds.): Development from Above or Below? John Wiley: Chichester.

Morris, A. S. (1978): "Urban growth patterns in Latin America with illustrations from Caracas", Urban Studies, 15, 299-312.

Moseley, M. J. (1973): "Growth centres - a shibboleth?", Area, 5, 143-50.

— (1974): Growth Centres in Spatial Planning. Pergamon Press: Oxford.

Moser, C. and Kalton, G. (1971): Survey Methods in Social Investigation. Heinemann: London.

Mountjoy, A. B. (1976): "Urbanization, the squatter and development in the Third World", Tijdschrift voor Economische en sociale Geografie, 67, 130-7.

— (1978): "Urbanization in the Third World", Chapter 10 in: Mountjoy, A. B. (Ed.) (1978): The Third World: Problems and Perspectives. Macmillan: London, pp. 102-11.

— (1980): "Worlds without end", Third World Quarterly, 2, 753-7.

— (1982): "Squatters in Hong Kong", Geographical Magazine, 53, 119-25.

Mulchansingh, V. C. (1983): "Use of natural gas in Trinidad and the Point Lisas pole de croissance", Caribbean Geography, 1, 133-9.

Muller, M. S. (1982): "Langa Langa: two site-and-service schemes in Kenya", Habitat International, 6, 89-107.

Murphy, P. E. (1978): "Preferences and perceptions of urban decision-making groups: congruence or conflict?", Regional Studies, 12, 749-59.

Murray, D. and Spencer, C. (1979): "Individual differences in the drawing of cognitive maps: the effects of geographical mobility, strength of mental imagery and basic graphic ability", Transactions of the Institute of British Geographers, New Series, 4, 385-91.

Myrdal, G. (1957): Economic Theory and Underdeveloped Areas. Duckworth: London.

Naipaul, S. (1973): The Chip-Chip Gatherers. Andre Deutsch: London.

Newsweek (1983): "An age of nightmare cities: flood tides of humanity will create mammoth urban problems for the Third World", Newsweek, October 31st, 26-31.

Northam, R. M. (1979): Urban Geography. Wiley: New York.

Nyerere, J. K. (1982): "On rural development", Habitat International, 6, 7-14.

O'Connor, A. M. (1976): "'Third World' or one world?", Area, 8, 269-71.

O'Connor, A. (1983): The African City. Hutchinson: London.

Oppenheim, A. N. (1966): Questionnaire Design and Attitude Measurement. Heinemann: London.

Osgood, C. E., Suci, G. J. and Tannenbaum, P. M. (1957): The Measurement of Meaning. University of Illinois Press: Urbana.

Pacione, M. (Ed.) (1981): Problems and Planning in Third World Cities. Croom Helm: London.

PAG (1965): The Future of Development Plans. Report of the Planning Advisory Group, H.M.S.O.: London.

Pahl, R. E. (1965): Urbs in Rure: the Metropolitan Fringe in Hertfordshire. London School of Economics and Political Science Geographical Papers, No. 2.

— (1970): "Whose City?". Longman: London.

Paquette, R. (1965): Concentration and Dispersal of settlements:Martinique and St Lucia. Unpublished M.A. dissertation, McGill University.

Parr, J. B. (1973): "Growth poles, regional development, and central place theory", Papers of the Regional Science Association, 31, 173-212.

— (1974): "Welfare differences within a nation: a comment", Papers of the Regional Science Association, 32, 83-91.

— (1978): "Models of the central place system: a more general approach", Urban Studies, 15, 35-49.

— (1980): "Health care facility planning: some developmental considerations", Socio-Economic Planning Sciences, 14, 121-7.

— (1981): "Temporal change in a central-place system", Environment and Planning, A, 13, 97-118.

Peattie, L. R. (1982): "Some second thoughts on sites-and-services", Habitat International, 6, 131-9.

Pedersen, P. O. (1970): "Innovation diffusion within and between national urban systems", Geographical Analysis, 2, 203-54.

Peek, P. and Standing, G. (1982): State Policies and Migration: Studies in Latin America and the Caribbean. Croom Helm.

Peet, R. (Ed.) (1977): Radical Geography. Methuen: London.

Penfold, A. H. (1970): "Caracas: urban growth and transportation", Town Planning Review, 41, 103-20.

Perloff, H. S. and Wingo, L. (1961): "Natural resource endowment and regional economic growth", in: Spengler, J. J. (Ed.): Natural Resources and Economic Growth. Resources for the Future: Washington D.C.

Perroux, F. (1950): "Economic space: theory and applications", Quarterly Journal of Economics, 64, 89-104.

— (1955): "Note sur la notion de 'pôle de croissance'", Économie Appliquée, 1-2, 307-20.

Pfister, F. (1982): "Housing improvement and popular participation in the Upper Volta", Habitat International, 6, 209-14.

Phillips, D. R. and Yeh, A. G. O. (1983): "Changing attitudes to housing provision: BLISS in the Philippines?", Geography, 68, 37-40.

Pocock, D. C. D. (1973): "Environmental perception: process and product", Tijdschrift voor Economische en Sociale Geografie, 64, 251-7.

— (1976): "Some characteristics of mental maps: an empirical study", Transactions of the Institute of British Geographers, New Series, 1, 493-512.

— (1979): "The novelist's image of the North", Transactions of the Institute of British Geographers, New Series, 4, 62-76.

— (1981): "Sight and knowledge", Transactions of the Institute of British Geographers, New Series, 6, 385-93.

Pocock, D. and Hudson, R. (1978): Images of the Urban Environment. Macmillan: London.

Pollard, H. J. (1984): "Oil in Trinidad: some gains and losses", Geography, 69, 72-5.

Porter, J., Hart, C. and Machin, J. (1975): "Cloze procedure tested in Hampshire", Area, 7, 196-8.

Potter, R. B. (1981): "Industrial development and urban planning in Barbados", Geography, 66, 225-8.
— (1982): The Urban Retailing System: Location, Cognition and Behavior. Gower: Aldershot.
— (1983a): "Tourism and development: the case of Barbados, West Indies", Geography, 68, 46-50.
— (1983b): "Urban development, planning and demographic change, 1970-80 in Barbados", Caribbean Geography, 1, 3-12.
— (1983c): "Spatial perceptions and physical development planning in Trinidad and Tobago: Preliminary results of a field research project", Bedford College, University of London, Papers in Geography, 15, 43 pp.
— (1983d): "Congruence between space preferences and socio-demographic structure in Barbados, West Indies: The use of cognitive studies in Third World urban planning and development", Geoforum, 14, 249-65.
— (1983e): "How Barbadians perceive their environment", The Bajan and South Caribbean, 357, 40-1.
— (1983f): "How Barbadians perceive planning issues", The Bajan and South Caribbean, 360, 30-1.
— (1984a): "Mental maps and spatial variations in residential desirability: a Barbados case study", Caribbean Geography, 1, 186-97.
— (1984b): "Perception research as input into urban planning in poor countries: a comparative project", Perception of Environment, 4, 4-5.
— (1984c): "Spatial perception and public involvement in Third World urban planning: the example of Barbados", Singapore Journal of Tropical Geography, 5, 30-44.
— (1985): "Perceived life domain satisfaction and social status", Journal of Social Psychology (in press).
Potter, R. B. and Coshall, J. T. (1984): "The hand analysis of repertory grids: an appropriate technique for Third World environmental studies", Area, 16, 315-22.
Potter, R. B. and Hunte, M. (1979): "Recent developments in planning the settlement hierarchy of Barbados: implications concerning the debate on urban primacy", Geoforum, 10, 355-62.
Potter, R. B. and Potter, V. (1978): "Urban development in the World Dryland regions: inventory and prospects", Geoforum, 9, 349-79.
Potter, R. B. and Wilson, M. G. (1983): "Age differences in the content and style of cognitive maps of Barbadian schoolchildren", Perceptual and Motor Skills, 57, 332.
Pounds, N. J. G. (1969): "The urbanization of the Classical World", Annals of the Association of American Geographers, 59, 135-57.
Prebish, R. (1950): The Economic Development of Latin America. United Nations: New York.

Pred, A. R. (1973): "The growth and development of systems of cities in advanced economies", in: Pred, A. and Törnqvist, G. (Eds.): Systems of Cities and Information Flows: Two Essays. Lund Studies in Geography, Series B, 38, 9-82.

--- (1977): City-Systems in Advanced Economies. Hutchinson:London.

Proshansky, H. M., Ittelson, W. H. and Rivlin, L. G. (Eds.) (1976):Environmental Psychology (Second Edition). Holt, Rinehart and Winston: New York.

Radford, J. P. (1979): "Testing the model of the pre-industrial city: the case of ante-bellum Charleston, South Carolina", Transactions of the Institute of British Geographers, New Series, 4, 392-410.

Renaud, B. (1981): National Urbanization Policy in Developing Countries. Published for the World Bank:Oxford University Press.

Richardson, H. W. (1972): "Optimality in city size, systems of cities and urban policy: a sceptics view", Urban Studies, 9, 29-43.

--- (1973a): "Theory of the distribution of city sizes: review and prospects", Regional Studies, 7, 239-51.

--- (1973b): The Economics of Urban Size. Saxon House: Farnborough.

---(1976): "The argument for very large cities reconsidered: a comment", Urban Studies, 13, 307-10.

--- (1977): City Size and National Spatial Strategies in Developing Countries. World Bank Staff Working Paper, No. 252.

--- (1978): Regional and Urban Economics. Penguin: Harmondsworth.

--- (1980): "Polarization reversal in developing countries", Papers of the Regional Science Association, 45, 67-85.

--- (1981): National urban development strategies in developing countries", Urban Studies, 18, 267-83.

Richardson, H. W. and Richardson, M. (1975): "The relevance of growth center strategies to Latin America", Economic Geography, 51, 163-78.

Riddell, J. B. (1970): The Spatial Dynamics of Modernization in Sierra Leone: Structure, Diffusion and Response. Evanston, Illinois.

Riddell, J. B. and Harvey, M. (1972): "The urban system in the migration process: an evaluation of step-wise migration in Sierra Leone", Economic Geography, 48, 270-83.

Rieser, R. L. (1972): "Urban spatial images: an appraisal of the choice of respondent and measurement situation", London School of Economics Graduate School of Geography Discussion Paper, No. 42.

Rieser, R. (1973): "The territorial illusion and behavioural sink: critical notes on behavioural geography", Antipode, 5, 52-7.

Roberts, B. R. (1978): Cities of Peasants: the Political Economy of Urbanization in the Third World. Arnold: London.

Robinson, M. E. (1974): "Cloze procedure and spatial comprehension tests", Area, 6, 137-42.

Robson, B. T. (1973): Urban Growth: an Approach. Methuen: London.

Rondinelli, D. A. (1982): "Intermediate cities in developing countries: a comparative analysis of their demographic, social and economic characteristics", Third World Planning Review, 4, 357-86

--- (1983a): "Dynamics of growth of secondary cities in developing countries", Geographical Review, 73, 42-57.

Rondinelli, D. A. (1983b): Secondary Cities in Developing Countries: Policies for Diffusing Urbanization. Sage: London.

Rostow, W. W. (1960): The Stages of Economic Growth: a non-communist manifesto. Cambridge University Press.

Saarinen, T. F. (1976): Environmental Planning: Perception and Behavior. Houghton Mifflin: Boston.

Safa, H. I. (Ed.) (1982): Towards a Political Economy of Urbanization in Third World Countries. Oxford University Press: Delhi.

Safier, M. (1969): "Towards the definition of patterns in the distribution of economic development over East Africa", East African Geographical Review, 7, 1-13.

Santos, M. (1979): The Shared Space: the two circuits of the urban economy in underdeveloped countries. Methuen: London.

Schenk, H. (1974): "Concepts behind urban and regional planning in China", Tijdschrift voor Economische en Sociale Geografie, 65, 381-9.

Schultz, T. W. (1953): The Economic Organization of Agriculture. McGraw-Hill: New York.

Schumpeter, J. A. (1911): Die Theorie des Wirtschaftlichen Entwicklung. Leipzig.

Sewell, W. R. D. and Coppock, J. T. (Eds.) (1977): Public Participation in Planning. John Wiley: Chichester.

Short, J. R. (1978): "Residential mobility", Progress in Human Geography, 2, 419-47.

Short, J. (1984): An Introduction to Urban Geography. Routledge and Kegan Paul: London.

Siebolds, P. and Steinberg, F. (1982): "Tanzania: sites-and-services", Habitat International, 6, 109-30.

Simon, H. R. (1957): Models of Man: Social and Rational. New York.

Simmons, J. W. (1968): "Changing residence in the city; a review of intra-urban mobility", Geographical Review, 58, 622-51.

Singh, J. (1979): "Central place hierarchy in a backward economy: Gorakhpur region (India)", Tijdschrift voor Economische en Sociale Geografie, 70, 300-6.

Sjoberg, G. (1960): The Preindustrial City: Past and Present. Free Press: New York.

--- (1965): "The origin and evolution of cities", Scientific American, September 1965; reprinted in Scientific American: Cities: Their Origin, Growth, and Human Impact. Freeman: San Francisco.

Skeffington Report (1969): People and Planning. H.M.S.O.: London.

Skinner, R. J. (1983): "Community participation: its scope and organization", Chapter 6 in: Skinner, R. J. and Rodell, M. J. (Eds.): People, Poverty and Shelter: Problems of Self-Help Housing in the Third World. Methuen: London.

Skinner, R. J. and Rodell, M. J. (1983): People, Poverty and Shelter: Problems of Self-Help Housing in the Third World. Methuen: London.

Skurnik, L. S. and George, F. (1964): Psychology for Everyman. Penguin: Harmondsworth.

Soja, E. W. (1968): The Geography of Modernization in Kenya: a Spatial Analysis of Social, Economic and Political Change. Syracuse University Press.
— (1974): "The geography of modernization: paths, patterns, and processes of spatial change in developing countries", in: Bruner, R. and Brewer, G. (Eds.): A Policy Approach to the Study of Political Development and Change. The Free Press.
Stadel, C. (1975): "The structure of squatter settlements in Medellín, Colombia", Area, 7, 249-54.
Sternstein, L. (1971): "The Image of Bangkok", Pacific Viewpoint, 12, 68-74.
Stevens, P. H. M. (1957): "Planning in the West Indies", Town and Country Planning, 25, 503-8.
Stohr, W. B. (1981): "Development from below: the bottom-up and periphery-inward development paradigm", Chapter 2 in Stohr, W. B. and Taylor, D. R. F. (Eds.): Development from Above or Below? Wiley, pp. 39-72.
Stohr, W. B. and Taylor, D. R. F. (1981): Development from Above or Below? The Dialectics of Regional Planning in Developing Countries. Wiley: Chichester.
Stohr, W. and Todtling, F. (1978): "Spatial equity - some antitheses to current regional development doctrine", Papers of the Regional Science Association, 38, 33-53.
Stokes, C. (1962):"A theory of slums",Land Economics,38,187-97.
Stretton, H. (1978): Urban Planning in Rich and Poor Countries. Oxford University Press.

Taaffe, E. J., Morrill, R. L. and Gould, P. R. (1963): "Transport expansion in underdeveloped countries: a comparative analysis", Geographical Review, 53, 503-29.
Taylor, J. L. and Williams, D. G. (Eds.) (1982): Urban Planning Practice in Developing Countries. Pergamon Press: Oxford.
Thapa, P. and Conway, D. (1983): "Internal migration in contemporary Nepal: models which internalize development policies", Papers of the Regional Science Association, 53, 27-42.
Tinbergen, J. (1967): Development Planning. Weidenfeld and Nicolson: London.
Town and Country Planning Division, Ministry of Finance and Planning (1973): Planning for Redevelopment: East Port of Spain. Trinidad and Tobago.
Town and Country Planning Division, Trinidad and Tobago (1978): National Physical Development Plan: Trinidad and Tobago. Trinidad and Tobago.
Town and Country Planning Office, Barbados (1970): Physical Development Plan for Barbados. Government Printing Office, Barbados.
Townsend, J. G. (1976): "Farm 'failures': the application of personal constructs in the tropical rainforest", Area, 8, 219-22.
--- (1977): "Perceived worlds of the colonists of tropical rainforest, Colombia", Transactions of the Institute of British Geographers, New Series, 2, 430-58.

Tuan, Y-F. (1974): "Review of Gould, P. and White, R., Mental Maps", Annals of the Association of American Geographers, 64, 59.

--- (1975): "Images and mental maps", Annals of the Association of American Geographers, 65, 205-13.

--- (1978): "The city: its distance from nature", Geographical Review, 68, 1-12.

Turner, A. (Ed.) (1980): The Cities of the Poor: Settlement Planning in Developing Countries. Croom Helm: London.

Turner, A. and Smulian, J. (1971): "New cities in Venezuela", Town Planning Review, 42, 3-18.

Turner, J. F. C. (1963): "Dwelling resources in South America", Architectural Design, 37, 360-93.

--- (1967): "Barriers and channels for housing development in modernizing countries", Journal of the American Institute of Planners, 33, 167-81.

--- (1968): "Housing priorities, settlement patterns and urban development in modernizing countries", Journal of the American Institute of Planners, 34, 354-63.

--- (1969): "Uncontrolled urban settlement: problems and policies", in Breese, G. (Ed.): The City in Newly Developing Countries. Prentice-Hall, pp. 507-35.

--- (1976): Housing by People: Towards Autonomy in Building Environments. Marion Boyars.

--- (1982): "Issues in self-help and self-managed housing", Chapter 4 in: Ward, P. M. (Ed.): Self-Help Housing: A Critique. Mansell: London, pp. 99-113.

--- (1983): "From central provision to local enablement: new directions for housing policies", Habitat International, 7, 207-10.

Turner, S. D. (1982): "Using sentence completions: a way into world-view?", Area, 14, 177-84.

Ulack, R. (1978): "The role of urban squatter settlements", Annals of the Association of American Geographers, 68, 535-50.

United Nations (1969): Growth of the World's Urban and Rural Population 1920-2000. United Nations: New York.

--- (1977): Demographic Yearbook 1976. United Nations: New York.

Unwin, T. (1983): "Perspectives on development - an introduction", Geoforum, 14, 235-41.

Urban Redevelopment Council (1972): East Port of Spain Social Survey Report. Trinidad and Tobago.

Vance, J. E. (1970): The Merchant's World: The Geography of Wholesaling. Prentice-Hall: Englewood Cliffs.

Vapnarsky, C. A. (1969): "On rank-size distributions of cities: an ecological approach", Economic Development and Cultural Change, 17, 584-95.

Walmsley, D. J. (1982): "Mass media and spatial awareness", Tijdschrift voor Economische en Sociale Geografie, 73, 32-42.

Walmsley, P. J. and Lewis, G. J. (1984): Human Geography: Behavioural Approaches. Longman: London.

Ward, B. (1980): "First, second, third and fourth worlds", The Economist, 18 May 1980, 65-73.
Ward, P. (1976): "The squatter settlement as slum or housing solution: some evidence from Mexico City", Land Economics, 52, 330-46.
Ward, P. M. (1978): "Social interaction patterns in squatter settlements in Mexico City", Geoforum, 9, 235-48.
— (1981): "Financing land acquisition for self-build housing schemes", Third World Planning Review, 3, 7-18.
— (Ed.) (1982): Self-Help Housing: a Critique. Mansell: London.
— (1982): "Introduction and purpose", Chapter 1 in: Ward, P. M. (Ed.) Self-Help Housing: a Critique. Mansell: London, pp. 1-13.
Warnes, A. M. (1978): "Registered change in urban geography research", Area 10, 4-7.
Webber, M. J. (1972): The Impact of Uncertainty on Location. M.I.T. Press.
Weinand, H. C. and Ward, R. G. (1979): "Area preferences in Papua New Guinea", Australian Geographical Studies, 17, 64-75.
Wells, A. (1969): "Low-cost housing in Casablanca", Architectural Association Quarterly, 1, 44-53.
Wheatley, P. (1967): "Some proleptic observations on the origins of urbanism", in: Steel, R. W. and Lawton, R. (Eds.): Liverpool Essays in Geography. Longmans: London, pp. 315-45.
— (1971): The Pivot of the Four Quarters. Chicago University Press.
Williams, P. (1978): "Urban managerialism: a concept of relevance?", Area, 10, 236-40.
Williamson, J. G. (1965): "Regional inequality and the process of national development: a description of the patterns", Economic Development and Cultural Change, 13,3-45.
Winters, C. (1982): "Urban morphogenesis in Francophone Black Africa", Geographical Review, 72, 139-54.
Wirth, L. (1938): "Urbanism as a way of life", American Journal of Sociology, 44, 1-24.
Wolf-Phillips, L. (1979): "Why Third World?", Third World Quarterly, 1, 105-13.
Wolpert, J. (1965): "Behavioral aspects of the decision to migrate", Papers of the Regional Science Association, 15, 159-69.
— (1966): "Migration as an adjustment to environmental stress", Journal of Social Issues, 22, 92-102.
Wong, K. Y. and Lee Fong, M. K. (1980): "Tai Po as a place of habitat - a perception study", Department of Geography and Geographical Research Centre, The Chinese University of Hong Kong, Occasional Paper, No. 11.
World Bank (1980): World Development Report 1980. The World Bank.
Wu, C. T. and Ip, D. F. (1981): "China: rural development - alternating combinations of top-down and bottom-up strategies", Chapter 6 in: Stohr, W. B. and Taylor, D. R. F. (Eds.): Development from Above or Below?. John Wiley: Chichester.

Yeates, M. and Garner, B. (1980): The North American City (Third Edition). Harper and Row: San Francisco.

Yeh, A. G. O. and Fong, P. K. W. (1984): "Public housing and urban development in Hong Kong", Third World Planning Review, 6, 79-94.

You, N. (1981): "Alternative strategies in urban development: some Chinese experiments in a quest for agropolitan space", Third World Planning Review, 3, 77-93.

Zelinsky, W. (1971): "The hypothesis of the mobility transition", Geographical Review, 61, 219-49.

Zetter, R. (1981): "Imported or indigenous planning education?: Some observations on the needs of developing countries", Third World Planning Review, 3, 21-42.

Zipf, G. K. (1949): Human Behaviour and the Principle of Least Effort. Cambridge, Mass.

Index

For Product Safety Concerns and Information please contact our EU representative GPSR@taylorandfrancis.com
Taylor & Francis Verlag GmbH, Kaufingerstraße 24, 80331 München, Germany

www.ingramcontent.com/pod-product-compliance
Ingram Content Group UK Ltd.
Pitfield, Milton Keynes, MK11 3LW, UK
UKHW022013180926
475701UK00003B/334
9780415853279